INTRODUCTION TO BANACH SPACES AND THEIR GEOMETRY

NORTH-HOLLAND
MATHEMATICS STUDIES **68**

Notas de Matemática (86)
Editor: Leopoldo Nachbin
Universidade Federal do Rio de Janeiro
and University of Rochester

Introduction to Banach Spaces and their Geometry

BERNARD BEAUZAMY

Département de Mathématiques
Université de Lyon I
Villeurbanne, France

1982

NORTH-HOLLAND PUBLISHING COMPANY – AMSTERDAM · NEW YORK · OXFORD

ISBN: 0 444 86416 4

Publishers
NORTH-HOLLAND PUBLISHING COMPANY
AMSTERDAM · OXFORD · NEW YORK

Sole distributors for the U.S.A. and Canada
ELSEVIER SCIENCE PUBLISHING COMPANY, INC.
52 VANDERBILT AVENUE
NEW YORK, N.Y. 10017

Library of Congress Cataloging in Publication Data

Beauzamy, Bernard, 1949-
Introduction to Banach spaces and their geometry.

(North-Holland mathematics studies ; 68)
(Notas de matemática ; 86)
Bibliography: p.
Includes index.
1. Banach spaces. I. Title. II. Series.
III. Series: Notas de matemática (Amsterdam, Netherlands) ; 86.
QA1.N86 no. 86 [QA322.2] 510s [515.7'32] 82-6463
ISBN 0-444-86416-4 (Elsevierscience) AACR2

PRINTED IN THE NETHERLANDS

INTRODUCTION

The study of Banach Spaces for themselves is a rather recent idea, since this branch of Functional Analysis came up, roughly speaking, after the work of S. Banach himself, in the 1930's. Soon, Banach Spaces Theory, as it is now called, received a quick development, considering the number of people working in this area, as well as the importance of the results they obtained. In some sense, it may appear as the natural continuation of the study of locally convex spaces, which culminated in the years 1950's. But, as will be seen, the spirit and the methods are quite different.

Meanwhile, courses are taught on Banach spaces in many Universities, at elementary or at high level. Several books have already been written on this topic. Let us mention first those of M.M. Day [13], Lindenstrauss-Tzafriri [34], E. Lacey [32], and many others which are devoted to some specialized feature of Banach Spaces.

All these books are intended for specialists of the field, and students who are barely acquainted with the basic knowledge in Analysis (that is : classical general Topology, classical Measure Theory, elementary properties of Hilbert spaces) may find them difficult (or even almost impossible) to read. The present book is intended to fill this gap and give an opportunity to students to become familiar with an elementary theory of Banach spaces, and to reach a level which will allow them to understand more specialized topics and read research papers. Rather than reaching the research level on some narrow field, we prefer to present the basic facts of the theory and give a general view of the subject.

Since we start at a rather low level, we have to present first the classical results of Functional Analysis, such as the Closed Graph Theorem, the Hahn-Banach Theorem, etc. But, since we have in mind the study of Banach spaces, we have avoided, as far as possible, the "Bourbaki style" of exposition, which develops a general theory of locally convex topological vector spaces. For example, we have completely neglected the general study of topologies compatible with a duality, and we have only presented the results which are of further use for our purpose : they are numerous enough! Also, in the context of normed spaces (endowed with their norm or with a weak topology) many classical theorems (connected, for example, with the Baire Property) have a simpler and more concrete proof : we have worked in this spirit as often as possible. But of course, when it makes no difference, we have no reason to restrict ourselves.

In contrast, we have given an important place, in this first part, to the study of reflexivity, with, of course, the classical results about duality and weak compactness, but also with R.C. James' conditions for reflexivity. These conditions are used several times in the sequel, and, from the start, they introduce the reader to the spirit of the geometric study of Banach spaces. The properties of separable Banach spaces are also investigated, and we end this first part with a study of the delicate notion of weak sequential compactness : we prove a weak form (weak, but sufficient for what follows) of the Eberlein-Smulian Theorem.

The second part enters really the subject. We start with a short chapter about Hilbert spaces. The proofs are given, but we think that most of the results should be familiar to the readers : we include these results in order to prepare the contrast with general Banach spaces.

In a second chapter, we investigate the notion of Schauder basis in a Banach space : up to a certain extent, this notion plays the role of an Hilbertian basis in a Hilbert space. We also develop a little the theory of unconditional bases, and prove some results, due to R.C. James, concerning reflexivity in Banach spaces with Schauder bases.

We then turn to the study of the most common Banach spaces : $\ell_p (1 \leq p \leq +\infty)$, c_0, $\mathscr{C}(K)$, $L_p (1 \leq p \leq +\infty)$. For all these spaces, we have chosen one theme : the classification of subspaces and of complemented subspaces (this theme is very common in any attempt of classification of Banach spaces). First we prove that Banach spaces contain uncomplemented subspaces. We then study the subspaces of ℓ_p and c_0. To investigate the Banach spaces of continuous functions requires the notion of extreme point of a compact convex set, which we have to develop first ; we can then give some results about the classification of the $\mathscr{C}(K)$ spaces. The last chapter is devoted to the L_p-spaces and also requires first some developments about measure theory (equi-integrability, and so on).

This second part has a double aim : first, to present some common Banach spaces, which will serve as examples in the sequel, and, second, to introduce already some geometric notions linked, for example, with the existence of subspaces of a given type, or with bases, or with extreme points.

In the third part, which is the shortest, we consider the "metric" properties of Banach spaces. We call "metric", in opposition to "topological", the properties which depend on the distance in the space, and need not be conserved if the norm is replaced by an equivalent one. Such properties can be local, like strict convexity and smoothness, or uniform, like uniform convexity and smoothness. We study the connections between them, their behaviour with respect to duality, and their relations with reflexivity and with the differentiability properties of the norm : this is now pure geometry of Banach spaces.

The fourth part is more specialized, and deals with the property of super-reflexity. In the last ten years, the super-reflexive Banach spaces have received a special interest, and they certainly constitute the best-known class of Banach spaces. We start with the definitions of finite-representability and super-properties (due to R.C. James). We investigate the tree-property, give the James'estimates for basic sequences in super-reflexive spaces, study two other characterizations : J - convexity and uniform non-squareness, and, finally, give a new and simple proof (due to B. Maurey) of the renorming theorem for super-reflexive spaces (originally due to P. Enflo, and, with better estimates,to G. Pisier).

So, as it can be seen, we present a rather general view of Banach spaces. All proofs are given in detail, and we mention with references the recent developments which may interest the reader. To help the beginner, we have also included exercises at the end of each chapter. We have tried to avoid the classical ones, and to stay in accordance with the general spirit of the book. Each chapter ends with a short paragraph, in which we give a brief history of the results presented, and some references. A general bibliography, as well as an index, are at the end of the volume.

For some parts of our chapters, we have followed, closely or not, various authors. Let us mention : N. Bourbaki : Espaces Vectoriels Topologiques (Chap. I à V) [11], M.M. Day : Normed Linear Spaces [13], G. Choquet : Lectures on Analysis (t. II) [12], G. Köthe : Topological

Vector Spaces [31], J. Lindenstrauss - L. Tzafriri : Classical Banach Spaces [34], J. Neveu : Notions élémentaires de la Théorie des Probabilités [38], W. Rudin : Real and Complex Analysis [43], K. Yoshida : Functional Analysis [50], A. Zygmund : Trigonometric series [51].

The readers should of course consult these books to obtain further information in the direction they wish. But, above all, it seems to us that S. Banach's book : Théorie des Opération linéaires [5], though it is not up-to-date, contains the best preparation to these matters.

To end this introduction, let us recall to the reader the words of Lavater : "Dieu préserve ceux qu'il chérit des lectures inutiles".

Finally, we wish to thank Mrs Reveillard, from the University of Lyon, for her prompt and excellent typing.

Paris, August 1981.

TABLE OF CONTENTS

CHAPTER 0

NOTATIONS AND PRELIMINARIES

In this chapter, we recall some elementary definitions and fix some terminology.

• A *topological space* is a set endowed with a topology.

• A *topological vector space* E is a vector space, equipped with a topology, such that the mappings $(x,y) \longrightarrow x + y$ (from $E \times E$ into E) and $(x,\lambda) \longrightarrow \lambda x$ (from $E \times \mathbb{K}$ into E) are continuous.

• If E is a vector space, its field of scalars will be denoted by $\mathbb{K}$. It will always be $\mathbb{R}$ or $\mathbb{C}$.

• A *locally convex topological vector space* E is a topological space such that the origin has a fundamental system of convex neighbourhoods. All the locally convex spaces that we shall meet are moreover *Hausdorff*, that is, two distinct points have neighbourhoods which do not meet.

• A *metric linear space* is a topological vector space in which the topology is given by a distance.

• A *semi-norm* on a vector space E is a mapping p from E into $\mathbb{R}^+$, which satisfies

$$\begin{cases} p(x + y) \leqslant p(x) + p(y) & \text{for all } x,y \in E \\ \quad \text{(sub-additivity)} & \\ p(\lambda x) = |\lambda| p(x) & \text{for all } \lambda \in \mathbb{K}, \text{ all } x \in E . \end{cases}$$

• A *norm* on E, denoted by $\|.\|$, satisfies moreover :

$\|x\| = 0$ implies $x = 0$.

A normed space is a locally convex topological vector space.

• A *Banach space* is a normed space which is complete for the norm topology.

• If E a normed space, $\mathscr{B}_E$ will be its closed unit ball, and $\mathscr{S}_E$ its unit sphere :

$$\mathscr{B}_E = \{x \in E \ ; \ \|x\| \leqslant 1\} \ , \qquad \mathscr{S}_E = \{x \in E \ ; \ \|x\| = 1\} \ .$$

• If E and F are two normed spaces, $\mathscr{L}(E,F)$ will be the space of all continuous linear mappings from E into F. A continuous linear mapping will just be called an *operator*. The *graph* of a mapping u is denoted by $\mathrm{Gr}\, u$, and is given by

$$\mathrm{Gr}\, u = \{(x, u(x)), x \in E\} \ .$$

• If A and B are two sets, the complement of B in $A = \{x \ ; \ x \in A, x \notin B\}$ is denoted by $A \setminus B$. The complement of A is $\complement A$. The symmetric difference $A \triangle B$ is $(A \cup B) \setminus (A \cap B)$.

• If $(x_n)_{n \in \mathbb{N}}$ are points in a topological vector space, we call :

$\mathrm{conv}\{(x_n)_{n \in \mathbb{N}}\}$ the convex hull of the x_n 's, that is

$$\mathrm{conv}\{(x_n)_{n \in \mathbb{N}}\} = \{\sum_0^N a_i x_i \ ; \ N \geqslant 0, \ a_0, \ldots, a_N \geqslant 0 \ \text{ and } \ \sum_0^N a_i = 1\}$$

$\mathrm{span}\{(x_n)_{n \in \mathbb{N}}\}$ the vector space spanned by the x_n 's, that is

$$\mathrm{span}\{(x_n)_{n \in \mathbb{N}}\} = \{ \sum_0^N a_i x_i \ ; \ N \geqslant 0, \ a_0, \ldots, a_N \in \mathbb{K} \}.$$

$\overline{\mathrm{conv}}\{(x_n)_{n \in \mathbb{N}}\}$, $\overline{\mathrm{span}}\{(x_n)_{n \in \mathbb{N}}\}$ are their closures.

• If E is a normed space, and K a compact in it, we shall say that a sequence $x_1, \ldots, x_n$ forms an ε-net for K if the balls centered on the x_i 's and with radius ε cover K.

• The space $\mathbb{K}^{(\mathbb{N})}$ is the set of *finite* sequences of elements of $\mathbb{K}$ (or, more exactly, the set of sequences which have only finitely many non-zero terms). The *canonical basis* of $\mathbb{K}^{(\mathbb{N})}$ will be called $(e_n)_{n \in \mathbb{N}}$; e_n is the sequence with 1 at the nth rank, 0 otherwise.

• The *diameter* of a set A in a normed space E is $\sup_{x,y \in A} \|x - y\|$.

• Filters.

When one deals (as we shall) with non-metrizable topologies, one cannot use only convergent sequences to define the topology : one has to use filters. Therefore, we recall briefly the necessary definitions (see Bourbaki [10] for details).

If I is a set, a set $\mathscr{F}$ of subsets of I is called a *filter* if :

a) if $X \in \mathscr{F}$ and $Y \supset X$, then $Y \in \mathscr{F}$.

b) if $X, Y \in \mathscr{F}$, then $X \cap Y \in \mathscr{F}$.

c) $I \in \mathscr{F}$.

d) $\phi \notin \mathscr{F}$.

If $I = \mathbb{N}$ and $\mathscr{F}$ is the set of $X \subset \mathbb{N}$ such that $\complement X$ is finite, then $\mathscr{F}$ is called the *Fréchet filter* on $\mathbb{N}$.

The set of filters on a set I can be ordered by inclusion : $\mathscr{F}$ is thinner than $\mathscr{F}'$ if $\mathscr{F}' \subset \mathscr{F}$. An *ultrafilter* is a maximal element for this ordering : no filter is strictly thinner.

If E is a topological space, the set of all neighbourhoods of a given point x_0 is a filter, denoted by $\mathscr{F}V_{x_0}$. In any set I, the set of all subsets containing a given element i_0 is an ultrafilter. Such an ultrafilter is called a trivial one.

On a topological space E, a filter $\mathscr{F}$ is said to *converge* to $x_0 \in E$ if $\mathscr{F}$ is thinner than $\mathscr{F}V_{x_0}$. This means that for every V, neighbourhood of x_0, there is $X \in \mathscr{F}$ with $X \subset V$. One says that x_0 is a *cluster* point for $\mathscr{F}$ if x_0 is a cluster point of all the elements X of $\mathscr{F}$.

If E is a metric space (for example, a normed space), $\mathscr{F}$ is called a *Cauchy filter* if, for every $\epsilon > 0$, there are elements $X \in \mathscr{F}$ such that the diameter of X is at most ϵ.

On a compact set E, every filter has at least a cluster point, and every ultrafilter is convergent. If E is a complete metric space (for example, a Banach space), every Cauchy filter converges.

A *filter base* $\mathscr{B}$ on a set I is a set of subsets of I with the following two properties :

1) If $B_1, B_2 \in \mathscr{B}$, $B_1 \cap B_2$ contains an element of $\mathscr{B}$.

2) $\mathscr{B}$ is not empty, and $\phi \notin \mathscr{B}$.

Then the set of subsets of I which contain an element of $\mathscr{B}$ is a filter, called *filter with base* $\mathscr{B}$.

In a topological space E, one says that a filter base $\mathscr{B}$ converges to a point $x_0 \in E$ if the filter with base $\mathscr{B}$ converges to x_0.

Let I be a set, with a filter $\mathscr{F}$; let E be a topological space, and f an application from I into E . One says that a point $y \in E$ is the limit of f for the filter $\mathscr{F}$ if y is the limit of the filter base $f(\mathscr{F})$. One writes $y = \lim_{\mathscr{F}} f$. This means also that, for any neighbourhood V of y , there is an $X \in \mathscr{F}$ such that $f(X) \subset V$.

For example, a family $(x_i)_{i \in I}$ of points of E converges to y for the filter $\mathscr{F}$ if y is the limit of the application $i \longrightarrow x_i$ for the filter $\mathscr{F}$. This means that for any neighbourhood V of y , there is an $X \in \mathscr{F}$, such that $\{x_i \ ; \ i \in X\} \subset V$. We then write $\lim_{\mathscr{F}} x_i = y$, or simply $x_i \underset{\mathscr{F}}{\longrightarrow} y$.

In particular, y is the limit of a sequence $(x_n)_{n \in \mathbb{N}}$ if y is the limit of the application $n \longrightarrow x_n$ for the Fréchet filter on $\mathbb{N}$. One checks that this definition coïncides with the usual one.

Let E be a topological space, and F a dense subset in E . For every $y \in E$, let $\mathscr{F}_1$ be the trace of $\mathscr{F}V_y$ on F (that is, the set of all $X \cap F$, $X \in \mathscr{F}V_y$). Then $\mathscr{F}_1$ is a filter on F , and a filter base on E . Let I be any set such that there is a surjection from I onto F ; call $f : i \longrightarrow x_i$ this surjection. Then $f^{-1}(\mathscr{F}_1)$ is a filter base $\mathscr{B}$ on I ; we call $\mathscr{F}$ the filter generated by $\mathscr{B}$. Then, the points $(x_i)_{i \in I}$ are also points of E , and $x_i \underset{\mathscr{F}}{\longrightarrow} y$.

Another way of obtaining the same conclusion is the following : let I be the set of all neighbourhoods of y . For every fixed $V \in I$, consider the set B_V of the neighbourhoods of y which are contained in V . Then $\mathscr{B} = \{B_V \ , \ V \in I\}$ is a filter base on I. Let $\mathscr{F}$ be the corresponding filter. For each $V \in I$, take $x_V \in F \cap V$. Then $x_V \underset{\mathscr{F}}{\longrightarrow} y$.

PART 1

FUNCTIONAL ANALYSIS

CHAPTER I

BAIRE'S PROPERTY AND ITS CONSEQUENCES

DEFINITION. - *We say that a topological space* E *has* Baire's Property *if every countable intersection of dense open sets in* E *is a dense set in* E.

The intersection needs not be open, since an infinite intersection of open sets is not open in general. For example, let $(q_n)_{n\geqslant 1}$ be an enumeration of the rational numbers in $\mathbb{R}$ (one knows that $\mathbb{Q}$ is countable), and let $O_n = \mathbb{R} - \{q_n\}$. Then O_n is open, for each $n \geqslant 1$, and $\bigcap_{n\geqslant 1} O_n = \mathbb{R} \setminus \mathbb{Q}$ is dense, but not open (since $\mathbb{Q}$ itself is also dense in $\mathbb{R}$).

In the definition, the word "countable" is essential, and there is no hope that any space could have the same property, for intersections indexed by $\mathbb{R}$: if $x \in \mathbb{R}$, put $O_x = \mathbb{R} \setminus \{x\}$, then O_x is open and dense in $\mathbb{R}$, but $\bigcap_{x\in\mathbb{R}} O_x$ is empty.

The definition is also equivalent to the following (just taking complements) : any countable union of closed sets with empty interior has empty interior.

The following theorem will provide two large classes of topological spaces having Baire's Property. Such a topological space will be called a *Baire space*.

THEOREM 1. - *Every complete metric space and every locally compact space are Baire spaces.*

PROOF. - We give it only for complete metric spaces.

Let $U_1,\dots,U_n,\dots$ be dense open sets in E, and $U = \bigcap_{n\geqslant 1} U_n$. We want to show that U is dense. For this, let W be any open set in E; we have to prove that $U \cap W$ is not empty.

We call d the distance on E, and $\mathscr{B}_0(x,r) = \{y \in E, d(x,y) < r\}$ the open ball of center x and radius r. $\overline{\mathscr{B}}_0(x,r)$ will be its closure ($\overline{\mathscr{B}}_0(x,r)$ is contained in $\mathscr{B}(x,r) = \{y \in E, d(x,y) \leqslant r\}$, but the inclusion can be strict).

Since U_1 is dense, $U_1 \cap W$ is open and non-empty, and so there are $x_1 \in E$ and $r_1 > 0$, $r_1 < 1$ such that $\overline{\mathscr{B}}_0(x_1, r_1) \subset U_1 \cap W$. Assume we have built two sequences $x_1, \dots, x_n$ in E, $r_1, \dots, r_n > 0$, with $0 < r_k < \frac{1}{k}$ $k = 1, \dots, n$, such that $\overline{\mathscr{B}}_0(x_k, r_k)$ is contained in $U_k \cap \overline{\mathscr{B}}_0(x_{k-1}, r_{k-1})$, for $k = 2, \dots, n$. Then $U_{n+1} \cap \mathscr{B}_0(x_n, r_n)$ is open and non-empty, since U_n is dense, and therefore contains a ball $\overline{\mathscr{B}}_0(x_{n+1}, r_{n+1})$, where r_{n+1} can be chosen smaller that $\frac{1}{n+1}$.

The sequence $(x_n)_{n \geqslant 1}$ thus built is a Cauchy sequence : for all $n \geqslant 1$, if $i,j > n$, $d(x_i, x_j) \leqslant \frac{2}{n}$, since x_i and x_j belong to $\overline{\mathscr{B}}_0(x_n, r_n)$, which is contained in $\overline{\mathscr{B}}_0(x_n, \frac{1}{n})$. Since E is complete, the sequence $(x_n)_{n \geqslant 1}$ converges to a point x. Since $\{(x_k)_{k \geqslant n}\}$ is contained in $\overline{\mathscr{B}}_0(x_n, r_n)$, x belongs to $\overline{\mathscr{B}}_0(x_n, r_n)$, for all n, and therefore belongs to all U_n's and to W. This proves the theorem, for complete metric spaces. The proof for locally compact spaces deals with the same type of arguments, but, instead of balls, one uses compact sets, and the fact that the intersection of a decreasing family of non-empty compact sets is not-empty. The details are left to the reader.

Theorem 1 has many applications in Functional Analysis and will be used directly several times throughout this book (see for example Second Part, Chapter VI). But is greatest achievements are realized by the intermediate of two very important theorems, Banach-Steinhaus Theorem and the Closed Graph Theorem, to which we shall now turn. But before, we shall recall a few definitions, concerning vector spaces :

- a *semi-norm*, on a vector space E, is a mapping from E into $\mathbb{R}^+$, denoted by p, satisfying the following two conditions :

 ★ $p(\lambda x) = |\lambda| p(x)$, for all $x \in E$, and for all $\lambda \in \mathbb{K}$ ($\mathbb{K} = \mathbb{R}$ or $\mathbb{C}$, is the field of scalars of the vector space E).

 ★ $p(x + y) \leqslant p(x) + p(y)$ for all $x,y \in E$.

- a *norm*, denoted by $\|.\|$, satisfies moreover :

★ $\|x\| = 0$ implies $x = 0$.

- a *Banach space* is a complete normed space.

- If E and F are normed linear spaces, and u is a continuous linear mapping from E into F , we set :

$$\|u\| = \sup_{\|x\|_E \leqslant 1} \|u(x)\|_F ,$$

and this formula defines a norm on the vector space $\mathcal{L}(E, F)$ of continuous linear mappings from E into F , since we know that u is continuous if and only if there is a constant $C > 0$ such that, for every $x \in E$, $\|u(x)\|_F \leqslant C\|x\|_E$. One checks easily that $\mathcal{L}(E, F)$ is complete as soon as F is complete (the completeness of E is not needed).

THEOREM 2. (Banach-Steinhaus) . - *Let* E *be a Banach space,* F *a normed space, and* $(u_i)_{i \in I}$ *a family of continuous linear mappings from* E *into* F *(the set* I *may have any cardinality). Then, two cases can occur :*

a) *either there exists in* E *a set* X *, which is a countable intersection of open sets (we say that* X *is a* G_δ *), dense in* E *, such that*

$$\sup_{i \in I} \|u_i(x)\|_F = + \infty , \quad \textit{for all } x \in X .$$

b) *or there exists a positive number* M *such that*

$$\|u_i\| \leqslant M , \quad \textit{for all } i \in I ,$$

and, of course, both terms of the alternative exclude each other.

PROOF. - We define, for $x \in X$, $q(x) = \sup_{i \in I} \|u_i(x)\|$. The values of q are in $\overline{\mathbb{R}}^+$. We put $U_n = \{x \in E, q(x) > n\}$.

For each $i \in I$, the function $x \longrightarrow \|u_i(x)\|_F$ is continuous on E , and therefore q is lower semi-continuous : this implies that each U_n is open (this fact can also be checked directly).

- If one of them, say U_N , is not dense in E , its complement has non-empty interior. Therefore there are $x_0 \in E$ and $r > 0$ such that, for all x in E with $\|x\| < r$, $x_0 + x \notin U_N$, that is $q(x_0 + x) \leqslant N$.

Consequently, for all $i \in I$, $\|u_i(x_0 + x)\| \leqslant N$, and we obtain, for all x with $\|x\| < r$,

$$\|u_i(x)\| \leq \|u_i(x+x_o)\| + \|u_i(x_o)\| \leq 2N .$$

If now $\|x\| \leq 1$, then $\|u_i(x)\| \leq \frac{2N}{r}$, from which we deduce b), with $M = \frac{2N}{r}$.

- If all the U_n's are dense in E , so is their intersection, by Theorem 1. This intersection is the set $\{x \in E, q(x) = +\infty\}$, and we obtain a).

COROLLARY 3. - *Let* E *be a Banach space,* F *be a normed space, and* $(u_n)_{n\in \mathbb{N}}$ *linear continuous mappings from* E *into* F *. We assume that, for all* $x \in E$ *, the sequence* $(u_n(x))_{n\in\mathbb{N}}$ *converges, when* $n \to +\infty$ *, to a limit which we call* $u(x)$ *. Then* u *is a continuous linear mapping,* $\sup_{n\in\mathbb{N}} \|u_n\| < +\infty$ *, and* $\|u\| \leq \sup_{n\in\mathbb{N}} \|u_n\|$.

PROOF. - The limit mapping u is obviously linear. If, for all $x \in E$, $u_n(x) \xrightarrow[n\to+\infty]{} u(x)$, then $\|u_n(x)\| \xrightarrow[n\to+\infty]{} \|u(x)\|$, and therefore, for all $x \in E$, $\sup_{n\in\mathbb{N}} \|u_n(x)\| < +\infty$.

By Theorem 2, there is a number M such that $\|u_n\| \leq M$ for all $n \in \mathbb{N}$. For every x in the unit ball of E , we have $\|u_n(x)\| \leq M$, and thus $\|u(x)\| \leq M$, which proves that u is continuous, and that $\|u\| \leq M$.

As a first application of Theorem 1, we shall give a result concerning algebraic bases in Banach spaces.

We shall call algebraic basis in a Banach space E a set of vectors $(e_i)_{i\in I}$ such that any element x in E has a unique decomposition $x = \Sigma\, \alpha_i e_i$ as a *finite* linear combination of the e_i's .

The existence of algebraic bases can be established very easily using Zorn's Axiom. We shall come back later (Second Part, Chapter II) on the notion of basis in a Banach space.

COROLLARY 4. - *In a Banach space the cardinality (that is, the number of the elements) of an algebraic basis is either finite or uncountable.*

PROOF. - Let $(e_i)_{i\in I}$ be an algebraic basis, and let us assume, on the contrary, that I is infinite and countable. Then we can assume $I = \mathbb{N}$. We call F_n the vector space spanned by $(e_o,\ldots,e_n)$. Since it is finite-dimensional, each F_n is closed, and, since $(e_i)_{i \in \mathbb{N}}$ is an algebraic basis, $E = \bigcup_{n \geqslant 0} F_n$. Since E is a Baire Space, one of the F_n's, say F_N, as non-empty interior. But F_N is a vector space, and if it contains an open ball, it contains an open ball centered at 0 : F_N is a neighbourhood of 0. Therefore $E = F_N$, and E is finite dimensional, which contradicts our assumption.

We shall now proceed to the second main application of Theorem 1 : the Closed Graph Theorem and its consequences.

Let us recall that the graph of a mapping u, from E into F, is the set (contained in $E \times F$) of all couples $(x,u(x))$, $x \in E$.

THEOREM 5. (Closed Graph Theorem). - *Let* E, F *be complete metric linear spaces. Let* u *be a linear mapping from* E *into* F. *Then* u *is continuous if and only if the graph of* u *is closed in* $E \times F$, *endowed with the product topology.*

REMARKS.

1) If the hypotheses on E and F are fulfilled, to show that u is continuous, it is enough to establish that :

If $(x_n)_{n \in \mathbb{N}}$ *is a sequence of points in* E, *converging to* 0, *and if* $(u(x_n))_{n \in \mathbb{N}}$ *converges, in* F, *to a point* y, *then* $y = 0$.

2) We know that if F is a Hausdorff space (which is the case if it is metric) the graph of any continuous mapping is closed. We only have to prove the converse implication.

PROOF. - We assume that the graph of u (denoted by $\mathrm{Gr}\, u$) is closed. We shall first establish a Lemma. We call $\mathscr{B}_E(\epsilon)$, $\mathscr{B}_F(\epsilon)$ the closed balls, in E and F, centered at 0, and with radius ϵ :

$$\mathscr{B}_E(\epsilon) = \{x \in E ,\ d(x,0) \leqslant \epsilon\} .$$

LEMMA 6. - *For every* $\epsilon > 0$, *there is a number* $\delta > 0$ *such that*

$$\mathscr{B}_E(\delta) \subset \overline{u^{-1}(\mathscr{B}_F(\epsilon))} .$$

PROOF of LEMMA 6. - We fix $\epsilon > 0$, and consider the sets $G_n = nu^{-1}(\mathscr{B}_F(\epsilon))$. The reunion $\bigcup_{n \geqslant 1} G_n$ covers E, therefore also

$\bigcup_{n\geq 1} \overline{G_n}$. Since E is a Baire space, one of the $\overline{G_n}$'s , say $\overline{G_N}$, must have a non-empty interior : it must contain some open ball of radius $\eta > 0$, and center x_0 , that is :

$$\text{if } d(x,x_0) < \eta \text{ , then } u(\frac{x}{n}) \in \mathscr{B}_F(\epsilon) \text{ , if } n \geq N .$$

Now, take y with $d(y,0) < \eta$; then

$$u(\frac{y}{n}) = u(\frac{y + x_0}{n}) - u(\frac{x_0}{n}) ,$$

and there is a N_1 such that if $n \geq N_1$, both $u(\frac{y + x_0}{n})$ and $u(\frac{x_0}{n})$ belong to $\mathscr{B}_F(\frac{\epsilon}{2})$, and $u(\frac{y}{n}) \in \mathscr{B}_F(\epsilon)$. For somme $\delta > 0$, small enough, we have

$$\mathscr{B}_E(\delta) \subset \frac{1}{N_1} \mathscr{B}_E(\eta) \subset \overline{u^{-1}(\mathscr{B}_F(\epsilon))} ,$$

which proves the Lemma. This Lemma uses only the linearity of u , and not the fact that its graph is closed.

Let us observe that we can assume that $\delta(\epsilon) < \epsilon$, for every $\epsilon > 0$. We come back to the proof of Theorem 5. We shall show that, for every $\epsilon > 0$, we have $u(\mathscr{B}_E(\delta(\frac{\epsilon}{2}))) \subset \mathscr{B}_F(\epsilon)$: this will prove that u is continuous at 0 , and therefore everywhere.

So, we fix $\epsilon > 0$, and $x \in \mathscr{B}_E(\delta(\frac{\epsilon}{2}))$. We choose $x_1 \in u^{-1}(\mathscr{B}_F(\frac{\epsilon}{2}))$, with $x - x_1 \in \mathscr{B}_E(\delta(\frac{\epsilon}{4}))$, and then $x_2 \in u^{-1}(\mathscr{B}_F(\frac{\epsilon}{4}))$ with $x - x_1 - x_2 \in \mathscr{B}_E(\delta(\frac{\epsilon}{8}))$, and so on : we build by induction a sequence $(x_i)_{i\geq 1}$, with $x_i \in u^{-1}(\mathscr{B}_F(\epsilon/2^i))$ and $x - \sum_{j=1}^{i} x_j \in \mathscr{B}_E(\delta(\epsilon/2^{i+1}))$, for $i = 1,2,\ldots$. Then $u(x_i) \in \mathscr{B}_F(\epsilon/2^i)$, and therefore the sequence $z_n = \sum_1^n u(x_i)$ is a Cauchy sequence, which converges to a point z , and, since $z_n \in \mathscr{B}_F(\epsilon)$ for all n , we have $z \in \mathscr{B}_F(\epsilon)$.

Also, $x - \sum_1^n x_i \in \mathscr{B}_E(\delta(\epsilon/2^{n+1})) \subset \mathscr{B}_E(\epsilon/2^{n+1})$, and the sequence $X_n = x_1 + \ldots + x_n$ converges to x . But since $u(X_n) = z_n$, and since Gr u is closed, we have $u(x) = z$. Therefore, $x \in u^{-1}(\mathscr{B}_F(\epsilon))$, which proves that $\mathscr{B}_E(\delta(\epsilon/2)) \subset u^{-1}(\mathscr{B}_F(\epsilon))$, and ends the proof of the theorem.

REMARK. - We do not really need that E is complete, but only that E is not covered by a countable union of closed subsets with empty interior : this is precisely the case for any Baire Space. This leads to the following definition :

DEFINITION. - *If* E *is a topological space, a subset* A *will be called* meager *in* A , *or* of the first category, *if it is contained in a countable union of closed subsets with empty interior. In the converse case, it will be called* of second category.

COROLLARY 7. - *Let* E *be a complete metric linear space,* F *a linear space of second category in itself. Let* u *be a continuous linear bijection from* E *onto* F . *Then* u^{-1} *is continuous (and* u *is an isomorphism).*

PROOF. - If u is continuous, then $\mathrm{Gr}\,u$ is closed in $E \times F$. But $\mathrm{Gr}\,u^{-1}$ is obtained from $\mathrm{Gr}\,u$ by "symmetry", that is by exchanging variables : $(x,y) \longrightarrow (y,x)$, and therefore it is also closed. So u^{-1} is continuous by the previous theorem.

DEFINITION. - *We say that a mapping is* open *if the image of any open set by this application is an open set.*

Of course, if the mapping is injective, it is open if and only if the inverse mapping is continuous.

In the sequel, the word "*operator*" will mean "continuous linear mapping".

THEOREM 8. (Open-mapping Theorem). - *Let* E *be a complete metric linear space,* F *a Hausdorff topological linear space,* u *an operator from* E *into* F . *Then :*

- *Either* $u(E)$ *can be covered by a countable union of closed sets, in* F , *with empty interior (that is ,* $u(E)$ *is meager in* F) ,

- *Or* $u(E) = F$, u *is open, and* F *is a complete metric linear space.*

PROOF. - We put $V = u(E)$: this is a vector subspace of F . Assume that it is not meager in F : then, certainly $\overline{V}$ has non empty interior. Since $\overline{V}$ is a subspace, 0 is in its interior. This implies that $\overline{V} = F$, and V is dense in F .

Still assuming that V is not meager in F , we shall see that V is not meager in itself : if $V = \bigcup_{n\geqslant 1} F_n$, where F_n 's are closed in V , then each F_n can be written $F_n = V \cap G_n$, where G_n is closed in F , and $V \subset \bigcap_{n\geqslant 1} G_n$, and so one of the G_n 's , say G_N , has non-empty interior $\overset{\circ}{G}_N$ in F. Since V is dense in F, $V \cap \overset{\circ}{G}_N$ is a non-empty open set in V, contained in F_N : this proves that F_N has non-empty interior in V, and that V is not meager in itself.

Now, let E_1 be the quotient of E by $\operatorname{Ker} u$, θ the canonical mapping from E onto E_1, $\tilde{u}$ the quotient map from E_1 into F. One knows that θ is an open mapping, that $\tilde{u}$ is injective and continuous, and that, endowed with the quotient-distance, E_1 is a complete metric space (recall that, if $\dot{x}$ is the class of x, $d(\dot{x},\dot{y}) = \inf\{d(x,y)\ ;\ x \in \dot{x}\ ,\ y \in \dot{y}\}$).

Since V is not meager in itself, and since E_1 is a complete metric space, we can apply Corollary 7, and $\tilde{u}$ is an isomorphism. Therefore V can be equipped with the metric brought by $\tilde{u}$ (that is, $d(\tilde{u}(x),\tilde{u}(y)) = d(x,y)$) and is complete for this metric. Since it was dense in F, we have $V = F$, u is surjective and F is a complete metric space. Moreover, u is open since $u = \tilde{u} \circ \theta$, and both are open. This proves the theorem.

REMARK. - If, for example, E and F are Banach spaces, the previous theorem says that, if u is not surjective, $u(E)$ is meager in F. But still, $u(E)$ can be dense in F. This is the case, for example, of the canonical injection from $L_p([0,1], dt)$ into $L_q([0,1], dt)$, if $p > q$. A set can very well be meager and still be dense in the whole space : for example, $\mathbb{Q}$ in $\mathbb{R}$.

To illustrate the previous theorem, we shall establish the following proposition :

PROPOSITION 9. - *Let* p, q, *with* $1 \leqslant p,q \leqslant +\infty$, *and let* E *be a vector subspace of* $L_p(\mathbb{R},dt) \cap L_q(\mathbb{R},dt)$. *If* E *is closed in* $L_p(\mathbb{R},dt)$ *and in* $L_q(\mathbb{R},dt)$, *the topologies induced on* E *by these two spaces coïncide.*

PROOF. - We call E_p the space E endowed with the norm L_p, and E_q with the norm L_q. We call I the identity mapping from E_p into E_q. We shall see that I is continuous, and, for this, prove that its graph is closed.

Let $(f_n)_{n \geqslant 1}$ be a sequence of functions in E ; we assume that (f_n, f_n) converges in $L_p \times L_q$ to an element (f,g). Therefore $f_n \xrightarrow[n \to +\infty]{} f$ in L_p, $f_n \xrightarrow[n \to +\infty]{} g$ in L_q. So we can extract from the sequence $(f_n)_{n \geq 1}$ a subsequence $(f'_n)_{n \geq 1}$ which converges to f almost everywhere ; this sequence converges to g in L_q . Therefore $f = g$, and the graph of I is closed. Since I is continuous and surjective, the converse mapping is also continuous (Corollary 7) and I is an isomorphism : therefore, both topologies coïncide on E , and the proposition is proved.

We shall see in the Second Part, Chapter VI, some examples of subspaces E which satisfy the assumptions of this Proposition.

EXERCISES ON CHAPTER I.

EXERCISE 1. (Condensation of singularities). - Let E be a complete metric vector space, F a metric vector space, and $(u_{p,q})_{\substack{p \in \mathbb{N} \\ q \in \mathbb{N}}}$ a double sequence of linear mappings from E into F . We assume that, for every $p \in \mathbb{N}$, there is a point $x_p \in E$ such that the limit $\lim_{q \to +\infty} u_{p,q}(x_p)$ does not exist. Show that the set of points $x \in E$ such that $\lim_{q \to +\infty} u_{p,q}(x)$ exists for no $p \in \mathbb{N}$ is of second category in E .

EXERCISE 2. - Let E and F be Banach spaces, u a surjective operator from E onto F . Show that, for every sequence of points $(y_n)_{n \geq 1}$ in F converging to a point $y_0 = u(x_0)$, there is a sequence of points $(x_n)_{n \geq 1}$ in E , converging to x_0 , with $u(x_n) = y_n$ for all $n \geq 1$.

EXERCISE 3. (lifting property of ℓ_1). - Let E and F be Banach spaces, and A a surjective operator from E onto F . For every operator T from ℓ_1 into F , there is an operator $\tilde{T}$ from ℓ_1 into E , such that $A\tilde{T} = T$.

[Show first that, if $(e_n)_{n \in \mathbb{N}}$ is the canonical basis of ℓ_1 , there are points $(x_n)_{n \in \mathbb{N}}$ in E , such that $\sup_n \|x_n\| < +\infty$, and $Ax_n = Te_n$ for all $n \in \mathbb{N}$. For this, use the fact that A is an open mapping] .

EXERCISE 4. - Let E and F be Banach spaces, and T an operator from E into F , with closed range. Show that there exists a $K > 0$ such that for all $y \in \operatorname{Im} T$, one can find $x \in E$, with $Tx = y$ and $\|x\| \leq K\|y\|$.

REFERENCES ON CHAPTER I.

Most results which are presented in this first chapter are due to S. BANACH himself, or, at least, appear in his book [5]. The proof of the Closed Graph Theorem which we give here is very close to the one given by W. RUDIN [43].

CHAPTER II

INFINITE-DIMENSIONAL NORMED SPACES

INTRODUCTION.

In this chapter, we shall develop the most important results concerning normed linear spaces. In fact, most of them make sense also in the more general setting of locally convex topological vector spaces, and we could have made a more general - and more abstract - study. We have chosen to restrict ourselves, both because we have in mind the study of Banach spaces (endowed with their norm, or with a weak topology, see below), and because this will allow us, in many cases, to present simpler and more concrete proofs. There is one exception, however : the Hahn-Banach Theorem, in its two forms (analytic and geometric), is given in its largest setting ; to do so does not require more work, and will be useful. Except for this result, we have avoided as much as we could to develop a general theory of locally convex spaces ; the reader interested by such a theory is referred to Bourbaki "Espaces Vectoriels Topologiques", Chapitres I à IV, [11].

A vector space E is *infinite-dimensional* if one can find in this space a sequence of points $(e_n)_{n \in \mathbb{N}}$, such that any finite subset of $\{(e_n)_{n \in \mathbb{N}}\}$ is made of linearly independent vectors. If the space is equipped with a norm, we may of course assume that, for all $n \in \mathbb{N}$, $\|e_n\| = 1$.

In an infinite-dimensional normed space, one can find a sequence of norm-one points which are far from each other :

PROPOSITION 1. - *Let* E *be an infinite-dimensional normed space. One can find in* E *a sequence of points* $(x_n)_{n \geq 1}$ *such that, for all* $n \geq 1$,

$$\|x_n\| = 1 ,$$

and

$$\text{dist}(x_{n+1} , \text{span}\{x_1,\ldots,x_n\}) \geq 1 .$$

This last condition implies of course $\|x_n - x_m\| \geq 1$ *for all* n,m, $n \neq m$.

PROOF. - Assume $x_1,\dots,x_n$ have been constructed. Put $F_n = \text{span}\{x_1,\dots,x_n\}$, and let $a \in E$, $a \notin F_n$. Put $G_n = \text{span}\{a,x_1,\dots,x_n\}$. Then F_n is an hyperplane of G_n, and so there is a linear functional f_n, defined on G_n (which is a finite-dimensional space), such that $f_n = 0$ on F_n. One may of course assume $\|f_n\| = 1$. Let now x_{n+1} be such that $\|x_{n+1}\| = 1$ and $f_n(x_{n+1}) = 1$ (such a point exists, since the unit ball of G_n is compact). Then, for every sequence $a_1,\dots,a_n$ of scalars:

$$\|x_{n+1} - \sum_1^n a_i x_i\| \geqslant f_n(x_{n+1} - \sum_1^n a_i x_i) = 1 ,$$

which means that

$$\text{dist}(x_{n+1} , \text{span}(x_1,\dots,x_n)) \geqslant 1 ,$$

and the Proposition is proved.

A linear functional on an infinite-dimensional space E is a linear application from E into the field of scalars of E. For us, the field of scalars, as we already mentioned, is denoted by $\mathbb{K}$, and is $\mathbb{R}$, or $\mathbb{C}$.

A topological vector space is a vector space endowed with a topology, for which the addition of vectors, and the multiplication of a vector by a scalar are continuous operations. If this topology is given by a metric, we have a metric vector space. A special case is given by the normed spaces. The metric vector spaces were already considered in the first chapter. All our topological spaces will always be Hausdorff.

If E is a topological vector space, we call $E^\star$ the dual of E : $E^\star$ is the set of continuous linear functionals on E (if E is finite-dimensional, any linear functional is automatically continuous ; this is not so if E is infinite-dimensional). If E is a normed space, we define a norm on $E^\star$ by the formula :

$$(1) \qquad \|\xi\|_{E^\star} = \sup_{\|x\|_E \leqslant 1} |\xi(x)| ,$$

and we know that $E^\star$ is always complete, even if E is not (this was recalled in Chapter I, p. 9 , for $\mathscr{L}(E,F)$, where F is a Banach space : here, $F = \mathbb{K}$).

The fact that there exist, on an infinite-dimensional normed space, non-trivial continuous linear functionals (that is, different from the zero linear functional) is not obvious : it is a consequence of the following theorem, which plays an essential role in Analysis.

§ 1. HAHN-BANACH THEOREM : ANALYTIC FORM.

THEOREM 1. (Hahn-Banach). - *Let* E *be a real vector space, and* p *a sub-additive and positively homogeneous function on* E *, that is :*

$$p(x+y) \leqslant p(x) + p(y) \qquad \text{for all} \quad x,y \in E ,$$

$$p(\lambda x) = \lambda p(x) \qquad \text{for all} \quad \lambda > 0 , \text{ all } \quad x \in E .$$

Let E *be a vector subspace of* E *,* f *a linear functional, defined on* F *and dominated by* p *:*

$$f(x) \leqslant p(x) \qquad \text{for all} \quad x \in F .$$

Then there is an extension $\widetilde{f}$ *of* f *to* E *, still dominated by* p *:*

$$\widetilde{f}(x) = f(x) \qquad \text{for all} \quad x \in F ,$$

$$\widetilde{f}(y) \leqslant p(y) \qquad \text{for all} \quad y \in E .$$

PROOF. - We first show that we can extend f to a vector subspace of E which contains F and has "one dimension more" than F .

Let $e \in E$, $e \notin F$. Let G be the vector subspace of E spanned by F and e . Then every point $y \in G$ has a unique decomposition $y = x + \lambda e$, where $x \in F$, $\lambda \in \mathbb{R}$. We define $\widetilde{f}(y) = f(x) + \lambda c$, for every $y \in G$, and we shall choose $c = \widetilde{f}(e)$ in order that $\widetilde{f} \leqslant p$ on G , that is

$$f(x) + \lambda c \leqslant p(x + \lambda e) , \qquad \text{for all} \quad x \in F , \quad \text{all} \quad \lambda \in \mathbb{R} .$$

This inequality, by assumption, is valid for $\lambda = 0$. For $\lambda > 0$, it can be written :

$$f\left(\frac{x}{\lambda}\right) + c \leqslant p\left(\frac{x}{\lambda} + e\right) ,$$

and therefore, we must have $c \leqslant p\left(\frac{x}{\lambda} + e\right) - f\left(\frac{x}{\lambda}\right)$, that is :

$$c \leqslant \inf\{p(x_1 + e) - f(x_1) \ ; \ x_1 \in F\} .$$

For $\lambda < 0$, we obtain, dividing by $-\lambda$:

$$f(x_2) - c \leqslant p(x_2 - e) , \quad \text{with} \quad x_2 = \frac{-x}{\lambda} ,$$

and therefore :

$$c \geqslant \sup\{f(x_2) - p(x_2 - e) \ ; \ x_2 \in F\} .$$

The choice of c will be possible if and only if :

$$\sup\{f(x_2) - p(x_2 - e) \ ; \ x_2 \in F\} \leqslant \inf\{p(x_1 + e) - f(x_1) \ ; \ x_1 \in F\} ,$$

that is, if and only if, for every $x_1, x_2 \in F$:

$$f(x_2) - p(x_2 - e) \leqslant p(x_1 + e) - f(x_1) ,$$

or $f(x_1) + f(x_2) \leqslant p(x_1 + e) + p(x_2 - e)$.

But we know that :

$$f(x_1) + f(x_2) = f(x_1 + x_2) \leqslant p(x_1 + x_2) = p[(x_1 + e) + (x_2 - e)]$$
$$\leqslant p(x_1 + e) + p(x_2 - e) ,$$

and therefore the choice of c is possible, and we can extend the functional f to G . Note the choice of c is not unique in general, and so the extension is not unique, either.

We shall now use Zorn's Axiom. Let us first recall the statement.

ZORN'S AXIOM. - *Let* A *be an ordered set, in which every totally ordered subset has a majorant (we say that* A *is inductive). Then any element of* A *is majorized by a maximal element (an element* a *is maximal if* $a \leqslant b \Rightarrow a = b$).

Let us consider the set of couples (H, f_H) , where H is a vector subspace of E , containing F , and f_H is a linear functional defined on H , extending f , and dominated by p on H . We order the set of these couples by the relation :

$$(H_1 , f_{H_1}) \leqslant (H_2 , f_{H_2})$$

if $H_2 \supset H_1$ and f_{H_2} is an extension of f_{H_1} .

Under this ordering, this set is inductive : let $(H_i , f_{H_i})_{i \in I}$ be a totally ordered subset. The element $(\mathscr{H}, f_{\mathscr{H}})$ defined by :

$$\begin{cases} \mathscr{H} = \bigcup_i H_i \\ f_{\mathscr{H}}(x) = f_{H_i}(x) \quad \text{if} \quad x \in H_i \end{cases}$$

is the majorant we were looking for.

Consequently, the element (F,f) is majorized by a maximal element, called (L , f_L). If L was not the whole space E, we could, by the first part of the proof, extend f_L to a subspace strictly larger than L, and L would not be maximal. Therefore $L = E$, and $f_L = \tilde{f}$ is the extension of f to the whole space E. This proves our theorem. It should be observed that no topology was introduced, and, in particular, F is not assumed to be closed or f to be continuous.

The previous theorem makes sense only for real vector spaces (since we assume $f \leqslant p$). We shall now pass to a corollary which is valid for complex vector spaces as well.

THEOREM 2. - *Let* E *be a vector space on* $\mathbb{R}$ *or* $\mathbb{C}$, p *a semi-norm on* E , F *a vector subspace of* E *and* f *a linear functional on* F , *dominated by* p , *in modulus* :

$$|f(x)| \leqslant p(x) \qquad \textit{for all} \quad x \in F .$$

Then there exists an extension $\tilde{f}$ *of* f *to* E , *still dominated by* p :

$$\tilde{f}(x) = f(x) \qquad \textit{for all} \quad x \in F ,$$

$$|\tilde{f}(y)| \leqslant p(y) \qquad \textit{for all} \quad y \in E .$$

PROOF. - Assume first that E is a real vector space. By Theorem 1, there is $\tilde{f}$, extending f , with $\tilde{f}(y) \leqslant p(y)$ for all $y \in E$. This implies (changing y into $-y$) : $\tilde{f}(-y) \leqslant p(-y)$, that is $-\tilde{f}(y) \leqslant p(y)$, and $|\tilde{f}(y)| \leqslant p(y)$.

Now, if E is a complex vector space, it is, a fortiori, a real vector space. So, if we put $g(x) = \mathcal{R}e\, f(x)$, we obtain a real functional, which we can extend into $\tilde{g}$, with $|\tilde{g}(y)| \leqslant p(y)$ for all $y \in E$, by the previous argument. But we have

$$Im(x) = -g(ix), \qquad \text{for all} \quad x \in F ,$$

and so, if we set

$$\tilde{f}(y) = \tilde{g}(y) - i\tilde{g}(iy) ,$$

we obtain a complex linear functional $\tilde{f}$ which extends f . We still have to show that p dominates $|\tilde{f}|$. For this, if $y \in E$, put $\theta = \text{Arg}\,\tilde{f}(y)$. We have :

$$|\tilde{f}(y)| = e^{-i\theta}\tilde{f}(y) = \mathcal{R}e\,\tilde{f}(e^{-i\theta}y) = \tilde{g}(e^{-i\theta}y) \leqslant p(e^{-i\theta}y) = p(y) ,$$

and this proves the theorem.

From this theorem results that the dual of a normed space contains non-zero elements. More precisely :

COROLLARY 3. - *Let* E *be a normed space,* $x_0 \in E$. *There is a continuous linear functional* f , *with* $\|f\| = 1$, *such that* $f(x_0) = \|x_0\|$.

PROOF. - If $x_0 = 0$, there is nothing to prove. Otherwise, we define f on the subspace F_{x_0} spanned by x_0 : $F_{x_0} = \{\lambda x_0 , \lambda \in \mathbb{K}\}$, by

$$f(\lambda x_0) = \lambda \|x_0\| ,$$

and so, for all $x \in F$, we have $|f(x)| = \|x\|$. We extend f to the whole space E , using theorem 2, taking the norm $\|.\|$ as semi-norm p : the extension $\tilde{f}$ satisfies

$$|\tilde{f}(y)| \leqslant \|y\| \qquad \text{for all} \quad y \in E .$$

This last formula shows that $\|\tilde{f}\| \leqslant 1$, and, since $f(x_0) = \|x_0\|$, we have $\|\tilde{f}\| = 1$, and the corollary is proved.

The linear functional thus constructed will be called "linear functional norming the point x_0 ", and will be denoted by f_{x_0} . This linear functional needs not be unique in general (since extensions are not unique, as we already mentioned) ; we shall come back on this problem in the Third Part, Chapter I.

On the same way, we obtain :

COROLLARY 4. - *Let* E *be a normed space,* F *a vector subspace,* f *a continuous linear functional on* F . *There is an extension* $\tilde{f}$ *with* $\|\tilde{f}\| = \|f\|$.

REMARK. - The formula, symmetric to formula (1) :

$$(2) \qquad \|x\|_E = \sup_{\|\xi\|_{E^\star} \leqslant 1} |\xi(x)|$$

is also a consequence of Hahn-Banach Theorem :

We have : $|\xi(x)| \leqslant \|\xi\| \,.\, \|x\|$, and thus :

$$\sup_{\|\xi\|_{E^\star}} |\xi(x)| \leqslant \|x\| \,,$$

but, if we choose $\xi = f_x$, we obtain

$$\sup_{\|\xi\|_{E^\star}} |\xi(x)| \leqslant |f_x(x)| = \|x\| \,,$$

and the formula is proved.

We shall now investigate several topologies which can be introduced on a normed space or on its dual.

§ 2. WEAK TOPOLOGY ON THE DUAL $E^\star$.

On $E^\star$, we have already introduced a norm, by the formula (1) of the previous paragraph. But there is another topology which can be defined : this is the topology of point-wise convergence on the elements of E .

A fundamental system of neighbourhoods of an element $\xi_0 \in E^\star$ for this topology is given by the following sets :

$$(1) \quad V_{\epsilon;x_1,\dots,x_n}(\xi_0) = \{\xi \in E^\star ; |\xi(x_1) - \xi_0(x_1)| < \epsilon, \dots, |\xi(x_n) - \xi_0(x_n)| < \epsilon\},$$

for all $\epsilon > 0$, all $n \geqslant 1$, all $x_1,\dots,x_n \in E$.

We call $\sigma(E^\star,E)$ this topology, which is completely defined by the previous fundamental system of neighbourhoods. If $(\xi_n)_{n \in \mathbb{N}}$ is a *sequence* of elements of $E^\star$, it converges to $\xi_0 \in E^\star$ if and only if, for all $x \in E$, $\xi_n(x) \xrightarrow[n \to +\infty]{} \xi_0(x)$ (to see this, just take the neighbourhoods $V_{\epsilon,x}(\xi_0)$, using only one point x). This is the reason why this topology is called "point-wise convergence on the elements of E ". But it should be noted that this definition with sequences does not suffice to characterize the topology, since it is not metrizable if E is infinite-dimensional.

The topology $\sigma(E^\star, E)$ is weaker than the norm-topology : a neighbourhood of the form (1) contains the ball centered at ξ_o, with radius $\epsilon / \max_{i=1,\dots,n} \|x_i\|$ (for a sequence ξ_n, if $\xi_n \xrightarrow[n \to +\infty]{} \xi_o$ for the norm, then $\|\xi_n - \xi_o\| \xrightarrow[n \to +\infty]{} 0$, and thus, for all $x \in E$, $|\xi_n(x) - \xi_o(x)| \leqslant \|\xi_n - \xi\| . \|x\| \xrightarrow[n \to +\infty]{} 0$). The topology $\sigma(E^\star, E)$ is in general strictly weaker, as shows the example of the vectors of the canonical basis of ℓ_2, which are of norm 1, but tend to zero, for $\sigma(\ell_2, \ell_2)$. The topology $\sigma(E^\star, E)$ is often called the weak $\star$ topology on $E^\star$.

The main interest of this topology is that it will allow us to obtain compactness results. One knows that, if a normed space E is infinite dimensional, its unit ball (for us, "unit ball" will always mean "closed unit ball") is not compact for the norm topology. This fact is the Riesz's Theorem : the unit ball of a normed space is norm-compact if and only if the space is finite-dimensional. Up to some extend, the weak topology will allow us to avoid this drawback.

THEOREM 1. (Alaoglu). - *The unit ball of* $E^\star$ *is compact for* $\sigma(E^\star, E)$.

PROOF. - Let us observe first that the statement involves two things : the unit ball of $E^\star$, which is defined with respect to the norm of $E^\star$, by formula 1 (1), and the weak topology $\sigma(E^\star, E)$. The unit ball $\mathscr{B}_{E^\star}$ is not a neighbourhood of 0 of $\sigma(E^\star, E)$, so the Riesz's theorem (which is true , in fact, for any neighbourhood of 0, in a locally convex space) does not apply to it, and $\mathscr{B}_{E^\star}$ can be $\sigma(E^\star, E)$ compact when $E^\star$ is infinite-dimensional.

Let us first observe that we can identify $E^\star$ with a subset of the product $\mathbb{K}^E$, by associating to each $\xi \in E^\star$ the set of its values : $\{\xi(x), x \in E\}$. One checks immediately on the definition of neighbourhoods that the topology $\sigma(E^\star, E)$ identifies itself with the topology induced on $E^\star$ by the product topology on $\mathbb{K}^E$. Moreover, the unit ball of $E^\star$ becomes a subset of the product :

$$I = \prod_{x \in E} \{\lambda \in \mathbb{K} \; ; \; |\lambda| \leqslant \|x\|\}$$

which is a product of compacts, and therefore is compact by Tychonoff's theorem. All we have to show is therefore that $\mathcal{B}_{E^\star}$ is closed in I.

For each $x,y \in E$, each $\lambda \in \mathbb{K}$, let us consider the functions $\varphi_{x,y}(\xi) = \xi(x) + \xi(y) - \xi(x+y)$ and $\psi_{x,\lambda}(\xi) = \xi(\lambda x) - \lambda\xi(x)$.

These functions are continuous on $\mathbb{K}^E$, and the unit ball of $E^\star$ is the intersection of I with

$$\bigcap_{x,y\in E} \varphi_{x,y}^{-1}(\{0\}) \cap \bigcap_{\substack{x\in E\\ \lambda\in E}} \psi_{x,\lambda}^{-1}(\{0\}),$$

which is a closed subset of I : our theorem is proved.

§ 3. THE BIDUAL $E^{\star\star}$ AND THE WEAK TOPOLOGY ON E.

We have seen that $E^\star$, endowed with the norm defined by the formula 1, (1), was a Banach space. So we can, again, consider its dual : it is a Banach space which is called the bidual of E, and is denoted by $E^{\star\star}$.

There are, in $E^{\star\star}$, some special elements, which can be immediately identified. These are the elements of the form :

$$(1) \qquad x_0 \in E \longrightarrow \{\xi \in E^\star \longrightarrow \xi(x_0)\}$$

(that is : to each element of $E^\star$, we associate its value at a fixed point $x_0 \in E$). This application will be denoted by $< x_0 , \xi >$. By formula 1, (2), we have :

$$\sup_{\|\xi\|_{E^\star} \leqslant 1} |< x_0 , \xi >| = \|x_0\|,$$

which says that the norm of the element of $E^{\star\star}$ defined by x_0 (by formula 1) is equal to the norm of x_0, in E. Therefore, it is legitimate to identifiy x_0 with the element of $E^{\star\star}$ which it defines, that it : to identify E with a subspace of $E^{\star\star}$, and the norm induced on E by $E^{\star\star}$ is the norm of E.

If E is normed, but not complete, $E^{\star\star}$ is certainly larger than E, since it is complete. But even if E is a Banach space $E^{\star\star}$ is generally larger than E : we shall come back on this question in the next chapter.

On $E^{\star\star}$, just repeating the definitions of the previous paragraph, we introduce a weak topology, called $\sigma(E^{\star\star},E^\star)$: it is the topology of point-wise convergence on the elements of $E^\star$.

Calling z the elements of E^{**}, and taking the notation $< z , \xi >$ instead of $z(\xi)$ to indicate the action of $z \in E^{**}$ on $\xi \in E^{*}$, we can give the form of a fundamental system of neighbourhoods of 0 for $\sigma(E^{**}, E^{*})$: these are the sets of the form :

(2) $V_{\epsilon;\xi_1,\ldots,\xi_n} = \{z \in E^{**} ; |< z, \xi_1 >| < \epsilon, \ldots, |< z, \xi_n >| < \epsilon\}$,

for all $\epsilon > 0$, all $n \geqslant 1$, all $\xi_1,\ldots,\xi_n \in E^{*}$.

A *sequence* of elements $(z_n)_{n \in \mathbb{N}}$ converges to an element z_0 for $\sigma(E^{**}, E^{*})$ if and only if, for all $\xi \in E^{*}$, $< z_n , \xi > \xrightarrow[n \to +\infty]{} < z_0 , \xi >$. Here again, this definition using sequences does not suffice to characterize the topology.

Let us now look for the trace on E of the topology $\sigma(E^{**}, E^{*})$: this has a meaning, since E is a subspace of E^{**} . A fundamental system of neighbourhoods of the origin is, by definition, constituted by the intersection with E of the sets (2). Therefore, this fundamental system is given by the sets :

(3) $W_{\epsilon;\xi_1,\ldots,\xi_n} = \{x \in E ; |< x, \xi_1 >| < \epsilon, \ldots, |< x, \xi_n >| < \epsilon\}$

for all $\epsilon > 0$, all $n \geqslant 1$, all $\xi_1,\ldots,\xi_n \in E^{*}$.

We shall call $\sigma(E, E^{*})$ this topology : it is the weak topology on E . From the form of the neighbourhoods (3), one deduces that it is the topology of point-wise convergence on the elements of E^{*} . In particular, a *sequence* $(x_n)_{n \in \mathbb{N}}$ of elements of E converges to a point x_0 for $\sigma(E, E^{*})$ if and only if, for every $\xi \in E^{*}$, $\xi(x_n) \xrightarrow[n \to +\infty]{} \xi(x_0)$ (or, with the new notations, $< x_n , \xi > \xrightarrow[n \to +\infty]{} < x_0 , \xi >$).

E^{**} is equipped with a norm, by a formula analogous to 1, (1). By Alaoglu's theorem, its unit ball $\mathscr{B}_{E^{**}}$ is compact for $\sigma(E^{**}, E^{*})$. But the unit ball $\mathscr{B}_E$ of E has no reason to be $\sigma(E, E^{*})$ compact : we shall come back on this point in the next chapter.

We shall now turn our interest to the spaces E and E^{*} , equipped with their weak topologies : $\sigma(E, E^{*})$ and $\sigma(E^{*}, E)$.

PROPOSITION 1. - *The topology* $\sigma(E^{*}, E)$ *is the weakest topology on* E^{*} *which renders continuous all the linear functionals* $\xi \in E^{*} \to < x , \xi >$, *for* $x \in E$.

PROOF. - First, it is obvious that if $E^\star$ is endowed with $\sigma(E^\star, E)$, these linear functionals are continuous : the inverse image of a neighbourhood of 0, in $\mathbb{K}$, $\{|\lambda| < \epsilon\}$, is the set $\{\xi \in E^\star, |< x, \xi >| < \epsilon\}$, which is of the type 2, (1).

Conversely, let $\mathcal{T}$ be a topology on $E^\star$ such that all the linear functionals $\xi \to < x, \xi >$, $x \in E$, are continuous. Let $x_1,\ldots,x_n$ be points in E, $\epsilon > 0$, and $V_{\epsilon ; x_1,\ldots,x_n}$ a neighbourhood of 0 for $\sigma(E^\star, E)$. Since, for $i = 1,\ldots,n$, the mapping $\xi \to < x_i, \xi >$ is continuous for $\mathcal{T}$, the set $\{\xi \in E^\star ; |< x_i, \xi >| < \epsilon\}$ is open for $\mathcal{T}$; we call it 0_i. The intersection $\bigcap_{i=1,\ldots,n} 0_i$ of these open sets is still open for $\mathcal{T}$, and it is $V_{\epsilon ; x_1,\ldots,x_n}$: this proves that $\mathcal{T}$ is stronger than $\sigma(E^\star, E)$.

One proves the same way :

PROPOSITION 2. - *The topology* $\sigma(E, E^\star)$ *is the weakest topology on* E *which renders continuous all the linear functionals* $x \in E \to < x, \xi >$, *for* $\xi \in E^\star$.

PROPOSITION 3. - *If* $E^\star$ *is endowed with* $\sigma(E^\star, E)$, *the dual of* $E^\star$ *can be identified with* E *: in other terms, there are no other continuous linear functionals on* $E^\star$, *than the applications* $\xi \in E^\star \to < x, \xi >$, $x \in E$.

PROOF. - We have seen that the linear functionals $\xi \to < x, \xi >$ were continuous, we shall see that they are the only ones. Let f be a linear functional on $E^\star$, continuous for $\sigma(E^\star, E)$. By definition, for every $\epsilon > 0$, there is $\eta > 0$, and $x_1,\ldots,x_n \in E$ such that, if $\xi \in V_{\eta ; x_1,\ldots,x_n}$, then $|f(\xi)| < \epsilon$. That is to say :

$$|\xi(x_1)| < \eta, \ldots, |\xi(x_n)| < \eta \Rightarrow |f(\xi)| < \epsilon ;$$

or

$$\operatorname*{Max}_{1 \leqslant k \leqslant n} |\xi(x_k)| < \eta \quad \Rightarrow \quad |f(\xi)| < \epsilon .$$

Using the fact that f is homogeneous, we obtain the existence of a constant $C > 0$ such that, for every $\xi \in E^\star$,

$$|f(\xi)| \leqslant C \operatorname*{Max}_{1 \leqslant k \leqslant n} |\xi(x_k)| .$$

Let us call F the vector subspace of E spanned by the points $x_1,\dots,x_n$. Let us assume that $x_1,\dots,x_r$ $(r \leq n)$ form an algebraic basis of F . If the r relations $\xi(x_i) = 0$, $i = 1,\dots,r$ are satisfied, then $f(\xi) = 0$.

For each $i = 1,\dots,r$, let us choose a linear functional ξ_i on F , with $\xi_i(x_i) = 1$, $\xi_i(x_j) = 0$ if $i \neq j$ $(j = 1,\dots,r)$, and extend it to a continuous linear functional, $\tilde{\xi}_i$, defined on the whole space E .

For each $\xi \in E^\star$, the linear functional $\xi' = \xi - \sum_{i=1}^{r} \xi(x_i)\tilde{\xi}_i$ is equal to zero on F : if $x \in F$, x can be written $x = \sum_1^r \alpha_i x_i$, with $\alpha_i = \tilde{\xi}_i(x)$, and

$$\xi'(x) = \sum_{i=1}^{r} \alpha_i \xi(x_i) - \sum_{i=1}^{r} \xi(x_i)\tilde{\xi}_i(x) = \sum_{i=1}^{r} \xi(x_i)\tilde{\xi}_i(x) - \sum_{i=1}^{r} \xi(x_i)\tilde{\xi}_i(x) = 0 .$$

Therefore, $f(\xi') = 0$, and

$$f(\xi) = \sum_{i=1}^{r} \xi(x_i) f(\tilde{\xi}_i) = \sum_{i=1}^{r} \beta_i \xi(x_i) = \xi\Big(\sum_{i=1}^{r} \beta_i x_i\Big) ,$$

with $\beta_i = f(\tilde{\xi}_i)$, for $i = 1,\dots,r$. This proves our proposition, since f is given by the element $\sum_{i=1}^{r} \beta_i x_i \in E$.

PROPOSITION 4. - *If* E *is endowed with the topology* $\sigma(E, E^\star)$, *its dual is* $E^\star$.

PROOF. - The topology $\sigma(E, E^\star)$ is weaker than the norm topology, therefore, there are fewer continuous linear functionals. But the elements of $E^\star$ remain continuous : therefore, the proposition is proved.

We see, from these last two propositions, that the duality is perfectly symmetric between $(E, \sigma(E, E^\star))$ and $(E^\star, \sigma(E^\star, E))$. This is not the case, in general, if $E^\star$ is endowed with its norm : its dual is then $E^{\star\star}$, usually strictly greater than E . We shall study this question in detail in the next chapter, but let us already give an example of this situation :

We take $c_0 = \{(x_n)_{n \in \mathbb{N}} \;;\; x_n \in \mathbb{K} \;\; \forall n, \;\; x_n \xrightarrow[n \to +\infty]{} 0\}$,

with the norm

$$\|(x_n)\|_{c_0} = \sup_{n \in \mathbb{N}} |x_n| .$$

Then c_0 is a Banach space, and has for dual

$$\ell_1 = \{(y_n)_{n\in\mathbb{N}} \; ; \; y_n \in \mathbb{K} \;\; \forall n , \; \sum_{n\in\mathbb{N}} |y_n| < +\infty \}$$

with the norm :

$$\|(y_n)\|_1 = \sum_{n\in\mathbb{N}} |y_n|$$

(the duality $<(x_n),(y_n)>$ is $\sum_{n\in\mathbb{N}} x_n y_n$).

The dual of ℓ_1 is

$$\ell_\infty = \{(z_n)_{n\in\mathbb{N}} \; ; \; z_n \in \mathbb{K} \;\; \forall n , \; \sup_{n\in\mathbb{N}} |z_n| < +\infty \}$$

with the norm

$$\|(z_n)\|_\infty = \sup_{n\in\mathbb{N}} |z_n| .$$

The norm of $(c_0)^{**} = \ell_\infty$ has the same expression as the norm of c_0 , but ℓ_∞ is a larger space : it is the space of the bounded sequences of scalars, whereas c_0 was the space of sequences tending to zero at infinity.

These spaces will be studied in great detail in Second Part, chapter IV.

§ 4. THE SEPARATION OF CONVEX SUBSETS : THE GEOMETRIC FORM OF HAHN-BANACH THEOREM.

Let us first recall a few facts. If E is a vector space and F a vector subspace of E , we call *codimension* of F in E the dimension of the quotient $E/_F$ (it may of course be infinite). An hyperplane is a vector subspace of codimension 1. It is the kernel of a linear functional : the hyperplane is closed if the linear functional is continuous, it is dense in the converse case.

An *affine* subspace is the translate of a vector subspace. In particular, an affine hyperplane is the set of points at which a linear functional takes a prescribed value.

THEOREM 1. (Hahn-Banach Theorem, geometric form). - *Let* E *be a topological vector space,* Ω *a non-empty convex open set in* E , L *a non-empty affine subspace of* E *, which does not meet* Ω *. Then, there is a closed affine hyperplane* H *, which contains* L *and does not meet* Ω .

PROOF. - We may, making a translation if necessary, assume that Ω contains the origin.

We first consider the case when E is a real vector space ; L is then a real affine subspace.

We define the gauge j of the convex set Ω by the formula :

$$j(x) = \inf\{\lambda \in \mathbb{R}^+ \ ; \lambda\Omega \ni x\} .$$

Note that this definition makes sense, since Ω , being open, is a neighbourhood of 0 : therefore, for all x , there is a $\lambda > 0$ such that $\lambda\Omega \ni x$.

It is obvious on the definition that, if $\alpha > 0$, $j(\alpha x) = \alpha j(x)$, and so j is positively homogeneous. Since Ω is convex, if

$\frac{x}{\lambda} \in \Omega$, if $\frac{y}{\mu} \in \Omega$, then

$$\frac{\lambda(\frac{x}{\lambda}) + \mu(\frac{y}{\mu})}{\lambda + \mu} \in \Omega ,$$

that is $\frac{x + y}{\lambda + \mu} \in \Omega$ $(\lambda, \mu > 0)$. From this follows that j is subadditive, that is $j(x + y) \leqslant j(x) + j(y)$, for all $x,y \in E$.

We have one more property of the gauge j , due to the fact that Ω is open :

LEMMA 2.

$$\Omega = \{x \in E \ ; \ j(x) < 1\} .$$

PROOF OF LEMMA 2. - It it clear that $\{x \in E , j(x) < 1\}$ is contained in Ω . Conversely, let $y \in \Omega$. We can find, since Ω is open, an $\epsilon > 0$, small enough to have $(1 + \epsilon)y \in \Omega$; therefore $j(y) < 1$. The lemma is proved.

Let us consider now the vector subspace F spanned by L . Since, by assumption, 0 is not in L , every element of L is of the form λx , $\lambda \in \mathbb{K}$, $x \in L$, and L is an affine hyperplane of F . Therefore, there is a linear functional f on F such that $L = \{x \in F \ ; \ f(x) = 1\}$. Consequently, we have $f(x) \leqslant j(x)$, for every x in L , and thus for every x in F . We may apply the analytic form of Hahn-Banach Theorem (§ 1, theorem 1), since j satisfies the necessary conditions (positive homogeneous, and subadditive). So there exists a linear functional $\tilde{f}$, defined on the whole space E, extending f and dominated by p: $\tilde{f}(z) \leqslant p(z)$, for all $z \in E$.

Let H be the affine hyperplane $\{y \in E\ ,\ \tilde{f}(y) = 1\}$: it contains L ; H and Ω are disjoint, since $j(y) \geqslant \tilde{f}(y) = 1$ if $y \in H$, and therefore $y \notin \Omega$. Let us observe finally that H is closed : from the inclusions $H \subset \bar{H} \subset E$ follows that H is closed or everywhere dense in E : but it cannot be dense, since it does not meet the open set Ω.

If now E is a complex vector space (L being a complex affine subspace), by the previous argument, there is a real closed affine hyperplane H_0, containing L, disjoint from Ω. Say that $H_0 = a + H'_0$, where H'_0 is a real vector hyperplane. Then $H' = H'_0 \cap (i\,H'_0)$ is a complex hyperplane, and $H = a + H'$ is a complex affine closed hyperplane, which does not meet Ω, and which contains L, since $L = a + L'$, where L' is a complex vector subspace, thus satisfying $iL' = L'$. This proves our theorem.

We could argue directly, for complex Banach spaces, using theorem 2, § 1, instead of theorem 1, § 1, but this would require another definition of the gauge, namely the following :

$$j(x) = \inf\{|\lambda|\ ;\ \lambda \in \mathbb{K}\ ,\ \lambda\Omega \ni x\}\ ,$$

which would be a "symmetric" gauge.

We now give several applications of Theorem 1 to the separation of convex sets.

COROLLARY 3. - *Let* E *be a real topological vector space,* A *a non-empty open convex subset,* B *a non-empty convex subset, which does not meet* A. *Then there is a continuous linear functional* f *and a real number* α *such that :*

$$A \subset \{x \in E\ ;\ f(x) < \alpha\}\ , \qquad B \subset \{x\ ;\ f(x) \geqslant \alpha\}\ .$$

PROOF. - The set $C = \bigcup_{z \in B} (A - z) = A - B$ (set of differences of elements of A and elements of B, not $A \cap \complement B$!) is open, as a reunion of open sets, and it is convex, since it is the image of $A \times B$ by the application $(y,z) \rightarrow y - z$. It is non-empty, and does not contain 0. By the Theorem, there exists a closed hyperplane which does not meet C : one can find a continuous linear functional f such that $f(x) < 0$, for all $x \in C$. Therefore, we have $f(y) < f(z)$, for all $y \in A$, all $z \in B$. So, if we put $\alpha = \inf_{z \in B} f(z)$, we have

$$f(y) \leqslant \alpha \leqslant f(z) , \quad \text{for all } y \in A , \text{ all } z \in B .$$

This is not yet our statement, but since A is open, we have

$$A \subset \{y \in E ; f(y) < \alpha\} ,$$

and the corollary is proved.

If we have $A \subset \{x \in E ; f(x) \leqslant \alpha\}$ and $B \subset \{x \in E ; f(x) \geqslant \alpha\}$, we say that the hyperplane $\{x ; f(x) = \alpha\}$ *separates* A and B : each of them lies in one closed half-space. If we have $A \subset \{x \in E ; f(x) < \alpha\}$ and $B \subset \{x \in E ; f(x) > \alpha\}$, we say that this hyperplane *strictly separates* A and B .

We shall now establish the analogue of Corollary 3 for complex vector spaces. We shall say that a set A , contained in a complex vector space E , is *$\mathbb{C}$-symmetric* if, whenever $x \in A$, then, for all $\lambda \in \mathbb{C}$ with $|\lambda| = 1$, $\lambda x \in A$ (such sets are also called "balanced").

COROLLARY 4. - *Let* E *be a complex topological vector space,* A *a non-empty, open,* $\mathbb{C}$*-symmetric and convex set,* B *a non-empty convex set, which does not meet* A . *Then there is a (complex) continuous linear functional* f *and a number* $\alpha > 0$ *such that :*

$$A \subset \{x \in E ; |f(x)| < \alpha\} , \quad B \subset \{x \in E ; |f(x)| \geqslant \alpha\} .$$

PROOF. - By Corollary 3, there is a real continuous linear functional g on E and a number $\alpha \in \mathbb{R}$ such that

$$A \subset \{x \in E ; g(x) < \alpha\} , \quad B \subset \{x \in E ; g(x) \geqslant \alpha\}$$

Since $0 \in A$ (A is convex and $\mathbb{C}$-symmetric), we have $0 < \alpha$. Put $f(x) = g(x) - ig(ix)$; then $\mathcal{R}e\, f = g$, and so

$$|f(x)| \geqslant \mathcal{R}e\, f(x) \geqslant \alpha \quad \text{if } x \in B .$$

Take now $x \in A$, and θ such that

$$|f(x)| = \mathcal{R}e\,(e^{i\theta} f(x)) = g(e^{i\theta} x) .$$

Since A is $\mathbb{C}$-symmetric, $e^{i\theta} x \in A$ and $|f(x)| < \alpha$, which proves the corollary.

COROLLARY 5. - *Let* E *be a real, Hausdorff, locally convex topological vector space. Let* A *be a non-empty compact convex set,* B *a non-empty closed convex set, which does not meet* A. *Then, there is a closed affine hyperplane which strictly separates* A *and* B *: there is a continuous linear functional* f *and a real* α *such that :*

$$A \subset \{x \in E \; ; \; f(x) < \alpha\} \; , \quad B \subset \{x \in E \; ; \; f(x) > \alpha\} \; .$$

PROOF. - If A is a compact set and B a closed set in a locally convex topological vector space, which do not intersect, one can find an open convex neighbourhood of the origin, U, such that $A+U$ and $B+U$ do not intersect (see for instance Bourbaki [11]). We can apply corollary 3 to $A+U$ and $B+U$, and we obtain

$$A+U \subset \{x \in E \; ; \; f(x) < \alpha\}$$

$$B+U \subset \{x \in E \; ; \; f(x) > \alpha\} \; ,$$

the last inclusion since $B+U$ is open. A fortiori, the same holds for A and B.

The same way, we have :

COROLLARY 6. - *Let* E *be a complex Hausdorff locally convex topological vector space,* A *a non-empty compact convex set,* B *a non-empty* $\mathbb{C}$*-symmetric closed convex set, which does not meet* A. *There exists a complex continuous linear functional* f *and a number* $\alpha > 0$ *such that :*

$$A \subset \{x \in E \; ; \; |f(x)| > \alpha\} \; , \quad B \subset \{x \in E \; ; \; |f(x)| < \alpha\} \; .$$

PROOF. - We do the same as in the previous lemma, and apply corollary 4.

We shall now apply these tools to the convex sets, closed under the weak or under the strong topologies, in a Banach space.

§ 5. CLOSED CONVEX SETS AND BOUNDED SETS IN THE WEAK AND IN THE STRONG TOPOLOGIES.

We now come back to the setting of § 2 and 3. It is obvious on the form of the neighbourhoods given for $\sigma(E^*, E)$, $\sigma(E^{**}, E)$, $\sigma(E, E^*)$, that these neighbourhoods are convex. Also, if two points are distinct, one can find a continuous linear functional which does not take the same value on both : therefore, these topologies are Hausdorff. We have thus introduced Hausdorff locally convex topologies on E, E^* or E^{**}, to which we shall be able to apply the tools of § 4.

The next result is a special case of a theorem of G. Mackey :

PROPOSITION 1. - *Let* E *be a normed space. Every convex subset of* E *which is closed for the norm is also closed for* $\sigma(E, E^*)$.

(The converse is obvious, since the norm-topology is stronger than $\sigma(E, E^*)$).

PROOF. - Let C be a convex set, closed for the norm. Let $x_0 \notin C$. By corollary 5, § 4, there is a real linear functional f and a number $\alpha \in \mathbb{R}$, with $f(x_0) < \alpha$, $f(y) > \alpha$ for all $y \in C$. But a set of the form $\{y \in E ; f(y) \geq \alpha\}$ is closed for $\sigma(E, E^*)$ (since $\{y \in E ; f(y) < \alpha\}$ is open, by definition of the neighbourhoods) ; this set contains C, and does not contain x_0 : this proves that x_0 cannot be in the closure of C for $\sigma(E, E^*)$.

It follows from this proposition that, for a convex set, the closures for the norm and for $\sigma(E, E^*)$ are the same.

Recall that a real function φ, defined on E, is lower semi-continuous if, for every $\lambda \in \mathbb{R}$, the set $\{x \in E, \varphi(x) \leq \lambda\}$ is closed. If, moreover, the function φ is convex, then these sets are also convex, and the previous proposition says that if a convex function is lower semi-continuous for the norm, it is also lower semi-continuous for the weak topology.

A very important example is the norm itself, which is therefore lower semi-continuous for the weak topology $\sigma(E, E^*)$. This means that the balls of E are $\sigma(E, E^*)$-closed, or that the balls of E^* are $\sigma(E^*, E^{**})$-closed. It implies that if $(x_n)_{n \in \mathbb{N}}$ is a sequence in the unit ball of E, weakly converging to an element $x \in E$, then :

$$\|x\| \leqslant \liminf_{n \to +\infty} \|x_n\|$$

(and the same for a family $(x_i)_{i \in I}$, converging for a filter $\mathcal{F}$ on I).

REMARK. - In $E^\star$, a convex set may be closed for the norm without being closed for $\sigma(E^\star, E)$. For example, in $E^{\star\star}$, the unit ball $\mathcal{B}_E$ is convex, norm-closed, but not $\sigma(E^{\star\star}, E^\star)$-closed (see proposition 4, below).

But, if $(f_i)_{i \in I}$ is a family of elements in $\mathcal{B}_{E^\star}$, converging for a filter $\mathcal{F}$, in $\sigma(E^\star, E)$, to an element f, we also have :

$$\|f\| \leqslant \liminf_{\mathcal{F}} \|f_i\| ,$$

since, for every $x \in \mathcal{B}_E$,

$$|f(x)| = \lim_{\mathcal{F}} |f_i(x)| \leqslant \liminf_{\mathcal{F}} \|f_i\| .$$

The same relation holds also in $E^{\star\star}$ for $\sigma(E^{\star\star}, E^\star)$.

The following proposition is also due to G. Mackey :

PROPOSITION 2. - *The subsets of* E *which are bounded for* $\sigma(E, E^\star)$ *are also norm-bounded (and conversely).*

PROOF. - We recall first that, in a topological vector space, a set B is *bounded* if, for every neighbourhood V of the origin, there is a scalar λ such that $\lambda V \supset B$.

LEMMA 3. - A *set* B *in a normed space* E *is bounded for* $\sigma(E, E^\star)$ *if and only if every continuous linear functional is bounded on* B, *that is, if for every* $f \in E^\star$, there is a number M such that $|f(x)| \leqslant M$, *for all* $x \in B$.

PROOF OF LEMMA 3. - Let $f \in E^\star$, $\epsilon > 0$, and V the neighbourhood of 0, for $\sigma(E, E^\star)$, defined by $V = \{x \in E ; |f(x)| < \epsilon\}$. If B is bounded, there is a λ such that $\lambda V \supset B$, that is, for all $x \in B$, $|f(x)| \leqslant |\lambda| \epsilon$, which proves that every continuous linear functional is bounded on B.

Conversely, we assume that every continuous linear functional is bounded on B. Let $V_{\epsilon ; \xi_1, \ldots, \xi_n}$ be a neighbourhood of 0 for $\sigma(E, E^\star)$.

There exists a number $M > 0$ such that, for all $x \in B$,

$$|\xi_1(x)| \leqslant M, \quad |\xi_2(x)| \leqslant M, \ldots, |\xi_n(x)| \leqslant M,$$

and so $B \subset \frac{MV}{\epsilon}$, and the lemma is proved.

Let us come back to the proof of proposition 2. The unit ball $\mathscr{B}_{E^\star}$ is a metric space, for the induced norm, and complete, therefore a Baire space. We consider the closed subsets, for $n \geqslant 1$:

$$F_n = \{\xi \in \mathscr{B}_{E^\star} ; |\xi(x)| \leqslant n, \quad \text{for all } x \in B\}$$

The lemma says that the reunion of F_n's covers $\mathscr{B}_{E^\star}$, and so one of them must have a non-empty interior : there must exist a $\xi_o \in E^\star$, $n_o \in \mathbb{N}$, $\epsilon > 0$, such that F_{n_o} contains the ball centered at ξ_o, with radius ϵ. This means that :

$$\|\xi - \xi_o\| < \epsilon \Rightarrow |\xi(x)| \leqslant n_o, \quad \text{for all } x \in B.$$

Therefore, if $\|f\| < \epsilon$, putting $f = \xi - \xi_o$, we obtain :

$$|f(x)| \leqslant |\xi(x)| + |\xi_o(x)| \leqslant 2n_o,$$

and thus

$$\|f\| \leqslant 1 \Rightarrow |f(x)| \leqslant \frac{2n_o}{\epsilon}, \quad \text{for all } x \in B.$$

Consequently,

$$\|x\| = \sup_{\|f\| \leqslant 1} |f(x)| \leqslant \frac{2n_o}{\epsilon}, \quad \text{for all } x \in B,$$

and B is bounded in norm ; the converse being obvious, our proposition is proved.

We have seen that the space E could be viewed as a subspace of $E^{\star\star}$, with the induced norm. We shall now look more closely at this question.

PROPOSITION 4. - *Let* E *be a normed space,* $E^\star$ *its dual,* $E^{\star\star}$ *its bidual. For the topology* $\sigma(E^{\star\star}, E^\star)$, *the ball* $\mathscr{B}_E = \{x \in E ; \|x\| \leqslant 1\}$ *is dense in the ball* $\mathscr{B}_{E^{\star\star}} = \{x \in E^{\star\star} ; \|x\| \leqslant 1\}$.

REMARK. - In this proposition, we consider E as a subspace of $E^{\star\star}$, and therefore, $\mathscr{B}_E$ is a subset of $\mathscr{B}_{E^{\star\star}}$.

It follows obviously from the proposition that E is dense in $E^{\star\star}$ for $\sigma(E^{\star\star}, E^{\star})$, but, if E is a Banach space, it is closed in $E^{\star\star}$ for the norm-topology.

PROOF. - Since $\mathscr{B}_{E^{\star\star}}$ is $\sigma(E^{\star\star}, E^{\star})$-compact, the set $\overline{\mathscr{B}_E}$ (closure for $\sigma(E^{\star\star}, E^{\star})$) is contained in $\mathscr{B}_{E^{\star\star}}$. Conversely, let us assume that there is a point x_0 in $\mathscr{B}_{E^{\star\star}}$, not in $\overline{\mathscr{B}_E}$. We can then separate strictly x_0 and $\overline{\mathscr{B}_E}$ by a closed hyperplane : there is an element $f \in E^{\star}$, with $\|f\| = 1$, and a number $a > 0$ such that

$$|f(x_0)| > a , \quad |f(x)| < a \quad \text{for all} \quad x \in \overline{\mathscr{B}_E}$$

(by corollary 6, § 4). This holds a fortiori for all $x \in \mathscr{B}_E$. But, on the other side, $a < |f(x_0)| \leqslant \|f\|.\|x_0\| \leqslant 1$, and so $a < 1$. Therefore, $\sup_{x \in \mathscr{B}_E} |f(x)| \leqslant a < 1$, and this contradicts formula 1.(1), since $\|f\| = 1$. This proves the proposition.

We shall now extend these notions to the operators between normed spaces.

§ 6. WEAK AND STRONG CONTINUITY. TRANSPOSITION.

Let E and F be two normed spaces. We know that a linear application u from E into F is continuous if and only if there is a constant $C > 0$ such that $\|u(x)\|_F \leqslant C\|x\|_E$, and we set $\|u\| = \sup_{\|x\|_E \leqslant 1} \|u(x)\|_F$. We call *operator* a continuous linear application.

Let $E^{\star}$, $F^{\star}$ be the duals of E and F. We shall say that u is *weakly continuous* from E into F, if it is continuous from E, endowed with $\sigma(E, E^{\star})$, into F, endowed with $\sigma(F, F^{\star})$.

For linear mappings, strong continuity implies weak continuity :

PROPOSITION 1. - *Every operator from* E *into* F *is weakly continuous.*

PROOF. - Let $V_{\epsilon;\xi_1,\ldots,\xi_n}$ be a neighbourhood of 0 in the image $u(E)$, for $\sigma(F, F^{\star})$: it is the set of $y \in u(E)$ such that $|\xi_1(y)| < \epsilon$, ..., $|\xi_n(y)| < \epsilon$. We put $\eta_1 = \xi_1 \circ u, \ldots, \eta_n = \xi_n \circ u$. These are continuous linear functionals on E, since u is continuous. Therefore, the set

$$U_{\epsilon;\eta_1,\ldots,\eta_n} = \{x \in E \ ; \ |\eta_1(x)| < \epsilon \ , \ldots, \ |\eta_n(x)| < \epsilon\}$$

is a neighbourhood of 0 for $\sigma(E, E^\star)$, the image of which, by u, is contained in $V_{\epsilon;\xi_1,\dots,\xi_n}$: this proves the weak continuity of u.

The *transpose* of u, denoted by ${}^tu$, is an application from $F^\star$ into $E^\star$, defined by the formula :

$${}^tu(\xi)(x) = \xi(u(x)) \quad \text{for all} \quad x \in E, \quad \text{all} \quad \xi \in F^\star .$$

This application is obviously linear, and has the following interesting continuity properties :

PROPOSITION 2. - *Let* u *be an operator from* E *into* F, ${}^tu$ *its transpose. Then :*

a) ${}^tu$ *is continuous from* $F^\star$ *into* $E^\star$ *equipped with their norms (we shall say : strongly continuous), and* $\|{}^tu\| = \|u\|$.

b) ${}^tu$ *is continuous from* $F^\star$ *equipped with* $\sigma(F^\star, F)$ *into* $E^\star$ *equipped with* $\sigma(E^\star, E)$.

PROOF. - Let us show that ${}^tu$ is strongly continuous. Let $\eta \in F^\star$ be a continuous linear functional, with $\|\eta\| = 1$. Let $x \in E$, $\|x\| = 1$. We have

$$|{}^tu(\eta)(x)| = |\eta(u(x))| \leqslant \|u\| .$$

We obtain :

$$\|{}^tu\| = \sup_{\|\eta\|_{F^\star} \leqslant 1} \ \sup_{\|x\|_E \leqslant 1} |{}^tu(\xi)(x)| \leqslant \|x\| ;$$

this proves that ${}^tu$ is continuous, and shows that $\|{}^tu\| \leqslant \|u\|$. Let now $\epsilon > 0$, $x_0 \in E$ with $\|x_0\| = 1$, such that $\|u(x_0)\| \geqslant (1-\epsilon)\|u\|$, and let $\eta_0 \in F^\star$, with $\eta_0(u(x_0)) = \|u(x_0)\|$. We have :

$$\|{}^tu\| = \sup_{\|x\|_E \leqslant 1,\ \|\eta\|_{F^\star} \leqslant 1} |{}^tu(\eta)(x)| \geqslant |{}^tu(\eta_0)(x_0)| = |\eta_0(u(x_0))| \geqslant (1-\epsilon)\|x\| ,$$

and this proves that $\|{}^tu\| = \|u\|$.

The proof of the continuity of ${}^tu$ from $(F^\star, \sigma(F^\star, F))$ into $(E^\star, \sigma(E^\star, E))$ is similar to the proof of the previous proposition ; we leave it to the reader. Let us observe, moreover, that, by proposition 1, ${}^tu$ is also continuous from $(F^\star, \sigma(F^\star, F^{\star\star}))$ into $(E^\star, \sigma(E^\star, E^{\star\star}))$.

We shall now investigate the links between the injectivity or surjectivity of u and the corresponding properties for ${}^t u$. Let us first recall some notions.

Let M be a vector subspace of E. We call *orthogonal* of M, and denote by $M^\perp$, the subspace of $E^\star$ defined by :

$$M^\perp = \{\xi \in E^\star \ ; \ \xi(x) = 0 \quad \text{for all} \quad x \in M\} .$$

This is obviously a closed subspace of $E^\star$ (if $\xi_n \xrightarrow[n \to +\infty]{} \xi$ and if $\xi_n(x) = 0$, then $\xi(x) = 0$). Also, we can consider the orthogonal of $M^\perp$, in $E^\star$ endowed with $\sigma(E^\star, E)$. This is the set $M^{\perp\perp}$, contained in E, and defined by :

$$M^{\perp\perp} = \{x \in E \ ; \ \xi(x) = 0 \quad \text{for all} \quad \xi \in M^\perp\} .$$

This is a subspace of E, which is $\sigma(E, E^\star)$-closed. In fact, $M^{\perp\perp}$ is the closure of M for $\sigma(E, E^\star)$ or for the norm (both closures are identical, by § 5, proposition 1). Indeed, one checks immediately that $\overline{M} \subset M^{\perp\perp}$, and if a point x_0 does not belong to $\overline{M}$, there is, by § 4, corollary 6, a linear functional on E endowed with $\sigma(E, E^\star)$ (that is : an element of $E^\star$), equal to zero on M, and strictly positive at x_0 : therefore x_0 is not in $M^{\perp\perp}$.

The same way, if N is a subspace of $E^\star$, we consider

$$N^\perp = \{x \in E \ ; \ \xi(x) = 0 \ , \ \text{for all} \quad \xi \in N\}$$

and

$$N^{\perp\perp} = \{\xi \in E^\star \ ; \ \xi(x) = 0 \ , \ \text{for all} \quad x \in N^\perp\} .$$

The subspace $N^{\perp\perp}$ is $\sigma(E^\star, E)$-closed, just a $M^{\perp\perp}$ was $\sigma(E, E^\star)$-closed : this is not a new notion, but just a consequence of the fact that the duality is perfectly symmetric between $(E, \sigma(E, E^\star))$ and $(E^\star, \sigma(E^\star, E))$. But one should observe that $M^{\perp\perp}$ is the norm-closure of M in E, whereas $N^{\perp\perp}$ (though norm-closed in $E^\star$) needs not be the norm-closure of N in $E^\star$.

PROPOSITION 3. - *Let* u *be an operator from* E *into* F, ${}^t u$ *its transpose. Then :*

a) $(\operatorname{Im} u)^\perp = \operatorname{Ker} {}^t u$;

b) $\operatorname{Ker} u = (\operatorname{Im}({}^t u))^{\perp}$;

c) $\overline{\operatorname{Im} u} = (\operatorname{Ker}({}^t u))^{\perp}$;

d) $(\operatorname{Ker} u)^{\perp} = \overline{\operatorname{Im}({}^t u)}$ (closure for $\sigma(E^{\star}, E)$).

PROOF. - Let us show a) :

$\xi \in (\operatorname{Im} u)^{\perp} \iff \xi(y) = 0$ for all $y \in \operatorname{Im} u \iff \xi(u(x)) = 0$, for all $x \in E \iff {}^t u(\xi)(x) = 0$, for all x in $E \iff {}^t u(\xi) = 0 \iff \xi \in \operatorname{Ker} {}^t u$.

b) Is obtained the same way.

c) Is obtained from a) by taking orthogonals, and d) from b). The proposition is proved.

Consequently, it is equivalent to say that u (resp. ${}^t u$) is injective and that ${}^t u$ (resp. u) has a dense image. One will observe that this last property differs from surjectivity. We shall now describe the conditions on u to obtain the surjectivity of ${}^t u$, and conversely :

PROPOSITION 4. - *The operator* u *is surjective if and only if* ${}^t u$ *is a weak isomorphism, (that is, a bijection which is bicontinuous for the weak topologies), from* $F^{\star}$ *onto* ${}^t u(F^{\star})$. *Also,* u *is a weak isomorphism from* E *onto* $u(E)$ *if and only if* ${}^t u$ *is surjective.*

PROOF. - Let us show the first part of the proposition. We assume u to be surjective. Then, by a) of the previous proposition, ${}^t u$ is injective : it is a bijection, weakly continuous, from $F^{\star}$ onto ${}^t u(F^{\star})$. We have to show that the inverse bijection is also weakly continuous. For this, let $V_{\epsilon ; y_1, \dots, y_n}$ be a neighbourhood of 0 , in $F^{\star}$, for $\sigma(F^{\star}, F)$. We have

$$V_{\epsilon ; y_1, \dots, y_n} = \{\eta \in F^{\star} \ ; \ |\eta(y_1)| < \epsilon, \dots, |\eta(y_n)| < \epsilon\}$$

and thus

$${}^t u(V_{\epsilon ; y_1, \dots, y_n}) = \{{}^t u(\eta) \ ; \ |\eta(y_1)| < \epsilon, \dots, |\eta(y_n)| < \epsilon\}$$

$$= \{\xi \in {}^t u(F^{\star}) \ ; \ |\xi(x_1)| < \epsilon, \dots, |\xi(x_n)| < \epsilon\} ,$$

if we denote by $x_1, \dots, x_n$ the points of E such that $u(x_1) = y_1, \dots, u(x_n) = y_n$. The image ${}^t u(V_{\epsilon ; y_1, \dots, y_n})$ is therefore a neighbourhood of 0 for $\sigma(E^{\star}, E)$, and ${}^t u$ is a weak isomorphism.

Conversely, let us assume ${}^t u$ to be a weak isomorphism, and let $y \in F$. Then y defines a linear functional on $F^\star$, which is $\sigma(F^\star, F)$-continuous. If we consider $y \circ ({}^t u)^{-1}$, we obtain a linear functional on $E^\star$, which is $\sigma(E^\star, E)$-continuous : such a functional, as we know (§ 3, prop. 3), is given by an element $x \in E$. Therefore, we have, for every $\eta \in F^\star$:

$$< u(x), \eta > = < x, {}^t u(\eta) > = < y \circ ({}^t u)^{-1}, {}^t u(\eta) > = < y, \eta > ,$$

and $u(x) = y$. This proves the first part of the proposition ; the second is proved analogously, and is left to the reader.

§ 7. DUALITY BETWEEN SUBSPACES AND QUOTIENTS FOR WEAK AND STRONG TOPOLOGIES.

In this paragraph, we shall investigate the following question :

If E is a normed space and F a subspace of E, what are the duals of F and of the quotient $E/_F$? The answer is given in the following proposition :

PROPOSITION 1. - *Let* E *be a normed space,* F *subspace of* E, *endowed with the induced norm. Then :*

a) *The dual* $F^\star$ *of* F *can be isometrically identified with the quotient* $E^\star/_{F^\perp}$ *; the duality mapping is defined by :*

(1) $\dot{\xi}(x) = \xi(x)$, *if* $x \in F$, $\dot{\xi} \in E^\star/_{F^\perp}$, *and* ξ *is any element in the class* $\dot{\xi}$, *in* $E^\star$.

b) *The weak topology* $\sigma(F, E^\star/_{F^\perp})$ *is the topology induced on* F *by* $\sigma(E, E^\star)$.

PROOF.

a) Let $\dot{\xi} \in E^\star/_{F^\perp}$; it is obvious that formula (1) defines a linear functional ζ on F, and one has :

$$\|\zeta\|_{F^\star} = \sup_{\|x\|_F \leqslant 1} |\zeta(x)| = \sup_{\|x\|_F \leqslant 1} |\xi(x)|$$

(if ξ is any element in the class $\dot{\xi}$),

$$\leqslant \sup_{\|x\|_E \leqslant 1} |\xi(x)| = \|\xi\|_{E^\star} ;$$

therefore

$$\|\zeta\|_{F^\star} \leqslant \inf_{\xi \in \dot{\xi}} \|\xi\|_{E^\star} = \|\dot{\xi}\|_{E^\star/F^\perp} .$$

Conversely, if ζ is a continuous linear functional on F, we can, using Hahn-Banach Theorem, extend it to a linear functional $\tilde{\zeta}$ on E, with the same norm, and the value $\tilde{\zeta}(x)$, $x \in F$, depends only on the class of $\tilde{\zeta}$ modulo $F^\perp$. Calling $\dot{\xi}$ the class of $\tilde{\zeta}$, we obtain :

$$\|\zeta\|_{F^\star} = \|\tilde{\zeta}\|_{E^\star} \geqslant \|\dot{\xi}\|_{E^\star/F^\perp} ,$$

which proves that $\|\zeta\|_F = \|\dot{\xi}\|_{E^\star/F^\perp}$.

b) Let $V_{\epsilon;\dot{\xi}_1,\dots,\dot{\xi}_n} = \{x \in F ; |\dot{\xi}_1(x)| < \epsilon,\dots, |\dot{\xi}_n(x)| < \epsilon\}$ be a neighbourhood of 0 for $\sigma(F, E^\star/F^\perp)$. We choose $\xi_1,\dots,\xi_n$ any elements in the classes $\dot{\xi}_1,\dots,\dot{\xi}_n$. We can write :

$$V_{\epsilon;\dot{\xi}_1,\dots,\dot{\xi}_n} = \{x \in F ; |\xi_1(x)| < \epsilon,\dots,|\xi_n(x)| < \epsilon\}$$

$$= \{x \in E ; |\xi_1(x)| < \epsilon,\dots,|\xi_n(x)| < \epsilon\} \cap F .$$

Conversely, if $V_{\epsilon;\xi_1,\dots,\xi_n}$ is a neighbourhood of 0 for $\sigma(E, E^\star)$, we have $F \cap V_{\epsilon;\xi_1,\dots,\xi_n} = \{x \in F ; |\dot{\xi}_1(x)| < \epsilon,\dots,|\dot{\xi}_n(x)| < \epsilon\}$, where $\dot{\xi}_1,\dots,\dot{\xi}_n$ are classes of $\xi_1,\dots,\xi_n$ modulo $F^\perp$, and the proposition is proved.

The dual of a subspace of E is therefore not a *subspace* of $E^\star$ but a *quotient* of $E^\star$. There is, however, an important case (on which we shall come back later), where the dual of F can be viewed as a *subspace* of $E^\star$.

If E is a normed space and F a closed subspace of E, we shall say that F is *complemented* in E if there exists a continuous linear projection P from E onto F (we recall that a projection satisfies $P^2 = P$). From now on, we shall just say "projection", and omit the words "continuous linear".

PROPOSITION 2. - *Let* E *be a normed space,* F *a closed subspace of* E. *We assume that there is a projection* P *from* E *onto* F. *Then the transpose* ${}^tP$ *is an isomorphism from* $F^\star$ *onto a closed subspace of* $E^\star$.

PROOF. - Since P is surjective from E onto F, it has a fortiori a dense range, and therefore ${}^tP$, from $F^\star$ into $E^\star$, is injective

(§ 6, prop. 3). We put $Y = \mathrm{Im}\, {}^tP$, and look at ${}^tP$, from $F^\star$ into Y. We shall prove that Y is closed in $E^\star$ (for the norm) : this will prove that Y is a Banach space, and, by the open mapping theorem (chapter I, Theorem 8), will prove that ${}^tP$ is bicontinuous.

We call $\widetilde{P}$ the operator P considered as an operator from E into itself (that is, $\widetilde{P} = i \circ P$, where i is the canonical embedding from F into E). We also put $j = {}^ti$, and we have the following diagrams :

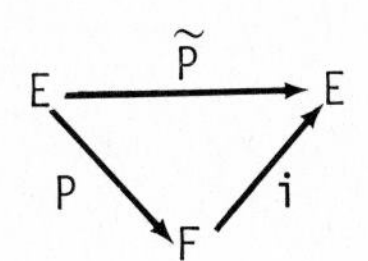

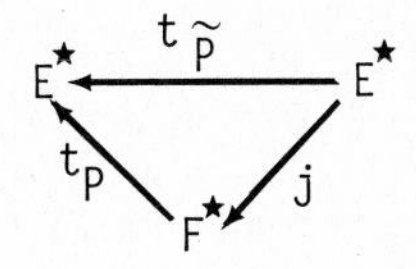

We have $\widetilde{P} \circ \widetilde{P} = \widetilde{P}$, therefore ${}^t\widetilde{P} \circ {}^t\widetilde{P} = {}^t\widetilde{P}$, and ${}^t\widetilde{P}$ is a projection from $E^\star$ onto a closed subspace of it. The embedding i is an isomorphism on its image, therefore a weak isomorphism (that is, $\sigma(F, F^\star)$ and $\sigma(E, E^\star)$ restricted to F coïncide ; see b) of the previous proposition), and j is surjective (§ 6, prop. 4). So $Y = \mathrm{Im}\, {}^tP = \mathrm{Im}\, {}^t\widetilde{P}$, and this last space is closed, which proves our assertion.

This proposition means also that $F^\star$ can be identified with a closed subspace of $E^\star$, by means of the formula :

$$\xi \in E^\star \longrightarrow \xi \circ P \in E^\star ,$$

but one should observe that ${}^t\widetilde{P}$ needs not be an isometry.

We shall now study the dual of quotients of E.

PROPOSITION 3. - *Let* E *be a normed space,* F *a subspace of* E, *endowed with the induced norm. Then :*

a) *The dual of* $E/_F$ *can be isometrically identified with* $F^\perp$, *the duality mapping being given by the following formula :*

(2) $< \dot{x}, \xi > = \xi(x)$, *if* $\dot{x} \in E/_F$, x *being any element in the class* $\dot{x}$, $x \in E$, *and* $\xi \in F^\perp$.

b) *The topology* $\sigma(E/_F , F^\perp)$ *is identical with the quotient topology of* $\sigma(E, E^\star)$ *by* F.

PROOF.

a) If $\xi \in F^{\perp}$, ξ defines a linear functional on $E/_F$ by $\xi(\dot{x}) = \xi(x)$, if x is any element in $\dot{x}$, and $\|\xi\|_{(E/_F)^\star} \leqslant \|\xi\|_{E^\star}$.

If $\xi \in (E/_F)^\star$, we define a linear functional on E by $\xi(x) = \xi(\dot{x})$, if $\dot{x}$ is the class of x modulo F. So $\xi \in F^{\perp}$, and $\|\xi\|_{E^\star} = \|\xi\|_{(E/_F)^\star}$.

b) We call p the canonical surjective mapping from E onto $E/_F$. Let $V_{\epsilon;\xi_1,\dots,\xi_n} = \{\dot{x} \in E/_F \; ; \; |\xi_1(x)| < \epsilon, \dots, |\xi_n(x)| < \epsilon\}$ $(\xi_1,\dots,\xi_n \in F^{\perp})$ be a neighbourhood of 0 for $\sigma(E/_F, F^{\perp})$. Then $V_{\epsilon;\xi_1,\dots,\xi_n}$ is the image by p of the neighbourhood

$$W_{\epsilon;\xi_1,\dots,\xi_n} = \{x \in E \; ; \; |\xi_1(x)| < \epsilon, \dots, |\xi_1(x)| < \epsilon\},$$

and this proves that the topology $\sigma(E, E^\star)/_F$ (that is : quotient by F) is stronger than $\sigma(E/_F, F^{\perp})$.

Conversely, let $W_{\epsilon;\xi_1,\dots,\xi_n}$ be a neighbourhood of 0 in E for $\sigma(E, E^\star)$ (with $\xi_1,\dots,\xi_n \in E^\star$). We shall show that we can find $f_1,\dots,f_n \in F^{\perp}$ such that

$$p(W_{\epsilon;f_1,\dots,f_n}) \subset p(W_{\epsilon;\xi_1,\dots,\xi_n}),$$

and this will prove that $\sigma(E/_F, F^{\perp})$ is stronger than $\sigma(E, E^\star)/_F$.

Let G be the vector subspace of $E^\star$ spanned by $F^{\perp}$ and $\xi_1,\dots,\xi_n$, and G_0 a supplement of $F^{\perp}$ in G, which is of finite dimension (at most n). Let $\eta_1,\dots,\eta_m$ $(m \leqslant n)$ be an algebraic basis of G_0. The restrictions to F of the linear functionals $\eta_1,\dots,\eta_m$ are linearly independent (if a linear combination $\alpha_1\eta_1 + \dots + \alpha_m\eta_m$ was equal to zero on F, it would belong to $F^{\perp}$, and this would contradict the definition of G_0).

We shall see that, for all $x \in E$, there is a $y \in F$, such that $\eta_j(x) = \eta_j(y)$, for $j = 1,\dots,m$. For this, let $e_1,\dots,e_m$ be points in F such that $\eta_i(e_j) = 0$ if $i \neq j$, $= 1$ if $i = j$ $(i,j = 1,\dots,m)$. For a given $x \in E$, we set $y = \sum_1^n \eta_j(x)e_j$: we have $y \in F$, and $\eta_j(y) = \eta_j(x)$, $j = 1,\dots,m$. Now, for $x \in E$, we set $x' = x - y$: we have $\eta(x') = 0$ for all $\eta \in G_0$, and therefore, we have a decomposition $x = x' + y$, where $x' \in G_0^{\perp}$ and $y \in F$. We also decompose, for $i = 1,\dots,n$:

$$\xi_i = f_i + f'_i \quad , \quad \text{where} \quad f_i \in F^{\perp} \quad \text{and} \quad f'_i \in G_o \ .$$

We obtain :

$$\xi_i(x') = f_i(x') + f'_i(x') = f_i(x') = f_i(x) \ .$$

Therefore, for all $x \in W_{\epsilon;f_1,\ldots,f_n}$, there is an $x' \in W_{\epsilon;\xi_1,\ldots,\xi_n}$ such that $x - x' \in F$: this proves that $p(W_{\epsilon;f_1,\ldots,f_n})$ is contained in $p(W_{\epsilon;\xi_1,\ldots,\xi_n})$, and ends the proof of our proposition.

COROLLARY 4. - *If* G *is a subspace of* $E^{\star}$ *, which is* $\sigma(E^{\star}, E)$*-closed, we have :*

$$\sigma(E/_{G^{\perp}} , G) = \sigma(E, E^{\star})\big|_{G^{\perp}} \ .$$

PROOF. - We come back to the situation of the previous proposition, putting $F = G^{\perp}$: we have $F^{\perp} = G$, since G is $\sigma(E^{\star}, E)$-closed.

EXERCISES ON CHAPTER II.

EXERCISE 1. - Show that if two Banach spaces E and F are isomorphic, their duals $E^{\star}$ and $F^{\star}$ are also isomorphic.

EXERCISE 2. (Banach's Limit). - Let E be the space of real bounded functions $x(s)$, defined on $[0, +\infty[$. For all $x \in E$, we put :

$$p(x) = \inf \{ \limsup_{s \to +\infty} \frac{1}{n} \sum_{k=1}^{n} x(s + \alpha_k) \ ; \ n \in \mathbb{N} , \ \alpha_1,\ldots,\alpha_n \in \mathbb{R}^+ \} \ .$$

a) Show that p is sub-additive and positively homogeneous.

b) We define on E a linear functional F , the following way : on the constant function 1, we put $F(1) = 1$, and we extend F to a linear functional dominated by p . We call $\operatorname{Lim}_{s \to +\infty} x(s)$ the value of this extension. Show that :

1) $\operatorname{Lim}(ax(s) + by(s)) = a \operatorname{Lim} x(s) + b \operatorname{Lim} y(s)$.

2) $\operatorname{Lim} x(s) \geqslant 0$, if $x(s) \geqslant 0$.

3) $\operatorname{Lim} x(s + s_o) = \operatorname{Lim} x(s)$ for every $s_o \geqslant 0$.

4) $\operatorname{Lim} 1 = 1$.

5) $\liminf_{s \to +\infty} x(s) \leqslant \text{Lim}\, x(s) \leqslant \limsup_{s \to +\infty} x(s)$

(for the 3°), one may choose the α_k's with $\alpha_1 = 0, \alpha_2 = s_0, \ldots, \alpha_n = (n-1)s_0$) .

c) Let $(\xi_n)_{n \in \mathbb{N}}$ be any bounded sequence of reals. We put :

$$x(s) = \xi_n \quad \text{if} \quad n - 1 < s \leqslant n \,, \quad \text{for} \quad n \geqslant 1 \,.$$

and $\text{Lim}_{n \to +\infty} \xi_n = \text{Lim}_{n \to +\infty} x(s)$. Show that :

1) $\text{Lim}(a\, \xi_n + b \eta_n) = a\, \text{Lim}\, \xi_n + b\, \text{Lim}\, \eta_n$.

2) $\text{Lim}\, \xi_n \geqslant 0$ if $\xi_n \geqslant 0$ for all $n \geqslant 0$.

3) $\text{Lim}\, \xi_{n+1} = \text{Lim}\, \xi_n$.

4) $\text{Lim}\, 1 = 1$.

5) $\liminf_{n \to +\infty} \xi_n \leqslant \text{Lim}_{n \to +\infty} \xi_n \leqslant \limsup_{n \to +\infty} \xi_n$.

EXERCISE 3. - Let $(f_n)_{n \in \mathbb{N}}$ be a sequence of linear functionals on ℓ_p $(1 \leqslant p < +\infty)$ (recall that f_n is given by an element of ℓ_q , $(\frac{1}{p} + \frac{1}{q} = 1)$. Show that $f_n \xrightarrow[n \to +\infty]{} f$ for $\sigma(\ell_q, \ell_p)$ if and only if one has simultaneously :

a) $\exists M > 0$, $\|f_n\|_{\ell^q} \leqslant M$, for all $n \in \mathbb{N}$

(that is : $\{(f_n)_{n \in \mathbb{N}}\}$ is bounded in ℓ_q).

b) For all $m \geqslant 0$, $f_n(m) \xrightarrow[n \to +\infty]{} f(m)$, where $f_n(m)$, $f(m)$ are the $m^{\underline{th}}$ coordinates of f_n and f .

EXERCISE 4. - Let $(f_n)_{n \in \mathbb{N}}$ be a sequence of linear functionals on L_p $(1 \leqslant p < +\infty)$ (recall that f_n is given by an element of L_q, $\frac{1}{p} + \frac{1}{q} = 1$).

Show that $f_n \xrightarrow[n \to +\infty]{} f$ for $\sigma(L_q, L_p)$ if and only if one has simultaneously :

a) $\exists M > 0$, $\|f_n\|_{L_q} \leqslant M$ for all $n \in \mathbb{N}$,

b) For all u , with $0 \leqslant u \leqslant 1$, $\int_0^u f_n(t)dt \longrightarrow \int_0^u f(t)dt$.

EXERCISE 5. - Let E be a Banach space, F a closed subspace of E. Show that $E/_F$, endowed with the quotient norm, is a Banach space.

EXERCISE 6. - Let E be a Banach space. Show that a linear functional z on $E^\star$ is continuous for $\sigma(E^\star, E)$ if and only if its restriction to $\mathscr{B}_{E^\star}$ is continuous for this topology.

REFERENCES ON CHAPTER II.

The elements of Functional Analysis which we give in this chapter can be found in every book dealing with these topics. Let us mention, in particular, G. CHOQUET [12], G. KÖTHE [31] and N. BOURBAKI [11]. Our presentations follows more closely these last two. The § 6 follows also I. EKELAND [17].

Exercises 2, 3, 4 come from Banach's book [5].

COMPLEMENTS ON CHAPTER II : HAHN-BANACH THEOREMS FOR OPERATORS.

The Hahn-Banach Theorem given in § 1 is valid only for linear functionals. One can wonder if such a theorem can exist for operators with values in another Banach space. One is led to the following definition :

DEFINITION. - *A Banach space* E *is said to be a* $(\mathscr{P}_\lambda)$*-space* $(\lambda \geqslant 1)$ *if one of the following equivalent properties holds :*

a) *For every Banach space* Z, *containing* E *as a subspace, there is a projection* P *from* Z *onto* E, *with* $\|P\| \leqslant \lambda$.

b) *For every Banach space* Z, *containing* E *as a subspace, every Banach space* Y, *every operator* T *from* E *into* Y, *there is an extension* $\tilde{T}$, *from* Z *into* Y, *with* $\|\tilde{T}\| \leqslant \lambda \|T\|$.

c) *For every Banach space* Y, *every operator* T *from* Y *into* E, *every Banach space* Z *containing* Y *as a subspace, there is an extension* $\tilde{T}$, *from* Z *into* E, *with* $\|\tilde{T}\| \leqslant \lambda \|T\|$.

The proof of the equivalence of these properties can be found in M.M. DAY's book [13].

The analogue of Hahn-Banch Theorem for operators corresponds to the case $\lambda = 1$, that is, to $(\mathscr{P}_1)$-spaces. These spaces were characterized by L. NACHBIN [37], in terms of intersection properties of balls.

We say that E has the binary intersection property (in short B.I.) if, for any collection of closed balls $(B_i)_{i \in I}$ in E , if $B_i \cap B_j$ is non-empty for all $i,j \in I$, then $\bigcap_{i \in I} B_i$ is non-empty. In [37], L. proved the following result :

THEOREM. - E *is a* $(\mathscr{P}_1)$*-space if and only if* E *has B.I.*

The spaces having this property were completely determined by L. NACHBIN, D.B. GOODER and J.L. KELLEY (see [13] for a proof) :

THEOREM. - E *has B.I. if and only if* E *is the space of continuous functions on an extremally disconnected compact set.*

(A set is extremally disconnected if the closure of any open set is open).

CHAPTER III

REFLEXIVE BANACH SPACES ; SEPARABLE BANACH SPACES

§ 1. REFLEXIVITY.

We have seen that the elements of E defined linear functionals on $E^\star$, by the formula $\xi \to \langle x, \xi \rangle$, $x \in E$. We say that a Banach space is *reflexive* if all the continuous linear functionals on $E^\star$ are obtained this way ; in other terms, if $E^{\star\star} = E$. Since $E^{\star\star}$ is always complete for the norm, this equality is possible only if E is complete, that is, if E is a Banach space. But not all the Banach spaces are reflexive : we have seen that if $E = c_0$, then $E^{\star\star} = \ell_\infty$. On the contrary, the spaces ℓ_p $(1 < p < +\infty)$ or L_p $(1 < p < +\infty)$ are reflexive.

In this paragraph, we shall study the properties of reflexive Banach spaces, and give "geometric" conditions for reflexivity.

First, if E is reflexive, the unit ball $\mathscr{B}_E$ is $\sigma(E, E^\star)$-compact, since we know that $\mathscr{B}_{E^{\star\star}}$ is $\sigma(E^{\star\star}, E^\star)$-compact, and $\mathscr{B}_E = \mathscr{B}_{E^{\star\star}}$. Conversely, this property characterizes reflexivity :

PROPOSITION 1. - E *is reflexive if and only if its unit ball* $\mathscr{B}_E$ *is* $\sigma(E, E^\star)$-*compact.*

PROOF. - We assume $\mathscr{B}_E$ to be $\sigma(E, E^\star)$-compact. Since the canonical embedding j of E into $E^{\star\star}$ is continuous from $\sigma(E, E^\star)$ into $\sigma(E^{\star\star}, E^\star)$ (by definition of $\sigma(E, E^\star)$), $\mathscr{B}_E$ is $\sigma(E^{\star\star}, E^\star)$-compact in $E^{\star\star}$. Since it is dense in $\mathscr{B}_{E^{\star\star}}$ for $\sigma(E^{\star\star}, E^\star)$ (chapter II, § 5, prop.4), we have $\mathscr{B}_{E^{\star\star}} = \mathscr{B}_E$, j is surjective, and E is reflexive.

COROLLARY 2. - *If* E *is reflexive, its dual* $E^\star$ *is also reflexive.*

PROOF. - We know that $\mathscr{B}_{E^\star}$ is $\sigma(E^\star, E)$-compact. If E is reflexive, $\mathscr{B}_{E^\star}$ is $\sigma(E^\star, E^{\star\star})$ compact, and by proposition 1, $E^\star$ is reflexive.

COROLLARY 3. - *If* E *is reflexive, so are all its closed subspaces.*

PROOF. - Let F be a closed subspace. The unit ball $\mathscr{B}_F$ is $\sigma(F, F^\star)$-compact, since $\sigma(F, F^\star) = \sigma(E, E^\star)|_F$ (chap. II, § 7, prop. 1), and since $\mathscr{B}_F$ is $\sigma(E, E^\star)$-closed.

COROLLARY 4. - *If* $E^\star$ *is reflexive, so is* E .

PROOF. - If $E^\star$ is reflexive, so is $E^{\star\star}$ (cor. 2), and so is E (cor. 3).

COROLLARY 5. - *If* E *is reflexive, for any continuous functional on* E *there is at least one point* x_0 *in the unit ball of* E *such that :*

$$(1) \qquad \xi(x_0) = \sup_{\|x\|_E \leqslant 1} |\xi(x)| .$$

PROOF. - Let $\xi \in E^\star$. Then ξ is continuous on $\mathscr{B}_E$ endowed with $\sigma(E, E^\star)$ (chap. II, § 3, prop. 4) ; this set is compact if E is reflexive. Therefore, there is a point $x_0 \in \mathscr{B}_E$ such that $|\xi(x_0)| = \sup_{\|x\|_E \leqslant 1} |\xi(x)|$. Replacing, if necessary, x_0 by $\pm x_0$ (if the space is real) or by $e^{i\theta} x_0$ (if the space is complex), we obtain (1), and, obviously, $\|x_0\| = 1$.

This point x_0 (which is not unique in general) is called a point at which ξ attains its norm.

Conversely, let us give an example of a Banach space and a linear functional on it which does not attain its norm. We consider the Banach space $c_0 = \{(x_n)_{n \in \mathbb{N}} \; ; \; x_n \in \mathbb{R} \text{ for all } n \, , \; x_n \xrightarrow[n \to +\infty]{} 0\}$, and the linear functional y given by the element of ℓ_1 : $y = (y_n)_{n \in \mathbb{N}}$, with $y_0 = 0$, $y_n = \frac{1}{2^n}$, for $n \geqslant 1$. If $x \in c_0$, $x = (x_n)_{n \in \mathbb{N}}$, then $< x , y > = \sum_n x_n y_n$. We have $\|y\|_{\ell_1} = 1$, but, for every $x = (x_n)_{n \in \mathbb{N}}$ such that $x_n \xrightarrow[n \to +\infty]{} 0$ and $\sup_n |x_n| \leqslant 1$, then $|\sum_n x_n y_n| < 1$. So y does not attain its norm on $\mathscr{B}_{c_0}$.

More generally, it has been proved by R.C. James [23] that, if every linear functional attains its supremum on the unit ball, the space must be reflexive.

As other examples of non-reflexive spaces, we shall meet L_1 , $\mathscr{C}([0,1])$ (continuous functions on [0,1]), L_∞ : we shall see later that these spaces have a subspace isomorphic to ℓ_1 , and so by corollary 3, they cannot be reflexive.

We shall now give a geometric characterization of non-reflexivity. We call it geometric, opposed to topological, because, whereas our previous characterization used weak compactness of the unit ball (which is pure topology), this one will use the structure of a sequence of points (or of two sequences, one of points, one of linear functionals). This result, which will be of great help for the sequel, is due to R.C. James [25].

THEOREM 6. - *For a Banach space* E , *the following conditions are equivalent* :

a) E *is not reflexive*

b) *For every* θ , $0<\theta<1$, *there is a sequence of points* $(x_k)_{k\in\mathbb{N}}$ *in* E , *with* $\|x_k\|=1$ *for all* k , *and a sequence of linear functionals* $(f_n)_{n\in\mathbb{N}}$, *with* $\|f_n\|=1$ *for all* n , *such that* :

$$f_n(x_k) = \theta \quad \text{if} \quad n\leqslant k$$
$$\qquad = 0 \quad \text{if} \quad n>k$$

c) *For every* θ , $0<\theta<1$, *there is a sequence* $(x_k)_{k\in\mathbb{N}}$ *of points in* E , *with* $\|x_k\|=1$ *for all* k , *such that* :

$$\left\{\begin{array}{l} \textit{For all } K\geqslant 1 \text{, all } k\leqslant K \text{, if } \alpha_1,\ldots,\alpha_k,\alpha_{k+1},\ldots,\alpha_K \\ \textit{are positive numbers satisfying } \sum_{i=1}^{k}\alpha_i = \sum_{i=k+1}^{K}\alpha_i = 1 \text{, then :} \\ \left\| \sum_1^k \alpha_i x_i - \sum_{k+1}^K \alpha_i x_i \right\| \geqslant \theta \ . \end{array}\right.$$

Since the set $\{\sum_1^k \alpha_i x_i \ ; \ k\geqslant 1 \ , \ \alpha_i\geqslant 0 \ , \ \sum_1^k \alpha_i = 1\}$ is the convex hull of $\{x_1,\ldots,x_k\}$ (which we call $\text{conv}(x_1,\ldots,x_k)$), this last condition can be written in short :

$$\left\{\begin{array}{l} \text{for all } K\geqslant 1 \text{, all } k\leqslant K \text{,} \\ \text{dist}(\text{conv}(x_1,\ldots,x_k) \ , \ \text{conv}(x_{k+1},\ldots)) \geqslant \theta \ , \end{array}\right.$$

and we shall call it "*condition* J " (for James). One interesting thing about this condition is that it allows to check that a space is non-reflexive without knowing its dual.

PROOF. - a) $\Rightarrow$ b). We assume that E is not reflexive. As before, we call j the canonical embedding of E into E^{**}. We know that $j(E)$ is a closed subspace of E^{**} (for the norm), which is not the whole space E^{**} : therefore, we can find a continuous linear functional on E^{**}, α, of norm 1, which is equal to zero on $j(E)$. Let θ, $0 < \theta < 1$. We choose $F \in E^{**}$, with $\|F\| < 1$, such that $\alpha(F) > \theta$.

We shall show by induction that there exist a sequence $(x_k)_{k \in \mathbb{N}}$ of points in E, a sequence $(f_n)_{n \in \mathbb{N}}$ of linear functionals, such that :

- $\|x_k\| = 1$, $\|f_n\| = 1$, for all $n, k \in \mathbb{N}$
- $F(f_n) = \theta$ for all $n \in \mathbb{N}$,
- $f_n(x_k) = \theta$ if $n \leqslant k$

  $= 0$ if $n > k$.

For this, we shall use the following lemma, which is called "Helly's Condition" :

LEMMA 7. (Helly's Condition). - *Let* E *be a normed space,* $f_1, \ldots, f_n$ *be continuous linear functionals,* $M > 0$, *and* $c_1, \ldots, c_n$ *scalars. Then, for every* $\epsilon > 0$, *one can find a point* $x \in E$, *with* $\|x\| = M + \epsilon$ *and* $f_k(x) = c_k$ *for* $k = 1, \ldots, n$, *if and only if, for every sequence* $a_1, \ldots, a_n$ *of scalars, one has :*

$$\text{(H)} \qquad \left| \sum_{i=1}^{n} a_i c_i \right| \leqslant M \left\| \sum_{i=1}^{n} a_i f_i \right\| .$$

PROOF OF LEMMA 7. - Assume that there is a point $x \in E$ with $\|x\| = M + \epsilon$, and $f_k(x) = c_k$, for $k = 1, \ldots, n$. Then, for every sequence of scalars $a_1, \ldots, a_n$, we have :

$$\left| \sum_1^n a_i c_i \right| = \left| \sum_1^n a_i f_i(x) \right| \leqslant \|x\| \left\| \sum_1^n a_i f_i \right\| = (M + \epsilon) \left\| \sum_1^n a_i f_i \right\| ,$$

from which (H) follows.

Conversely, assume (H) to be satisfied. We may assume that the f_i's are not all equal to zero (otherwise, we have $c_i = 0$ for all i, and we take any x with $\|x\| = M + \epsilon$). Among the n f_i's, we can find at most r $(r \leqslant n)$ which are linearly independent. We assume, for the sake of simplicity, that it is $f_1, \ldots, f_r$. We call $H_i = \{x \in E \ ; \ f_i(x) = x_i\}$, for $i = 1,\ldots,n$, and $F = \bigcap_{i=1,\ldots,r} H_i$ is an affine subspace of codimension r . From (H) follows also that $F = \bigcap_{i=1,\ldots,n} H_i$.

We put $a = \inf\{\|x\| \,;\, x \in F\}$. Then F does not meet the open ball, centered at 0 , of radius a . By the geometric form of Hahn-Banach Theorem (chap. II, § 4, th. 1), there is an affine hyperplane $\mathscr{H}$ containing F , which does not meet this open ball. This means that there is a continuous linear functional f and a scalar c such that $F \subset \mathscr{H} = \{x \in E \,;\, f(x) = c\}$, and we have :

$$\inf_{x \in \mathscr{H}} \|x\| = \frac{|c|}{\|f\|} \geqslant a .$$

But $F \subset \mathscr{H}$, and therefore the n conditions $f_i(x) = c_i$, $i = 1,\ldots,n$ imply $f(x) = c$. It follows that there exist scalars a_i , $i = 1,\ldots,n$, such that :

$$f = \sum_{i=1}^{n} a_i f_i \quad \text{and} \quad c = \sum_{i=1}^{n} a_i c_i .$$

By (H), we have :

$$\frac{|c|}{\|f\|} = \frac{\left|\sum_{i=1}^{n} a_i c_i\right|}{\left\|\sum_{1}^{n} a_i f_i\right\|} \leqslant M ,$$

and therefore $a \leqslant M$. This indicates that for every $\epsilon > 0$, we can find $x_0 \in F$ with $\|x_0\| \leqslant M + \epsilon$. Now, take $x' \in E$ such that $f_i(x') = 0$, $i = 1,\ldots,n$, and $\|x'\| = 1$. Then, for some $\lambda \in \mathbb{R}$, we may choose $x = \lambda x_0 + (1 - \lambda)x'$. This proves the lemma.

Let us come back to the proof of a) $\Rightarrow$ b) in the theorem. We observe first that, since $\text{dist}(F, j(E)) > \theta$, we have $\|F\| > \theta$. Therefore, there is an $f_1 \in E^\star$, with $\|f_1\| = 1$, such that $F(f_1) = \theta$. Thus $\|f_1\| > |F(f_1)| = \theta$. We then choose $x_1 \in E$, with $\|x_1\| = 1$, and $f_1(x_1) = \theta$. This is the first step of the induction. Let us now assume that x_k and f_k have been chosen, following the prescribed conditions up to $k = p - 1$. We now choose f_p , with $\|f_p\| = 1$, $F(f_p) = \theta$ and $f_p(x_k) = 0$ if $k < p$. This choice is possible, because the following Helly's Condition is satisfied :

$$\theta \leqslant \frac{\theta}{\text{dist}(F,j(E))} \left\| \sum_{i=1}^{p-1} a_i \; j(x_i) + F \right\|_{E^{\star\star}} , \quad \text{for all } a_1,\dots,a_{p-1} .$$

We must now choose x_p . We have :

$$\left| \sum_{i=1}^{p} a_i \, \theta \right| = \left| \sum_{i=1}^{p} a_i \; F(f_i) \right| \leqslant \|F\| \cdot \left\| \sum_{i=1}^{p} a_i f_i \right\| .$$

Since $\|F\| < 1$, it is possible to choose x_p with $\|x_p\| = 1$, and $f_i(x_p) = \theta$ if $i \leqslant p$: this gives by induction the construction of the sequences $(x_k)_{k \in \mathbb{N}}$ and $(y_n)_{n \in \mathbb{N}}$, and shows a) $\Rightarrow$ b).

b) $\Rightarrow$ c). Let $K \geqslant 1$, $1 \leqslant k \leqslant K$, $\alpha_1,\dots,\alpha_k , \alpha_{k+1},\dots,\alpha_K$ be positive numbers, with $\sum_1^k \alpha_i = \sum_{k+1}^K \alpha_i = 1$. We have :

$$\theta = f_{k+1}\left(\sum_1^k \alpha_i x_i - \sum_{k+1}^K \alpha_i x_i \right) \leqslant \|f_k\| \cdot \left\| \sum_1^k \alpha_i x_i - \sum_{k+1}^K \alpha_i x_i \right\| ,$$

and c) is proved.

c) $\Rightarrow$ a). Assume that c) holds for some θ , $0 < \theta < 1$. We shall prove that E is not reflexive. Assume on the contrary that E is reflexive. For $k \in \mathbb{N}$, we put $A_k = \{x_k , x_{k+1} , \dots\}$, and $B_k = \text{conv}\, A_k$. The sets $\overline{B_k}$ (closure for the norm, or for $\sigma(E, E^\star)$: both coïncide) form a decreasing sequence of compact sets, since the ball $\mathscr{B}_E$ is $\sigma(E, E^\star)$ compact. Therefore, $\bigcap_{k \in \mathbb{N}} \overline{B_k}$ must be non-empty. Let y_o be a point in it. In particular, x_o belongs to the norm-closure of B_o : there is a (finite) linear combination $\alpha_1 x_1 + \dots + \alpha_\ell x_\ell$, with α_i's $\geqslant 0$, $\sum_1^\ell \alpha_i = 1$, such that $\left\| \sum_{i=1}^\ell \alpha_i x_i - y_o \right\| \leqslant \frac{\theta}{3}$ (we have called ℓ the last index in the combination). Since y_o is also in the norm-closure of $B_{\ell+1}$, there is another linear combination $\alpha_{\ell+1} x_{\ell+1} + \dots + \alpha_L x_L$, with α_i's $\geqslant 0$, $\sum_{\ell+1}^L \alpha_i = 1$, such that $\left\| \sum_{i=\ell+1}^L \alpha_i x_i - y_o \right\| \leqslant \frac{\theta}{3}$. Finally, we obtain :

$$\left\| \sum_1^\ell \alpha_i x_i - \sum_{\ell+1}^L \alpha_i x_i \right\| \leqslant \left\| \sum_1^\ell \alpha_i x_i - y_o \right\| + \left\| y_o - \sum_{\ell+1}^L \alpha_i x_i \right\| \leqslant \frac{2\theta}{3} ,$$

and this contradicts (J), and proves the theorem.

REMARKS.

1) We only used, in this last proof, the fact that c) holds for some θ ,

$0 < \theta < 1$. Therefore, c) is also equivalent to :

(c') : For some θ , $0 < \theta < 1$, there is a normalized sequence $(x_k)_{k \in \mathbb{N}}$ such that (J) holds.

2) This last proof shows that a sequence satisfying condition (J) cannot be weakly convergent. Since (J) applies also to the subsequences of the given sequence, it shows in fact that a sequence satisfying condition (J) cannot have any weakly convergent subsequence. So we obtain :

COROLLARY 8. - *If, in a Banach space* E , *every bounded sequence has a weakly convergent subsequence, then* E *is reflexive.*

A very simple example of a sequence satisfying (J) is the canonical basis of ℓ_1 : so we have a new proof (slightly longer !) of the fact that ℓ_1 is not reflexive. But we shall have many opportunities of using James'Theorem in its full strength.

We shall now turn to the study of the Banach spaces which, though they are infinite-dimensional, do not have a too large cardinality.

§ 2. SEPARABLE BANACH SPACES.

We say that a normed space E is separable if there is a subset A of E which is countable and dense in E . It is the case, as one sees easily for the spaces ℓ_1 , ℓ_2 , ℓ_p $(1 \leqslant p < +\infty)$: the finite sequences, with rational coefficients, form a dense and countable set. It is not the case for ℓ_∞ (cf. exercise 4).

More generally, we shall say that a set M contained in a normed space E is separable if there exists a subset $A \subset M$, which is countable and dense in M (that is, $M \subset \overline{A}$) .

PROPOSITION 1. - *For a normed space* E , *the following assertions are equivalent :*

1) E *is separable,*
2) *The unit ball* $\mathcal{B}_E$ *is separable,*
3) *The unit sphere* $\mathcal{S}_E = \{x \in E \ ; \ \|x\| = 1\}$ *is separable.*

PROOF. - 1) $\Rightarrow$ 2). If A is dense in E , $A \cap \mathscr{B}_E$ is dense in $\mathscr{B}_E$.

2) $\Rightarrow$ 3). If $(u_n)_{n \in \mathbb{N}}$ is a dense sequence in $\mathscr{B}_E$ (the previous set, which was countable, and has been enumerated), then $\left(\frac{u_n}{\|u_n\|} \right)_{n \in \mathbb{N}}$ is dense in $\mathscr{S}_E$.

3) $\Rightarrow$ 1). Let $(u_n)_{n \in \mathbb{N}}$ be a dense sequence in $\mathscr{S}_E$. Then the set $A = \{\lambda u_n \; ; \; n \in \mathbb{N} , \lambda \in Q^+\}$ is countable and is dense in E .

PROPOSITION 2. - *If* E *is separable, so are all its subspaces.*

PROOF. - Let $(u_n)_{n \in \mathbb{N}}$ be a dense sequence in E . We consider the family of the balls centered at u_n's , with radii $\frac{y}{m}$, for $m \geqslant 1$: let $B(u_n , \frac{1}{m})$ be these balls. Let $A_{n,m} = F \cap B(u_n , \frac{1}{m})$, $n \in \mathbb{N}$, $m \geqslant 1$. The reunion of the sets $A_{n,m}$, $n \in \mathbb{N}$, $m \geqslant 1$, covers F (but some, of course, are empty). In each of the non-empty $A_{n,m}$, we choose a point $x_{n,m}$: we obtain this way a countable family which, as one checks immediately, is dense in F .

REMARK. - The definition we have given for separability extends, more generally, to topological vector spaces. But the previous proposition does not hold in this larger frame.

PROPOSITION 3. - *If* E *is separable and infinite dimensional, there is a dense and linearly independent sequence* $(v_n)_{n \in \mathbb{N}}$ *in* E .

PROOF. - First recall that a sequence is said to be linearly independent if every *finite* sub-family is linearly independent.

Let $(u_n)_{n \in \mathbb{N}}$ be a dense sequence in E . We take $v_0 = u_0$. Assume $v_1, \ldots, v_k$ have been chosen, in order to satisfy :
$\|v_1 - u_1\| < \frac{1}{2}$, $\|v_2 - u_2\| < \frac{1}{2^2}$, ... , $\|v_k - u_k\| < \frac{1}{2^k}$. Then we choose v_{k+1} , independent of $v_0, v_1, \ldots, v_k$, and satisfying $\|v_{k+1} - u_{k+1}\| < \frac{1}{2^{k+1}}$. This is possible, since the open ball of center u_{k+1} and radius $\frac{1}{2^{k+1}}$ has non-empty interior, and therefore cannot been contained in $\mathrm{span}\{v_0, v_1, \ldots, v_k\}$, which is finite dimensional. Therefore, we build the sequence $(v_n)_{n \in \mathbb{N}}$ by induction : it is independent, by construction, and it is dense because, for every $x \in E$, there is a subsequence $(u_{n_j})_{j \in \mathbb{N}}$ which converges to x , and the subsequence $(v_{n_j})_{j \in \mathbb{N}}$ also converges to x .

This proposition, which is very easy to prove (as the reader has seen) illustrates well the difference with the finite-dimensional case, where such a thing is obviously false : in a sense, one has much more freedom when the space is infinite-dimensional.

The main interest of separable spaces is that we shall be allowed, in them, to use sequences to test continuity or convergence (instead of filters, which are usually required). One knows that sequences suffice if the space is metrizable. This is of course so when we consider the norms, in E or $E^\star$, but is not so when we turn to weak topologies $\sigma(E, E^\star)$ or $\sigma(E^\star, E)$: these topologies, on the whole spaces E or $E^\star$, are not metrizable if the space is infinite dimensional. But, if we restrict ourselves to balls, the situation may be different :

PROPOSITION 4. - *If* E *is separable, the balls of* $E^\star$ *are metrizable for the topology* $\sigma(E^\star, E)$, *and there exists a countable set which is dense in* $E^\star$ *for* $\sigma(E^\star, E)$ *(one sometimes says that* $E^\star$ *is weakly separable).*

PROOF. - Let $(u_n)_{n\in\mathbb{N}}$ be a dense sequence in E. We shall show that the unit ball $\mathcal{B}_{E^\star}$ is metrizable. For this, we consider the following family of semi-distances, defined on $E^\star$ by :

$$d_n(f,g) = |f(u_n) - g(u_n)| \quad \text{for} \quad f,g \in E^\star$$

(this is only a semi-distance, since $d_n(f,g) = 0$ does not imply $f = g$). This family is countable, and therefore defines a metrizable topology, which we call $\mathcal{T}$. We shall now show that, on $\mathcal{B}_{E^\star}$, $\mathcal{T}$ coïncides with $\sigma(E^\star, E)$.

Let $V_{\epsilon;x_1,\dots,x_n}$ be a neighbourhood of 0 for $\sigma(E^\star, E)$, and $u_{m_1},\dots,u_{m_n}$ be points of the family $(u_j)_{j\in\mathbb{N}}$ such that $\|u_{m_i} - x_i\| < \frac{\epsilon}{2}$, for $i = 1,\dots,n$. We now define a neighbourhood of 0 for $\mathcal{T}$, by :

$$V' = \{f \in \mathcal{B}_{E^\star} \ ; \ d_{m_i}(f,0) < \frac{\epsilon}{2}, \quad \text{for} \quad i = 1,\dots,n\} .$$

Then $V' \subset V_{\epsilon;x_1,\dots,x_n}$: if $d_{m_i}(f,0) < \frac{\epsilon}{2}$, then, for $i = 1,\dots,n$, $|f(u_{m_i})| < \frac{\epsilon}{2}$, and thus :

$$|f(x_i)| \leqslant |f(u_{m_i})| + |f(x_i - u_{m_i})| \leqslant \frac{\epsilon}{2} + \|f\| \cdot \|x - u_{m_i}\| < \epsilon$$

Conversely, let W be a neighbourhood of 0 of $\mathcal{T}$; it contains a neighbourhood of 0 of the form :

$$W' = \{f \in \mathcal{B}_{E^\star} \; ; \; d_{m_1}(f,0) < \epsilon, \ldots, d_{m_n}(f,0) < \epsilon\} ,$$

for some $\epsilon > 0$ and some indices $m_1, \ldots, m_n$. Therefore, we have :

$$W' = \{f \in \mathcal{B}_{E^\star} \; ; \; |f(u_{m_i})| < \epsilon \quad \text{for} \quad i = 1, \ldots, n\} ,$$

and this is a neighbourhood of 0 for $\sigma(E^\star, E)$. This proves the first part of our proposition.

In order to prove the second, we shall first observe that the restriction of $\sigma(E^\star, E)$ to $\mathcal{B}_{E^\star}$ has a countable basis. Indeed, this is the case, since the sets of the form (if E is real)

$$\mathcal{O}_{m_1, \ldots, m_n \, ; \, I_1, \ldots, I_n} = \{f \in \mathcal{B}_{E^\star} \; ; \; f(u_{m_1}) \in I_1, \ldots, f(u_{m_n}) \in I_n\}$$

where $n \in \mathbb{N}$, $m_1, \ldots, m_n \in \mathbb{N}$, and $I_1, \ldots, I_n$ and open intervals with rational extremities, constitute a basis for this topology, and the family of these sets is countable. If E is complex, we replace this condition by $f(u_{m_1}) \in I_1 \times iI_1'$, and so on, where I and I' have rational end points.

So, in order to end the proof of our proposition, all we have to do is to prove the following lemma :

LEMMA 5. - *Let* G *a metrizable topological space. Its topology has a countable basis if and only if there exists a countable set in* G *, dense in* G *.*

PROOF OF THE LEMMA. - The condition is necessary, because if $(B_n)_{n \in \mathbb{N}}$ is a basis of the topology of G , and if one chooses $b_n \in B_n$ for all n , the set $\{(b_n)_{n \in \mathbb{N}}\}$ is dense. Conversely, if there exists in G a dense sequence $(u_n)_{n \in \mathbb{N}}$, if $\overset{\circ}{B}_{m,n}$ is the open ball of center u_n , radius $\frac{1}{m}$ $(n \in \mathbb{N}, m \geqslant 1)$, the balls $\overset{\circ}{B}_{m,n}$ are a basis of the topology.

PROPOSITION 6. - *If the dual* $E^\star$ *is separable, the space* E *itself is separable. (The converse is not true, as the example of* ℓ_1 *,* ℓ_∞ *, shows).*

PROOF. - If $E^\star$ is separable, by the previous proposition, the ball $\mathcal{B}_{E^{\star\star}}$ is metrizable for $\sigma(E^{\star\star}, E^\star)$, and this topology, restricted to $\mathcal{B}_{E^{\star\star}}$, has a countable basis. Therefore, the topology $\sigma(E, E^\star)$, restricted to $\mathscr{B}_E$ (which is the topology induced on $\mathscr{B}_E$ by $\sigma(E^{\star\star}, E^\star)$) has a countable basis. The previous lemma then says that there is a countable family $(u_n)_{n\in\mathbb{N}}$ which is dense in $\mathscr{B}_E$ for $\sigma(E, E^\star)$.

The convex hull $\mathrm{conv}\{(u_n)_{n\in\mathbb{N}}\}$ is therefore, a fortiori, dense in $\mathscr{B}_E$ for $\sigma(E, E^\star)$. By Chapter II, § 5, prop. 1, it is also dense for the norm-topology. If now we consider :

$$A = \{\Sigma\lambda_i u_i \ ; \ (\lambda_i) \text{ finite sequence of positive rationals, with } \Sigma\lambda_i \leqslant 1\},$$

we obtain a countable set. This set is dense in $\mathscr{B}_E$. To see this, it suffices to show that it is dense in $\mathrm{conv}\{(u_n)_{n\in\mathbb{N}}\}$. But, if

$x = \sum_{i=1}^{N} \alpha_i u_i$, $\alpha_i \in \mathbb{R}$, $\alpha_i > 0$, $\sum_{i=1}^{N} \alpha_i = 1$, we choose, for a given $\epsilon > 0$, rationals λ_i with $0 \leqslant \lambda_i \leqslant \alpha_i$ and $\alpha_i - \lambda_i < \frac{\epsilon}{N}$, for $i = 1,\dots,N$, and we have, if $x' = \Sigma\lambda_i u_i$,

$$\|x - x'\| \leqslant \sum^{N} (\alpha_i - \lambda_i) < \epsilon .$$

This proves that $\mathscr{B}_E$ is separable, and therefore E is, by proposition 1.

§ 3. THE UTILIZATION OF CONVERGENT SEQUENCES IN ORDER TO DEFINE A TOPOLOGY ; WEAK COMPACTNESS.

We have seen that, if E was separable, the unit ball $\mathscr{B}_{E^\star}$ was a metrizable compact for $\sigma(E^\star, E)$. On this ball, we shall therefore be allowed to use sequences, instead of filters, to define the topology. For example, if we want to show that a function φ, defined on $\mathscr{B}_{E^\star}$, and real-valued, is continuous at $x_0 \in \mathscr{B}_{E^\star}$ for $\sigma(E^\star, E)$, we only have to show that, for every sequence $(x_n)_{n\in\mathbb{N}}$ converging to x_0, for $\sigma(E^\star, E)$, $\varphi(x_n) \xrightarrow[n\to+\infty]{} \varphi(x)$.

If $E^\star$ is separable, the balls of $E^{\star\star}$ will be metrizable for $\sigma(E^{\star\star}, E^\star)$, and therefore the balls of E for $\sigma(E, E^\star)$: here again, but now in E, one can use sequences.

If one wants to study the weak compactness of a subset of E , on can also use sequences. This is a much deeper fact, which does not depend on the separability of E or $E^{\star}$, to which we shall now turn our attention.

PROPOSITION 1. - *Let* E *be a Banach space,* A *a subset of* E . *If* A *is relatively weakly compact (that is, if the closure* $\overline{A}$ *of* A *for* $\sigma(E, E^{\star})$ *is compact for this topology), every sequence of points in* A *contains a subsequence which is weakly convergent (to a point of the closure* $\overline{A}$ *).*

PROOF. - Let $(x_n)_{n\in\mathbb{N}}$ a sequence of points in A . Let E_0 be the closed subspace spanned by the x_n's . This vector space is separable, and consequently, its dual $E_0^{\star}$ is weakly separable. Let H be a countable set of linear functionals on E_0 , dense in $E_0^{\star}$ for $\sigma(E_0^{\star}, E_0)$. By Hahn-Banach Theorem, we can extend each element of H into a linear functional on E . We then enumerate H : $h_1, h_2, \dots, h_n, \dots$.

The set $\{h_1(x_n) \; ; \; n \in \mathbb{N}\}$ is bounded in $\mathbb{K}$ (since it is contained in $h_1(\overline{A})$, which is compact) ; therefore there is a subsequence $(x_n^{(1)})_{n\in\mathbb{N}}$, contained in $(x_n)_{n\in\mathbb{N}}$, such that $(h_1(x_n^{(1)}))_{n\in\mathbb{N}}$ converges in $\mathbb{K}$. The same way, there is a subsequence $(x_n^{(2)})_{n\in\mathbb{N}}$, contained in $(x_n^{(1)})_{n\in\mathbb{N}}$, such that $(h_2(x_n^{(2)}))_{n\in\mathbb{N}}$ converges in $\mathbb{K}$. By repeating this process, we extract, for all $m \geqslant 1$, a subsequence $(x_n^{(m)})_{n\in\mathbb{N}}$, such that $(h_m(x_n^{(m)}))_{n\in\mathbb{N}}$ converges in $\mathbb{K}$. We consider the "diagonal subsequence", $(y_n)_{n\in\mathbb{N}} = (x_n^{(n)})_{n\in\mathbb{N}}$. By construction, for each $m \geqslant 1$, $(h_m(y_n))_{n\in\mathbb{N}}$ converges.

Since A is weakly relatively compact, every sequence in A has an accumulation point for $\sigma(E, E^{\star})$. Let y_0 be an accumulation point of $(y_n)_{n\in\mathbb{N}}$. We shall prove that $(y_n)_{n\in\mathbb{N}}$ is weakly convergent to y_0 .

Assume that this is not the case. Then, we can find $\epsilon > 0$, $f \in E^{\star}$ and a subsequence $(y_{n_k})_{k\in\mathbb{N}}$ of $(y_n)_{n\in\mathbb{N}}$ such that :

$$(1) \quad |f(y_{n_k}) - f(y_0)| \geqslant \epsilon \qquad \text{for all} \quad k \geqslant 1 .$$

But the sequence $(y_{n_k})_{k \in \mathbb{N}}$ has itself an accumulation point, which we call y'_0. The point y'_0 belongs to E_0, and $h(y_{n_k}) \xrightarrow[k \to +\infty]{} h(y'_0)$, for all $h \in H$ (by definition of an accumulation point, for each $h \in H$, each $\epsilon > 0$, the set $\{y \in E \; ; \; |h(y) - h(y'_0)| < \epsilon\}$ contains infinitely many of the y_{n_k}'s, therefore the limit $\lim_{k \to +\infty} h(y_{n_k})$ satisfies $\left| \lim_{k \to +\infty} h(y_{n_k}) - h(y'_0) \right| < \epsilon$). Therefore, $h(y_0) = h(y'_0)$, for all H. But H is $\sigma(E^\star, E)$ dense in $E^\star$, and we obtain

$$f(y_0) = f(y'_0) \quad \text{for all} \quad f \in E \text{ , that is } \quad y_0 = y'_0 \text{ ,}$$

which contradicts (1), and proves our proposition.

COROLLARY 2. - *In a reflexive space, every bounded sequence has a weakly convergent subsequence.*

Conversely, we have seen as a corollary of James' Theorem (§ 1) that, *when* A *is the unit ball of the space*, if A is not weakly compact (that is, if the space is not reflexive), there is a sequence of points in A which has no weakly convergent subsequence.

This is true more generally, but is harder to prove : this fact is known as the Eberlein-Šmulian Theorem :

THEOREM (EBERLEIN-ŠMULIAN). - *Let* E *be a Banach space,* A *a subset of* E, $\bar{A}$ *its closure for* $\sigma(E, E^\star)$.

Then $\bar{A}$ *is weakly compact if and only if, from every sequence in* A *one can extract a subsequence which is weakly convergent (to a point of* $\bar{A}$*).*

We shall not have the opportunity, here, to use this result in its full strength : Proposition 1 and the Corollary of James' Theorem which we just recalled will be sufficient for us. Therefore, we shall not prove this result, and we refer the reader to (for example) Dunford-Schwartz [16] (p.430). See also exercise 10 below.

EXERCISES ON CHAPTER III.

EXERCISE 1. - Let E be a reflexive space and F a closed subspace of E. Prove that the quotient $E/_F$ is reflexive.

EXERCISE 2. - Find in the unit ball of c_0 a sequence satisfying "Condition J".

EXERCISE 3. - Let $(E_n)_{n\in\mathbb{N}}$ be a sequence of Banach spaces. We set, for $1 \leqslant p < +\infty$,

$$(\prod_{n\in\mathbb{N}} E_n)_p = \{(x_n)_{n\in\mathbb{N}} \ ;\ x_n \in E_n \ \text{ for all } n \text{, and } \sum_{n\in\mathbb{N}} \|x_n\|_{E_n}^p < +\infty\}$$

endowed with the norm

$$\|(x_n)_{n\in\mathbb{N}}\| = (\sum_{n\in\mathbb{N}} \|x_n\|_{E_n}^p)^{1/p} .$$

a) Show that if $1 \leqslant p < +\infty$, the dual of $(\prod_{n\in\mathbb{N}} E_n)_p$ can be identified with $(\prod E_n^\star)_q$, $\frac{1}{p} + \frac{1}{q} = 1$.

b) Show that if each of the E_n's is reflexive and $1 < p < +\infty$, then $(\prod E_n)_p$ is reflexive.

c) Show that if each of the E_n's is separable and $1 \leqslant p < +\infty$, then $(\prod E_n)_p$ is separable.

EXERCISE 4. - By considering the sequences made of ± 1, show that ℓ_∞ is not separable. Show that $\mathscr{C}([0,1])$ is separable.

EXERCISE 5. - Let E be a Banach space, $(A_n)_{n\in\mathbb{N}}$ a *sequence* of subspaces of E. Show that if all the A_n's are separable, then $\overline{\bigcup_n A_n}$ is also separable.

EXERCISE 6. - Show that, if E is separable, from any family $(f_i)_{i\in I}$ of norm-one linear functionals satisfying

$$\bigcap_{i\in I} \ker f_i = \{0\} ,$$

one can extract a countable subfamily which has the same property.

(One may introduce the set $C = \bigcap_{i \in I} \{x \in E \; ; \; |f_i(x)| \leqslant 1\}$, and consider the gauge of C).

EXERCISE 7. - An operator $T : E \to F$ is said to be weakly compact if $T(\mathscr{B}_E)$ is weakly relatively compact in F (i.e. : the closure of $T(\mathscr{B}_E)$ for $\sigma(F, F^\star)$ is compact for this topology).

1) Prove that T is weakly compact if and only if ${}^{tt}T(E^{\star\star}) \subset F$.

2) Prove that if $(T_n)_{n \geqslant 1}$ is a sequence of weakly compact operators, from E into F , converging to T in operator norm, then T is weakly compact.

3) Prove that T is weakly compact if and only if ${}^tT$ is continuous from $(F^\star, \sigma(F^\star, F))$ into $(E^\star, \sigma(E^\star, E^{\star\star}))$.

4) Prove that T is weakly compact if and only if ${}^tT$ is.

EXERCISE 8. - Let E be a Banach space, C a bounded set, $\sigma(E, E^\star)$-closed. We assume that C is not $\sigma(E, E^\star)$-compact. We put $\overline{C}$ = closure of C , in $E^{\star\star}$, for $\sigma(E^{\star\star}, E^\star)$.

1) Show that $\overline{C} \cap E = C$.

2) Show that, if $F \in E^{\star\star}$, $F \in \overline{C}$, but $F \notin C$, then $F \notin E$. Put $\Delta = \text{dist}(F,E)$. Let θ , $0 < \theta < \Delta$. Show that $\|F\| > \theta$.

3) Consider the following neighbourhood of F for $\sigma(E^{\star\star}, E^\star)$:

$$V_{f,\epsilon}(F) = \{z \in E^{\star\star} \; , \; |\langle F,f \rangle - \langle z,f \rangle| < \epsilon\} .$$

Choose $f_1 \in E^\star$ such that $\langle F,f_1 \rangle > \theta$, $\|f_1\| = 1$. Using the fact that $F \in \overline{C}$, show that, for every $\epsilon > 0$, $V_{f_1,\epsilon}(F)$ contains a point of C . Deduce that one can find a point $x_1 \in C$ such that $f_1(x_1) > \theta$.

4) Assume that $x_1,\dots,x_{p-1} \in C$, $f_1,\dots,f_{p-1} \in E^\star$ with $\|f_i\| = 1$ have been chosen so that, if $n,k \leqslant p-1$:

$$\begin{cases} f_n(x_k) > \theta & \text{if} \quad n \leqslant k \\ f_n(x_k) = 0 & \text{if} \quad n > k \\ F(f_n) > \theta & \end{cases}$$

Use Helly's condition, and the fact that $\text{dist}(F,E) > \theta$, to find

$f_p \in E^\star$, $\|f_p\| = 1$, with $f_p(x_k) = 0$, $k \leqslant p-1$, and $F(f_p) > \theta$

5) Use the fact that the neighbourhood of F :

$$V_{f_1,\dots,f_p;\epsilon}(F) = \{z \in E^{\star\star} \ ; \ |<z,f_1> - <F,f_1>| < \epsilon, \dots, \\ |<z,f_p> - <F,f_p>| < \epsilon\}$$

meets C to find a point x_p in C such that $f_n(x_p) > \theta$, $n \leqslant p$.

6) Deduce the following theorem :

If C is a bounded, $\sigma(E,E^\star)$-closed, subset of a Banach space, if C is not weakly compact, there exist a sequence $(x_k)_{k\geqslant 1}$ of points in C , a sequence $(f_n)_{n\geqslant 1}$ of norm-one linear functionals, such that

$$\begin{cases} f_n(x_k) > \theta & \text{if} \quad n \leqslant k \\ f_n(x_k) = 0 & \text{if} \quad n > k \end{cases}$$

(This theorem extends Th. 6, § 1, taking C as the unit ball).

7) Show that if the assumptions of the theorem are satisfied, the sequence $(x_k)_{k\geqslant 1}$ in C satisfies, for every $K \geqslant 1$:

$$\text{dist}(\text{conv}(x_1,\dots,x_K) , \text{conv}(x_{K+1},\dots,)) > \theta .$$

8) Conversely, show that if, for some $\theta > 0$, there exists such a sequence in C , then C cannot be weakly compact.

EXERCISE 9. - Let E , F be Banach spaces, and T an operator from E into F. Show that T is not weakly compact if and only if there exist, for some $\theta > 0$, a sequence $(x_k)_{k\geqslant 1}$ of norm -1 points in E , a sequence $(f_n)_{n\geqslant 1}$ of norm -1 points in $F^\star$, such that

$$\begin{cases} f_n(T_{x_k}) > \theta & \text{if} \quad n \leqslant k \\ f_n(T_{x_k}) = 0 & \text{if} \quad n > k \end{cases}$$

EXERCISE 10. - Using exercise 8, show that a closed bounded set C in E is weakly compact if and only if every sequence $(x_n)_{n\geqslant 1}$ in C contains a subsequence $(x'_n)_{n\geqslant 1}$ which converges, in C , for $\sigma(E, E^\star)$.

REFERENCES ON CHAPTER III.

The theorem of § 1 is due to R.C. JAMES [25] ; the other statements are classical (see, for instance, BOURBAKI [11]).

For the § 2, most of the results were known to S. BANACH [5] ; the proof we give are closer to BOURBAKI [11].

The Eberlein-Šmulian Theorem may be found in DUNFORD-SCHWARTZ [16].

Exercise 3 comes from S. BANACH [5] ; exercise 6 is from a paper of B. MAUREY and the author [9]. Exercise 8, comes from R.C. JAMES [25].

PART 2

THE STRUCTURE OF SOME COMMON BANACH SPACES

CHAPTER I

HILBERT SPACES

This chapter is devoted to the main results about Hilbert spaces. It may seem too elementary, but we preferred to start this way, because one will see more easily in what respect the examples of Banach spaces which we shall study later can differ from Hilbert spaces.

One calls *Hilbert space* a vector space (real or complex), equipped with a scalar product (denoted by $<\cdot,\cdot>$), which is complete for the norm deduced from this scalar product. This norm is defined by the formula :

$$\|x\| = [< x,x >]^{1/2} .$$

Obviously, this is a special case of Banach space. A Hilbert space H has the following property, usually called "parallelogram identity" :

$$(1) \qquad \forall x,y \in H , \quad \left\| \frac{x+y}{2} \right\|^2 + \left\| \frac{x-y}{2} \right\|^2 = \frac{1}{2} (\|x\|^2 + \|y\|^2) ,$$

which is immediately established, just going back to the definition of the norm. Conversely, this equality is characteristic of the spaces, the norm of which comes from a scalar product : it is not valid in general in Banach spaces, and a great part of the study of the Geometry of Banach spaces consists in looking for conditions, close to (1), but valid in a more general setting.

PROPOSITION 1. (Projection on closed convex sets). - *Let* C *be a closed convex set in a Hilbert space* H . *There exists a "best approximation" projection from* H *onto* C , *that is, for all* $x \in H$, *there is a unique point* $y = Px \in C$ *such that*

$$\|x - y\| = \inf_{z \in C} \|x - z\| .$$

PROOF. - Let $x \in H$, and put $d = \inf_{z \in C} \|x - z\|$. Let $(y_n)_{n \in \mathbb{N}}$ be a sequence of points in C such that $\|x - y_n\| \searrow d$. For $m,n \in \mathbb{N}$, the parallelogram identity gives :

$$\left\| x - \frac{y_n + y_m}{2} \right\|^2 + \left\| \frac{y_n - y_m}{2} \right\|^2 = \frac{1}{2} (\|x - y_n\|^2 + \|x - y_m\|^2) .$$

But $\|x - y_m\|^2 \xrightarrow[m \to +\infty]{} d^2$, $\|x - y_n\|^2 \xrightarrow[n \to +\infty]{} d^2$, and $\left\| x - \frac{y_n + y_m}{2} \right\| \geqslant d$, since $\frac{y_n + y_m}{2} \in C$, therefore $\limsup_{n,m \to +\infty} \|y_n - y_m\|^2 = 0$, and so $(y_n)_{n \in \mathbb{N}}$ is a Cauchy sequence, and converges to some point $y \in C$. This point satisfies $\|x - y\| = d$. There is only one point $y \in C$ satisfying such an equality : if there were y, y' with $y \neq y'$, then, by (1)

$$\left\| x - \frac{y + y'}{2} \right\| < \frac{1}{2} (\|x - y\|^2 + \|x - y'\|^2) < d^2 ,$$

which is contradictory.

This point $y = Px$ can be characterized by the following property :

PROPOSITION 2. - *The point* $y = Px$ *is the unique point satisfying*

$$(2) \qquad \mathcal{R}e < x - y, z - y > \leqslant 0 \qquad \text{for all } z \in C .$$

PROOF.

1) Assume (2). Then
$$\|x - z\|^2 = \|x - y\|^2 - 2\,\mathcal{R}e < x - y , z - y > + \|y - z\|^2 \geqslant \|x - y\|^2 , \quad \text{for all } z \in C .$$

2) Conversely, assume $\|x - z\|^2 \geqslant \|x - y\|^2 \ \forall z \in C$. Then, for all $z \in C$, all $t \in [0,1]$, set $z' = (1 - t)y + tz$... Then :

$$\begin{aligned} \|x - y\|^2 \leqslant \|x - z'\|^2 &= < x - y + t(y - z), x - y + t(y - z) > \\ &= \|x - y\|^2 + 2\,\mathcal{R}e < x - y, t(y - z) > + t^2\|y - z\|^2 , \end{aligned}$$

and therefore

$$2t\,\mathcal{R}e < x - y, y - z > + t^2\|y - z\|^2 \geqslant 0$$

and finally $\mathcal{R}e < x - y, y - z > \geqslant 0$.

Let us now consider the case when C is a closed subspace.

PROPOSITION 3. - *Let* F *be a closed vector subspace of* H . *The projection* P *on* F *is a linear application of norm one (it will be called "orthogonal projection").*

PROOF. - Let $x \in E$, y its projection onto F . For all $z \in F$, one has : $\mathcal{R}e < x - y, z - y > \leqslant 0$. Therefore, since F is a vector space, for all $z' \in F$, $\mathcal{R}e < x - y, z' > \leqslant 0$, and, changing z' into $-z'$, and into iz' , one obtains $< x - y, z' > = 0$, for all $z' \in F$. So $x - y = x - Px$ is orthogonal to F .

For all $\lambda \in \mathbb{C}$, one has $< \lambda x - \lambda y, z > = \lambda < x - y, z > = 0$, $\forall z \in F$, and so $P(\lambda x) = \lambda Px$. Finally :

$$< x_1 + x_2 - (Px_1 + Px_2), z > = < x_1 - Px_1, z > + < x_2 - Px_2, z > = 0 ,$$

and therefore $P(x_1 + x_2) = Px_1 + Px_2$.

Take $z = Px$ in $< x - Px, z > = 0$. Then

$$\|Px\|^2 = < x, Px > \leqslant \|x\| \cdot \|Px\| \text{ , and } \|Px\| \leqslant \|x\| \text{ , for all } x \in E .$$

We denote by $G = F^\perp$ the orthogonal of F , $F^\perp = \{z \in H, <z,x> = 0 \ \forall x \in F\}$. It is a closed subspace, and $F^{\perp\perp} = F$ (if F was not assumed to be closed, $F^{\perp\perp} = \overline{F}$). We obtain :

PROPOSITION 4. - *The orthogonal* $F^\perp$ *of a closed subspace* P *is the Kernel of the orthogonal projection* P *on this subspace ;* $I - P$ *is the orthogonal projection on* $F^\perp$, *and* H *is the direct sum of* F *and* $F^\perp$.

For subspaces and quotients, we have :

PROPOSITION 5. - *Every closed subspace of a Hilbert space is a Hilbert space.*

This is obvious, since it is equipped with the induced scalar product, and is complete for the norm.

PROPOSITION 6. - *Let* F *be a closed subspace of* H *; the quotient space* $H/_F$ *is a Hilbert space.*

This is clear, since it is isometric to the Kernel of the orthogonal projection onto F .

From now on, we denote by P_F the projection onto F .

PROPOSITION 7.

a) *If* F_1 *and* F_2 *are closed subspaces, with* $F_1 \subset F_2$ *, then* $P_{F_1} \circ P_{F_2} = P_{F_2} \circ P_{F_1} = P_{F_1}$

b) *If* $(F_n)_{n \in \mathbb{N}}$ *is an increasing [resp. decreasing] sequence of closed subspaces, and if* $F = \overline{\bigcup_n F_n}$ *[resp.* $\bigcap_n F_n$ *], then, for all* $x \in E$ *,* $P_{F_n} x \xrightarrow[n \to +\infty]{} P_F x$.

PROOF.

a) Is obvious. For b), it is enough to consider the case when the F_n's are increasing : the other will follow by taking orthogonals. If $x \in H$, $x = P_F x + (I - P_F)x$, and $(I - P_F)x$ is orthogonal to F , and so to all F_n's . Therefore $P_{F_n}(I - P_F)x = 0$, and we just have to prove the result for $P_F x$, that is, to assume $x \in F$. Put $L = \{x \in F, P_{F_n} x \xrightarrow[n \to +\infty]{} x\}$. This is a subspace, which is closed since, if $x_p \in L$ and $x_p \to x$:

$$\|P_{F_n} x - x\| \leqslant \|P_{F_n} x - P_{F_n} x_p\| + \|P_{F_n} x_p - x_p\| + \|x_p - x\| \quad \text{and} \quad \|P_{F_n}\| = 1 .$$

But L contains each F_m , because, if $n \geqslant m$, $P_{F_n} x = x$, if $x \in F_m$. So $L \supset \overline{\bigcup F_n} = F$.

DEFINITIONS. - *A family* $(e_i)_{i \in I}$ *of elements of a Hilbert space* H *is called* orthogonal *if, for all* i, j , $i \neq j$, $< e_i , e_j > = 0$. *It is* orthonormal *if it is orthogonal and* $\|e_i\| = 1$ *for all* i . *It is an* orthonormal basis *if, moreover,* H *is the closed linear space generated by* $(e_i)_{i \in I}$.

PROPOSITION 8. - *Let* $(e_n)_{(0 \leqslant n \leqslant N)}$ *be a finite or infinite* $(N = +\infty)$ *orthonormal sequence, and* $F = \overline{\text{span}}\{(e_n) ; 0 \leqslant n \leqslant N\}$. *Then the application* $(a_n)_{0 \leqslant n \leqslant N} \to \sum_0^N a_n e_n$ *defines an isometry from the Hilbert space*

$\ell^2[0,N]$ *onto* F , *that is* $\left\| \sum_o^N a_n e_n \right\| = \left(\sum_o^N |a_n|^2 \right)^{1/2}$. *Moreover, for all* $x \in E$, *the sequence* $(<x, e_n>)_{o \leqslant n \leqslant N}$ *is in* $\ell^2[0,N]$, *and* $P_F x = \sum_{o \leqslant n \leqslant N} <x, e_n> e_n$. *One also has* :

$$\sum_o^N |<x, e_n>|^2 = \|P_F x\|^2 \leqslant \|x\|^2 \qquad \textit{(Bessel-Parseval Inequality)}.$$

PROOF. - The case $N < +\infty$ is immediate. Assume $N = +\infty$. Let $(a_n)_{n \in \mathbb{N}} \in \ell^2$, that is $\sum_o^\infty |a_n|^2 < +\infty$. Then the series $\sum_n a_n e_n$ converges in H , because $\left\| \sum_{m \leqslant n} a_m e_m - \sum_{m \leqslant p} a_m e_m \right\| = \left\| \sum_{n+1}^p a_m e_m \right\| = \left(\sum_{n+1}^p |a_m|^2 \right)^{1/2} \xrightarrow[m,p \to +\infty]{} 0$

So the application $(a_n) \in \ell^2 \to \sum_n a_n e_n$ is defined ; it is an isometry, since $\left\| \sum_n a_n e_n \right\|^2 = \lim_{p \to +\infty} \left\| \sum_{n \leqslant p} a_n e_n \right\|^2 = \lim_{p \to +\infty} \sum_{n \leqslant p} |a_n|^2 = \sum_o^\infty |a_n|^2$.

For all $x \in H$, one has $\sum_{n \leqslant p} |<x, e_n>|^2 \leqslant \|x\|^2$, for all $p \in \mathbb{N}$, and so $\sum_n |<x, e_n>|^2 \leqslant \|x\|^2$, and $(<x, e_n>)_{n \in \mathbb{N}} \in \ell^2$. Therefore $\sum_n <x, e_n> e_n$ converges in H . Let $y = \sum_n <x, e_n> e_n$. For all k , $<x - y, e_k> = 0$, and so $x - y$ is orthogonal to F , and $y = P_F x$.

Also :

$$\sum_n |<x, e_n>|^2 = \|y\|^2 = \|P_F x\|^2 \leqslant \|x\|^2 .$$

Finally, the application $(a_n)_{n \in \mathbb{N}} \to \sum_n a_n e_n$ is surjective, from ℓ^2 onto F , since, if $x \in F$, $x = P_F x = \sum_n <x, e_n> e_n$.

COROLLARY 9. - *The orthonormal sequence* $(e_n)_{n \in \mathbb{N}}$ *is an orthonormal basis if and only if, for all* $x \in H$, $x = \sum_n <x, e_n> e_n$. *This happens if and only if, for all* $x \in H$, $\|x\|^2 = \sum_n |<x, e_n>|^2$.

It can be proved that every Hilbert space has an orthonormal basis. We shall prove it only for separable Hilbert spaces, but then the family $(e_i)_{i\in I}$ will be countable.

PROPOSITION 10. - *Every separable Hilbert space has a countable orthonormal basis.*

PROOF. - It will be obtained from the Gram-Schmidt orthogonalization process :

Let $(u_n)_{n\in\mathbb{N}}$ a dense sequence in H. Let n_1 be the first index such that $u_{n_1} \neq 0$. Put $e_1 = \frac{u_{n_1}}{\|u_{n_1}\|}$ and $u_n^{(1)} = u_n - \langle u_n, e_1\rangle e_1$, for $n \geq 1$. Then :

$$u_n^{(1)} = 0 \text{ if } n \leqslant n_1 , \text{ and } \langle u_n^{(1)}, e_1\rangle = 0 \text{ for all } n .$$

Now choose n_2 to be the first index n such that $u_n^{(1)} \neq 0$, put $e_2 = \frac{u_{n_2}^{(1)}}{\|u_{n_2}^{(1)}\|}$, and $u_n^{(2)} = u_n^{(1)} - \langle u_n^{(1)}, e_2\rangle e_2$. Then $u_n^{(2)} = 0$ if $n \leqslant n_2$, and $\langle u_n^{(2)}, e_2\rangle = 0$ for all n. And so on : one chooses as n_k the first index n such that $u_n^{(k-1)} \neq 0$, puts

$$e_k = \frac{u_{n_k}^{(k-1)}}{\|u_{n_k}^{(k-1)}\|} , \quad u_n^{(k)} = u_n^{(k-1)} - \langle u_n^{(k-1)}, e_k\rangle e_k ,$$

and so, $\langle e_k, u_n^{(k)}\rangle = 0$ for all n. For all $k \geqslant 1$, e_k and $(u_n^{(k)})_n$ are linear combinations of the $(u_n^{(k-1)})_n$. If $\ell > k$, e_ℓ is a linear combination of the $(u_n^{(k)})_n$, and so $\langle e_k, e_\ell\rangle = 0$. The sequence $(e_n)_{n\in\mathbb{N}}$ is therefore orthonormal. But the u_i's are themselves linear combinations of the e_n's (this is checked by induction), and so the span of the e_n's contains all the u_i's, and, since $(u_i)_{i\in\mathbb{N}}$ is a dense sequence, $\overline{\text{span}}\{(e_n)_{n\in\mathbb{N}}\} = H$.

Therefore, two separable Hilbert spaces are isometric : choose orthonormal bases $(e_n)_{n\in\mathbb{N}}$ of the first, $(f_n)_{n\in\mathbb{N}}$ of the second. Then the application $e_n \to f_n$ defines an isometry between these spaces. This is the case, for example, of the spaces ℓ^2 and L^2 : in ℓ^2, the sequences $(e_n)_{n\in\mathbb{N}}$ defined by $\begin{cases} e_n(k) = 0 & \text{if } k \neq n , \\ = 1 & \text{if } k = n , \end{cases}$ form an orthonormal basis ; in L^2 , this is the case for the complex exponentials $(e^{in\theta})_{n\in\mathbb{Z}}$ (which can be arranged in a sequence indexed by $\mathbb{N}$).

We turn now to the dual of a Hilbert space. It can be identified with the space itself :

PROPOSITION 11. - *For all* $a \in H$, *the mapping* $x \to \langle x, a \rangle$ *is a continuous linear form on* H , *called* $a^\star$. *The application* $a \to a^\star$ *is bijective, continuous, isometric, and antilinear (that is* $(\lambda a)^\star = \bar{\lambda} a^\star$ *) from* H *onto* $H^\star$.

PROOF. - The continuity of $a^\star$ follows from Cauchy-Schwarz inequality

$$|\langle x, y \rangle| \leqslant \|x\| \cdot \|y\| , \quad \text{for all } x,y \in H .$$

All the announced properties of $a \to a^\star$ are immediately checked. Let us just show that it is surjective. Let $\xi \in H^\star$, $\xi \neq 0$. Let $F = \operatorname{Ker} \xi$: this is a closed hyperplane. Choose a' orthogonal to F . Then $x \to \langle x, a' \rangle$ is a linear form on H , which has the same kernel as ξ . Write $x = \langle x, a' \rangle a' + y$, with $y = P_F x$. Then $\xi(x) = \langle x, a' \rangle \xi(a')$, and ξ comes from a point of H .

Therefore, a Hilbert space is reflexive, and its unit ball is compact for $\sigma(H, H^\star)$.

To end this chapter, let us see how finite-dimensional Hilbert spaces are realized :

PROPOSITION 12. - *Let* $N \geqslant 1$. *If* $a_1,\ldots,a_N$ *are positive numbers, the norm*

$$\|x\| = \sum_{i=1}^{N} a_i |x_i|^2 , \quad \text{if } x = (x_1,\ldots,x_N) \in \mathbb{K}^N$$

comes from the scalar product $\langle x, y \rangle = \sum_1^N a_i x_i \overline{y_i}$, *and therefore* $\mathbb{K}^N$ *equipped with this norm is a Hilbert space. The unit ball is the set* $\{(x_1,\ldots,x_N) \; ; \; a_1|x_1|^2 + \ldots + a_N|x_N|^2 \leqslant 1\}$, *which is an ellipsoïd in* $\mathbb{K}^N$.

So we have seen that the geometric structure of a Hilbert space appears rather simple : all subspaces and quotients are Hilbert spaces, all subspaces are complemented (i.e. there is a projection on them), and separable spaces have bases. Unfortunately none of these facts holds in general Banach spaces : the following chapters are devoted to the study of these questions.

EXERCISES ON CHAPTER I.

EXERCISE 1. - Let F be a closed convex subset, in a Hilbert space H, $a \in H$, $\xi \in H^{\star}$. Show that there is a unique point $b \in F$ which minimizes the function

$$x \in F \longrightarrow \|x - a\|^2 + < \xi , x > ,$$

and that this point b is characterized by :

$$\text{For all } y \text{ such } b + y \in F , \ \mathcal{R}e(2 < b - a, y > + \xi(y)) \geqslant 0 .$$

EXERCISE 2. - Find the best approximation projection of the function $x \to e^x$ on the subspace of polynomials of degree $\leqslant 2$ in $L_2([-1,+1], dx)$.

EXERCISE 3. - In ℓ_2, consider the sequences $x = (x(k))_{k \in \mathbb{N}^\star}$ such that, for all $k \geqslant 1$, $|x(k)| \leqslant \frac{1}{k}$. Show that the set of such sequences is compact in ℓ_2 for the norm (the "Hilbert Cube").

EXERCISE 4. - In a Hilbert space H, let $(x_n)_{n \in \mathbb{N}^\star}$ be a sequence of elements with $\|x_n\| \leqslant 1$ for all $n \geqslant 1$. For all $z \in H$, we define, for $n \geqslant 1$:

$$\varphi_n(z) = \frac{1}{n} \sum_1^n \|z - x_j\|^2 .$$

Show that φ_n is a convex function, which attains its minimum at a unique point s_n, and that $s_n = \frac{1}{n} \sum_1^n x_j$. Show that if $H = \ell_2$, the $k^{\underline{th}}$ coordinate of s_n, $s_n(k)$, minimizes

$$t \to \frac{1}{n} \sum_{j=1}^n |t - x_j(k)|^2 .$$

Study the same questions, when $H = \ell_2$, if $\varphi_n^{(p)}$ is defined by

$$\varphi_n^{(p)}(z) = \frac{1}{n} \sum_1^n \|z - x_j\|^p . \qquad 1 \leqslant p < +\infty , \quad p \neq 2 .$$

EXERCISE 5. - The notations are the same as in the previous exercise. Let $(z_n)_{n \geq 1}$ be a sequence in H. Show that if $\varphi_n(z_n) - \varphi_n(s_n) \xrightarrow[n \to +\infty]{} 0$, then $z_n - s_n \xrightarrow[n \to +\infty]{} 0$ in H. For this, establish the formula :

$$\varphi_n(z) \geq \varphi_n(s_n) + \frac{1}{2} \|z - s_n\|^2 , \qquad \forall z \in H .$$

EXERCISE 6. - Same notations. Assume that a subsequence $(s_{n_\ell})_{\ell \in \mathbb{N}}$ converges weakly to some point $a \in \ell^2$, and that another subsequence $(s_{n'_\ell})_{\ell \in \mathbb{N}}$ converges weakly to $b \in \ell^2$. Put $\alpha = \|a - b\|^2$. Prove that there exists $\ell_0 \geq 1$ such that, if $\ell \geq \ell_0$

$$\frac{1}{n_\ell} \sum_{j=1}^{n_\ell} \|b - x_j\|^2 \geq \frac{1}{n_\ell} \sum_{j=1}^{n_\ell} \|a - x_j\|^2 + \frac{\alpha}{8} .$$

For this, take $\epsilon > 0$ and choose K such that $\sum_{k>K} |a(k)|^2 < \epsilon^2$ and $\sum_{k>K} |b(k)|^2 < \epsilon^2$. If $k \leq K$, prove that

$$\frac{1}{n_\ell} \sum_{1}^{n_\ell} |b(k) - x_j(k)|^2 \geq \frac{1}{n_\ell} \sum_{1}^{n_\ell} |a(k) - x_j(k)|^2 + \frac{1}{4} |a(k) - b(k)|^2 .$$

EXERCISE 7. - Let C be a closed convex subset of the unit ball of ℓ_2. Let A be a (non-necessarily linear) contraction from C into C, that is, an application satisfying $\|Ax - Ay\| \leq \|x - y\|$, for all $x,y \in C$.

1°) Let $a,y \in C$. Show that $< Aa - Ay, a - y > \leq < a - y, a - y >$.

Deduce that $< Aa - a + y - Ay, a - y > \leq 0$.

2°) Assume that $s_n \in C$ converges weakly to a point $f (\in C)$, and that $As_n - s_n \xrightarrow[n \to +\infty]{} 0$. Prove that, for all $y \in C$,

$$< y - Ay, f - y > \leq 0 .$$

3°) For any $z \in C$, take $y = tz + (1 - t)f$, $0 < t \leq 1$. Show that

$$< f + t(z - f) - A(f + t(z - f)) , f - z > \leq 0 .$$

Deduce that

$$< f - Af, f - z > \leq 0 .$$

4°) Prove that f is a fixed point for A, that is $Af = f$.

EXERCISE 8. - Let A be as in the previous exercise. Fix $e \in C$, and take $x_j = A^j e$. Define s_n as in exercise 4.

Use exercise 5 to show that $As_n - s_n \xrightarrow[n \to +\infty]{} 0$. Use exercise 7 to show that any weak limit of (s_n) is a fixed point of A . Use exercise 6 to show that (s_n) is weakly convergent. So one obtains the following theorem :

THEOREM. (J.B. BAILLON) - *Let* C *be a closed convex bounded set in a Hilbert space, and* A *a (non-linear) contraction from* C *into* C . *Then, for any* $e \in C$, *the Cesaro averages* $\frac{1}{n} \sum_1^n A^j e$ *converge weakly to a fixed point of* A .

REFERENCES ON CHAPTER I.

Almost every book devoted to Functional Analysis contains a more or less extensive study of Hilbert space. For a more detailed study than ours, we refer the reader to the book by L. SCHWARTZ [45].

The theorem obtained in exercise 8 was proved by J.B. Baillon. The proof given here (via the functions φ_n) is simpler than the original proof, and comes from a paper by P. Enflo and the author ("Théorèmes de points fixes et d'approximation", to appear), which applies also to ℓ_p , $1 < p < +\infty$. These functions correspond to a special case of a geometric notion, called "minimal point" of a set. See [9] for a general study of this notion.

CHAPTER II

SCHAUDER BASES IN BANACH SPACES

§ 1. SCHAUDER BASES.

Obviously, the notions of algebraic basis, in a finite dimensional vector space, or of orthonormal basis, in a Hilbert space, are essential tools for the study of these spaces. Therefore, if we want to investigate the structure of general Banach spaces, it is natural to try to find a corresponding notion.

In the first part (chapter I) we have introduced the concept of an algebraic basis for a Banach space, and we have proved a result concerning the cardinality of such a basis. One can always - using Zorn's axiom - establish the existence of an algebraic basis in a given Banach space, but such a basis does not provide satisfactory informations, because it has no links with the topology of the space. Let us explain this : let $(e_i)_{i \in I}$ be an algebraic basis in a Banach space E , let $(z_n)_{n \in \mathbb{N}}$ be a sequence of points in E , converging to some point z . Each of the z_n's and z have (finite) decompositions : $z_n = \sum_i \alpha_i^{(n)} e_i$, $z = \sum_i \alpha_i e_i$. But there is no reason why for all i , $\alpha_i^{(n)}$ should converge to α_i , and this not so in general. But this convergence of "coordinates" is certainly the least one can ask. So, in order to obtain it, we shall replace the notion of algebraic basis, where only finite decompositions are used, by another one, where each point will be represented as the sum of a series. Also, nothing will prevent, in this case, the cardinal of such a basis to be countable.

Let us start with some definitions concerning the convergence of series in a Banach space E :

DEFINITIONS.

1) *Let* $(x_n)_{n \in \mathbb{N}}$ *be a sequence of points in* E . *We say that the series* $\sum_{n \in \mathbb{N}} x_n$ converges *to a point* x , *and we put* $x = \sum_{n \in \mathbb{N}} x_n$, *if*

$$\| x - \sum_{n \leq p} x_n \| \xrightarrow[p \to +\infty]{} 0 .$$

It is essential to observe that this definition depends on the order in which the points $x_0, x_1, \ldots, x_n, \ldots$ *are put : the series may converge to another point, or may not converge at all (in* $\mathbb{R} : x_n = \frac{(-1)^n}{n}$ *) if the* x_n's *are ordered differently.*

2) *We say that the series* $\sum_{n \in \mathbb{N}} x_n$ converges unconditionally *to* x *if, for all permutations* π *of* $\mathbb{N}$ *, the series* $\sum_{n \in \mathbb{N}} x_{\pi(n)}$ *converges to* x .

3) *We say that the series* $\sum_{n \in \mathbb{N}} x_n$ converges absolutely, *or* normally, *to* x *, if it converges to* x *and if the series of positive numbers* $\sum_{n \in \mathbb{N}} \|x_n\|$ *is convergent.*

The unconditional convergence obviously implies the convergence. We shall see that the normal convergence implies the unconditional convergence.

For this, let $(x_n)_{n \in \mathbb{N}}$ be a sequence of points ; assume that $\sum_n x_n$ converges normally to a point x . Let π be a permutation of the integers.

For any $\epsilon > 0$, there is an $n_0 \geq 0$ such that, if $n \geq n_0$, $\left\| \sum_0^n x_k - x \right\| < \epsilon$ and $\sum_{n > n_0} \|x_n\| < \epsilon$. Take $N_0 = \sup_{n \leq n_0} \pi(n)$. If $N > N_0$, put $\sum_1^N x_{\pi(n)} = \sum_{n \leq n_0} x_n + y$, and $\|y\| < \epsilon$. We obtain :

$$\left\| \sum_1^N x_{\pi(n)} - x \right\| = \left\| \sum_{n \leq n_0} x_n + y - x \right\| < 2\epsilon ,$$

and this proves our assertion.

If $(x_n)_{n \in \mathbb{N}}$ is a sequence of points in a Banach space E , we denote by $\text{span}\{(x_n)_{n \in \mathbb{N}}\}$ the vector space spanned by the x_n's , and by $\overline{\text{span}}\{(x_n)_{n \in \mathbb{N}}\}$ its closure.

Let us turn now to definitions concerning basicity :

DEFINITIONS.

4) *The sequence* $(x_n)_{n \in \mathbb{N}}$ *will be called a* basic sequence *in* E *if,*

for all points $x \in \overline{\text{span}}\{(x_n)_{n \in \mathbb{N}}\}$, *one can find a unique sequence* $(a_n)_{n \in \mathbb{N}}$ *of scalars, such that the series* $\sum_n a_n x_n$ *converges to* x .

5) *The sequence* $(x_n)_{n \in \mathbb{N}}$ *is a* Schauder basis *of* E *if it is a basic sequence and if* $\overline{\text{span}}\{(x_n)_{n \in \mathbb{N}}\} = E$.

In this case, every point x *of* E *has a unique decomposition* $x = \sum_n a_n x_n$. *The scalars* $(a_n)_{n \in \mathbb{N}}$ *will be called the* coordinates *of* x *on the basis* $(x_n)_{n \in \mathbb{N}}$. *We shall often assume the sequence* $(x_n)_{n \in \mathbb{N}}$ *normalized, that is* $\|x_n\| = 1$ *for all* $n \in \mathbb{N}$.

The basicity of a sequence $(x_n)_{n \in \mathbb{N}}$ can be characterized by a condition on the projections on finite subsets of the coordinates :

PROPOSITION 1. - *A sequence* $(x_n)_{n \in \mathbb{N}}$ *is basic if and only if there exists a number* $K > 0$ *such that, for all integers* p *and* q , *with* $p \leqslant q$, *and all scalars* $a_0 \dots a_q$, *one has* :

$$(1) \qquad \left\| \sum_0^p a_n x_n \right\| \leqslant K \left\| \sum_0^q a_n x_n \right\| .$$

REMARK. - This condition means that the projections $P_{q,p}$, from $\text{span}\{x_0,\dots,x_q\}$ onto $\text{span}\{x_0,\dots,x_p\}$, have norms uniformly bounded by K.

PROOF.

a) Let us assume first that $(x_n)_{n \in \mathbb{N}}$ is basic. We shall use the following lemma, the proof of which is elementary and is left to the reader :

LEMMA 2. - *Let* $(x_n)_{n \in \mathbb{N}}$ *a sequence of points in a Banach space* E , *with* $x_n \neq 0$ *for all* n . *Let* $F = \{(a_n)_{n \in \mathbb{N}}$ *(scalars), such that* $\sum a_n x_n$ *converges*$\}$, *equipped with the norm* :

$$\|(a_n)_{n \in \mathbb{N}}\|_F = \sup_{N \in \mathbb{N}} \left\| \sum_0^N a_n x_n \right\| .$$

Then F *is a Banach space.*

Now, if $x \in E_1 = \overline{\text{span}}\{(x_n)_{n \in \mathbb{N}}\}$, x has, by definition, a unique decomposition $x = \sum_n a_n x_n$, and so the application $(a_n)_{n \in \mathbb{N}} \in F \to x = \sum_n a_n x_n$ is a linear bijective mapping from F onto E_1 . But this mapping is continuous, since :

$$\left\| \sum_n a_n x_n \right\| \leqslant \sup_{N \in \mathbb{N}} \left\| \sum_0^N a_n x_n \right\| = \left\| (a_n)_{n \in \mathbb{N}} \right\|_F .$$

By the open mapping theorem (Ist part, chapter I, theorem 8), the inverse mapping is continuous : there is a constant $K > 0$ such that

$$\sup_{N \in \mathbb{N}} \left\| \sum_0^N a_n x_n \right\| \leqslant K \left\| \sum_n a_n x_n \right\| ,$$

for all $(a_n)_{n \in \mathbb{N}} \in F$. Therefore, for all p, q with $p \leqslant q$,

$$\left\| \sum_{n \leqslant p} a_n x_n \right\| \leqslant K \left\| \sum_{n \leqslant q} a_n x_n \right\| ,$$

and (1) is proved.

b) Conversely, assume (1) to be satisfied. Take $x \in \overline{\text{span}}\{(x_n)_{n \in \mathbb{N}}\}$. Then there exists a sequence $(y_k)_{k \in \mathbb{N}}$, each y_k belonging to $\text{span}\{(x_n)_{n \in \mathbb{N}}\}$, with $y_k \xrightarrow[k \to +\infty]{} x$. Each y_k can be written :

$$y_k = \sum_{\ell=0}^{N_k} \alpha_\ell^{(k)} x_\ell .$$

From (1) we deduce, letting $q \to +\infty$, that if a series $\sum_n a_n x_n$ converges, one has, for all p

$$(2) \qquad \left\| \sum_{n \leqslant p} a_n x_n \right\| \leqslant K \left\| \sum_n a_n x_n \right\| .$$

Let us call P_p the projection defined by

$$P_p\left(\sum_n a_n x_n \right) = \sum_{n \leqslant p} a_n x_n .$$

For all $p \in \mathbb{N}$, we obtain

$$(3) \qquad \| P_p(y_k - y_{k'}) \| \leqslant K \| y_k - y_{k'} \| \xrightarrow[k,k' \to +\infty]{} 0 ,$$

and it follows that, for any fixed p , $P_p y_k$ has a limit in E when $k \to +\infty$: let us call it X_p . Taking successively $p = 0,1,2,\dots,$ we deduce that the coefficients $\alpha_0^{(k)}$ have a limit α_0 , that the coefficients $\alpha_1^{(k)}$ have a limit $\alpha_1, \dots$ when $k \to +\infty$. Therefore $X_p = \sum_{\ell \leqslant p} \alpha_\ell x_\ell$.

From (3), we obtain, for all k and p :

$$\|P_p\, y_k - X_p\| \leqslant K\|y_k - x\| .$$

Let $\epsilon > 0$. We can find k_0 such that, if $k \geqslant k_0$, $\|y_k - x\| < \epsilon$. Choose p_0 such that, if $p \geqslant p_0$, $\|P_p\, y_{k_0} - u_{k_0}\| < \epsilon$. Then :

$$\|X_p - x\| \leqslant \|X_p - P_p\, y_{k_0}\| + \|P_p\, y_{k_0} - y_{k_0}\| + \|y_{k_0} - x\|$$

$$\leqslant K\|y_{k_0} - x\| + \epsilon + \|y_{k_0} - x\| \leqslant (2 + K)\,\epsilon ,$$

which proves our assertion. The uniqueness of the decomposition is clear : if $x = \sum_\ell \alpha_\ell x_\ell$ and $x = \sum_\ell \beta_\ell x_\ell$, then $\sum_1^n (\alpha_\ell - \beta_\ell)x_\ell \to 0$, and so, by (1), for all p , $\sum_{\ell \leqslant p} (\alpha_\ell - \beta_\ell)x_\ell = 0$, and $\alpha_\ell = \beta_\ell$ for all ℓ . This ends the proof of the proposition.

The smallest constant K satisfying (1) will be called the *basis-constant* of the sequence $(x_n)_{n\in\mathbb{N}}$. The basis will be called *monotone* if $K = 1$.

If the sequence $(a_n)_{n\in\mathbb{N}}$ is basic and normalized, we can define the *coordinate functionals* $(f_k)_{k\in\mathbb{N}}$, by

$$f_k(x) = a_k , \quad \text{if} \quad x = \sum_n a_n x_n .$$

These functionals are defined on $E_1 = \overline{\operatorname{span}}\{(x_n)_{n\in\mathbb{N}}\}$ and are continuous, since :

$$|a_k| = \|a_k x_k\| \leqslant \Big\|\sum_{\ell\leqslant k} a_\ell x_\ell\Big\| + \Big\|\sum_{\ell\leqslant k-1} a_\ell x_\ell\Big\| \leqslant 2K\|x\| ,$$

and so, by the Hahn-Banach theorem, they can be extended to E , as functionals, still called $(f_k)_{k\in\mathbb{N}}$, with $\|f_k\| \leqslant 2K$. The sequence $(f_k)_{k\in\mathbb{N}}$ is also sometimes called the biorthogonal sequence associated to the sequence $(x_n)_{n\in\mathbb{N}}$ (this is because $f_k(x_n) = 0$ if $n \neq k$, $= 1$ if $n = k$). Since these functionals are continuous when $(x_n)_{n\in\mathbb{N}}$ is a Schauder basis, if $(z_n)_{n\in\mathbb{N}}$ is a sequence of points converging to some point z , then, for all k , $f_k(z_n) \xrightarrow[n\to+\infty]{} f_k(z)$. This new notion of a basis is already more satisfactory than the algebraic one. But then, the question is : what Banach spaces have a Schauder basis ?

The first result concerning the existence of basic sequences is due to Banach himself ; it is the following :

PROPOSITION 3. - *Every infinite-dimensional Banach space contains a basic sequence.*

PROOF. - We shall build this sequence by induction. For this purpose, it is clearly enough to prove the following lemma :

LEMMA 4. - *Let* $(x_1,\dots,x_n)$ *be a finite sequence, of basis constant* A . *For all* $\epsilon > 0$, *one can find a point* x_{n+1} *such that* $(x_1,\dots,x_n, x_{n+1})$ *has basis constant at most* $A(1+\epsilon)$.

PROOF OF LEMMA 4. - Put $E_n = \operatorname{span}\{x_1,\dots,x_n\}$, $\epsilon' = \frac{\epsilon}{1+\epsilon}$, and let $(z_j)_{j=1\dots N}$ be an ϵ'-net in the unit sphere of E_n : the balls of radius ϵ' , centered at $z_1,\dots,z_N$ cover this unit sphere. Let $(\varphi_j)_{j=1\dots N}$ be continuous linear functionals on E , with $\varphi_j(z_j) = 1$ for $j = 1\dots N$, and choose $x_{n+1} \in \bigcap_{j=1\dots N} \operatorname{Ker} \varphi_j$, with $\|x_{n+1}\| = 1$. This is possible since E is infinite-dimensional. To show that $(x_1,\dots,x_{n+1})$ has basis constant at most $A(1+\epsilon)$, it is enough to see that, for all scalars $\alpha_1,\dots,\alpha_{n+1}$,

$$\left\|\sum_1^n \alpha_j x_j\right\| \leqslant (1+\epsilon)\left\|\sum_1^{n+1} \alpha_j x_j\right\| .$$

By homogeneity, we may assume $\left\|\sum_1^n \alpha_j x_j\right\| = 1$. Let j_0 be the index (between 1 and N) such that $\left\|z_{j_0} - \sum_1^n \alpha_j x_j\right\| < \epsilon'$. Then :

$$1 = \varphi_{j_0}(z_{j_0} + \alpha_{n+1} x_{n+1}) \leqslant \left\|z_{j_0} + \alpha_{n+1} x_{n+1}\right\|$$

$$\leqslant \left\|z_{j_0} - \sum_1^n \alpha_j x_j\right\| + \left\|\sum_1^{n+1} \alpha_j x_j\right\| \leqslant \epsilon + \left\|\sum_1^{n+1} \alpha_j x_j\right\| ,$$

and

$$\left\|\sum_1^{n+1} \alpha_j x_j\right\| \geqslant 1 - \epsilon' = \frac{1}{1+\epsilon} .$$

This proves the lemma.

The problem of the existence of a Schauder basis in a given Banach space is not so simple. First, obviously, the space must be separable, since the linear combinations, with rational coefficients (or rational real and imaginary parts of the coefficients), of the elements of the basis constitute a dense, countable, set. The most common Banach spaces all have Schauder bases. For example, one checks easily that in ℓ^p $(1 \leq p < +\infty)$, the canonical basis of $\mathbb{K}^{(\mathbb{N})}$ is a Schauder basis (which will be called the *canonical basis* of ℓ^p). The same is true for c_0 . We shall see in the sequel that the Haar system is a Schauder basis for $L^p([0,1], dt)$. This system is defined by : $h_1(t) = 1 \quad 0 \leq t \leq 1$, and

$$h_{2^k+\ell}(t) = 1 \quad \text{if} \quad t \in \left[\frac{2\ell - 2}{2^{k+1}}, \frac{2\ell - 1}{2^{k+1}}\right[$$

$$= -1 \quad \text{if} \quad t \in \left[\frac{2\ell - 1}{2^{k+1}}, \frac{2\ell}{2^{k+1}}\right[$$

$$= 0 \quad \text{otherwise,}$$

for $k = 0,1,2,\ldots,$ and $\ell = 1,2,\ldots,2^k$.

The problem of the existence of a Schauder basis in every separable Banach space remained open for a long time ; it was solved negatively by P. ENFLO [19] (1974). Then it was proved that, for $p \neq 2$, $1 \leq p < +\infty$, each $L^p([0,1], dt)$ had a subspace without basis (though the whole space has one). The most recent and striking result in this direction is due to A. SZANKOWSKI [49] : the space $\mathscr{L}(H)$ of all bounded linear operators on a Hilbert space (equipped with the norm of operators) does not have the approximation property (a weaker notion, see the "complements", p. 98).

Despite these facts, the notion of Schauder basis still keeps some importance, since the geometry of spaces having a basis is relatively well known. Let us mention that no criterion is known allowing to say that a space does or does not have a basis.

To illustrate how a Schauder basis can be used, let us see how reflexivity of a space can be characterized on the basis. This result is due to R.C. JAMES [22].

THEOREM 5. - *Let* E *be a Banach space having a Schauder basis* $(e_n)_{n\in\mathbb{N}}$. *Then* E *is reflexive if and only if* $(e_n)_{n\in\mathbb{N}}$ *satisfies simultaneously the following two properties :*

a) *For every sequence* $(\alpha_n)_{n\in\mathbb{N}}$ *of scalars such that*

$$\sup_n \left\| \sum_0^n \alpha_j e_j \right\| < +\infty ,$$

there exists $x \in E$ *with* $x = \sum_0^\infty \alpha_j e_j$, *(the basis will be called* "boundedly complete"*)*.

b) *For every linear functional* $\xi \in E^\star$,

$$\lim_{n\to+\infty} \sup\{|\xi(x)| \; ; \; x = \sum_{j\geq n} \alpha_j e_j , \quad \|x\| = 1\} = 0$$

(the basis will be called "shrinking"*)*.

PROOF.

1) Assume first E reflexive. Let $(\alpha_n)_{n\in\mathbb{N}}$ be scalars such that $\sup_n \left\| \sum_0^n \alpha_i e_i \right\| < +\infty$. Put $u_n = \sum_0^n \alpha_j e_j$. Then $(u_n)_{n\in\mathbb{N}}$ is a bounded sequence in a weakly compact set. Then we know (first part, chap. III, § 3, prop. 1) that $(u_n)_{n\in\mathbb{N}}$ contains a subsequence $(u_{n_k})_{k\in\mathbb{N}}$, weakly convergent to a point u . Since $(e_n)_{n\in\mathbb{N}}$ is a basis, u has a decomposition $u = \sum_0^\infty \beta_i e_i$, and, if $(f_k)_{k\in\mathbb{N}}$ are the coordinate functionals, we have $f_m(u) = \beta_m$, for all $m \in \mathbb{N}$, and $f_m(u_n) = \alpha_m$ if $m \leq n$. But $f_m(u_{n_k}) \xrightarrow[k\to+\infty]{} f_m(u)$, for all m , and so $\alpha_m = \beta_m$ for all m , and $\sum_i \alpha_i e_i = u$. This proves a). Assume b) to be false. Then there is a linear functional ξ , a number $\epsilon > 0$, a strictly increasing sequence $(p_n)_{n\in\mathbb{N}}$ of integers, and a sequence $(v_n)_{n\in\mathbb{N}}$, $v_n = \sum_{i=p_n}^\infty a_i^{(n)} e_i$, with $\|v_n\| = 1$ and $|\xi(v_n)| > \epsilon$, for all n . Here again, the v_n's must have a weakly converging subsequence. Since all the coordinates of v_n with index $\leq n$ are equal to zero, the limit must have all its coordinates equal to zero, and therefore must be 0 . But this contradicts $|\xi(v_n)| \geq \epsilon$ for all n , and proves the first part of the theorem.

2) Conversely, assume a) and b) to be satisfied. Let K be the basis constant of $(e_n)_{n\in\mathbb{N}}$. Let us observe first that if

$$x = \sum_{0}^{\infty} \alpha_i e_i \text{ , then, for all } n,p \in \mathbb{N} \text{ , } \left\| \sum_{n}^{n+p} \alpha_i e_i \right\| \leqslant 2K\|x\| \text{ .}$$

Let $(y_n)_{n \in \mathbb{N}}$ a sequence of points in E, with, for all n, $\|y_n\| \leqslant 1$. We shall show that $(y_n)_{n \in \mathbb{N}}$ contains a subsequence $(y_{n_k})_{k \in \mathbb{N}}$ converging for $\sigma(E, E')$. This will show that the unit ball of E is weakly compact (first part, chap. III, § 3, prop. 1), and E is reflexive.

Each y_n has a decomposition $y_n = \sum_{i=0}^{\infty} \alpha_i^{(n)} e_i$, with $|\alpha_i^{(n)}| \leqslant \frac{2K}{\|e_i\|}$, for all i and n.

For all $i \in \mathbb{N}$, we may choose a sequence $(q_n)_{n \in \mathbb{N}}$ of integers such that $\lim_{n \to +\infty} \alpha_i^{(q_n)}$ exists, and, by a diagonal procedure, a sequence $(p_n)_{n \in \mathbb{N}}$ of integers such that $\lim_{n \to +\infty} \alpha_i^{(p_n)}$ exists for all $i \in \mathbb{N}$. Put $\alpha_i = \lim_{n \to +\infty} \alpha_i^{(p_n)}$. For all $N \in \mathbb{N}$, we have :

$$\left\| \sum_{i=0}^{N} \alpha_i^{(p_n)} e_i \right\| \leqslant K \left\| y_{p_n} \right\| \leqslant K \text{ ,}$$

and thus

$$\left\| \sum_{i=0}^{N} \alpha_i e_i \right\| \leqslant K \text{ .}$$

By condition a), we can find a point $y \in E$ such that $y = \sum_{i=0}^{\infty} \alpha_i e_i$. We shall see that y is the weak limit of the y_{p_n}'s.

Let $\eta > 0$, $\xi \in E^{\star}$. Choose N such that

$$\sup\{|\xi(x)| \text{ ; } x = \sum_{i \geqslant N} \alpha_i e_i \text{ , } \|x\| = 1\} < \frac{\eta}{8K} \text{ ;}$$

this is possible by b). Then choose M such that, if $n \geqslant M$,

$$\left\| \sum_{i=1}^{N-1} (\alpha_i - \alpha_i^{(p_n)}) e_i \right\| < \frac{\eta}{2\|\xi\|} \text{ .}$$

We know that for all $n,p \in \mathbb{N}$:

$$\left\| \sum_{i=N}^{N+p} \alpha_i^{(p_n)} e_i \right\| \leqslant 2K \text{ , and thus, for all } p \in \mathbb{N}$$

$$\left\|\sum_{i=N}^{N+p} \alpha_i e_i\right\| \leqslant 2K \ , \quad \text{and} \quad \left\|\sum_{i \geqslant N} \alpha_i e_i\right\| \leqslant 2K \ .$$

Therefore, if $n \geqslant M$:

$$\left|\xi(y) - \xi(y_{p_n})\right| \leqslant \left|\xi\left(\sum_{i=0}^{N-1} (\alpha_i - \alpha_i^{(p_n)})(e_i)\right| + \left|\xi\left(\sum_{i \geqslant N} \alpha_i e_i\right)\right| + \left|\xi\left(\sum_{i \geqslant N} \alpha_i^{(p_n)} e_i\right)\right|$$

$$\leqslant \frac{\eta}{2} + \frac{\eta}{4} + \frac{\eta}{4} = \eta \ ,$$

and $\xi(y_{p_n}) \xrightarrow[n \to +\infty]{} \xi(y)$, which proves the theorem.

We shall now turn to a more restrictive class of Schauder bases, for which we shall obtain more precise results.

§ 2. UNCONDITIONAL SCHAUDER BASES.

We say that a Schauder basis $(e_n)_{n \in \mathbb{N}}$ is *unconditional* if whenever the series $\sum_n a_n e_n$ converges, it converges unconditionally.

PROPOSITION 1. - *For a Schauder basis* $(e_n)_{n \in \mathbb{N}}$, *the following properties are equivalent :*

a) $(e_n)_{n \in \mathbb{N}}$ *is an unconditional basis.*

b) *For every convergent series* $\sum_{n \in \mathbb{N}} a_n e_n$, *and every sequence* $(\epsilon_n)_{n \in \mathbb{N}}$ *with* $\epsilon_n = \pm 1 \quad \forall n \in \mathbb{N}$, *the series* $\sum_n \epsilon_n a_n e_n$ *converges.*

c) *For every convergent series* $\sum_n a_n e_n$ *and every strictly increasing sequence* $(n_i)_{i \in \mathbb{N}}$ *of integers, the series* $\sum_i a_{n_i} e_{n_i}$ *converges.*

d) *For every convergent series* $\sum_n a_n e_n$ *and every sequence of scalars* $(b_n)_{n \in \mathbb{N}}$ *such that, for all* n , $|b_n| \leqslant |a_n|$, *the series* $\sum_n b_n e_n$ *converges.*

e) *For every* $\epsilon > 0$, *every convergent series* $\sum_n a_n e_n = x$, *there is a finite subset* A_0 *of* $\mathbb{N}$ *such that, for every finite* $A \supset A_0$, $\left\|\sum_{i \in A} a_i e_i - x\right\| < \epsilon$.

f) *There exists a constant* $K \geq 1$ *such that if* A *and* B *are finite subsets of* $\mathbb{N}$ *, with* $A \subset B$ *, then for any sequence* $(a_n)_{n \in \mathbb{N}}$ *of scalars :*

$$\left\| \sum_{n \in A} a_n e_n \right\| \leq K \left\| \sum_{n \in B} a_n e_n \right\| .$$

We shall call *unconditional basis constant* of $(e_n)_{n \in \mathbb{N}}$ the smallest K satisfying this property. The basis will be called *unconditonally monotone* if $K = 1$.

REMARK. - Obviously, the sums of $\sum_n a_n e_n$, $\sum_n \epsilon_n a_n e_n$, $\sum_i a_{n_i} e_{n_i}$, $\sum_n b_n e_n$ may be all different.

PROOF. - a) $\Rightarrow$ e). Assume that e) is false. Then there is an $\epsilon > 0$ such that, for every A_0 finite, we can find $A \supset A_0$, with $\left\| \sum_{i \in A} a_i e_i - x \right\| \geq \epsilon$. By induction, we build a sequence $(A_n)_{n \geq 1}$ with $A_n \supset A_{n-1} \cup \{1,\ldots,n\}$, such that $\left\| \sum_{i \in A_n} a_i e_i - x \right\| \geq \epsilon$. Let π the permutation of $\mathbb{N}$ obtained by enumerating first A_1 , then $A_2 \setminus A_1$, $A_3 \setminus A_2, \ldots$; the series $\sum_i a_{\pi(i)} e_{\pi(i)}$ cannot converge to x .

e) $\Rightarrow$ f). If A is a finite subset of $\mathbb{N}$, define the projection P_A by $P_A(\sum \alpha_n e_n) = \sum_{n \in A} \alpha_n e_n$. Since $(e_n)_{n \in \mathbb{N}}$ is a basis, each P_A is continuous, and f) says that the P_A's are uniformly bounded. In order to prove that $\sup_{A \subset \mathbb{N}} \|P_A\| < +\infty$, it is enough, by Banach-Steinhaus Theorem (first part, chap. I), to show that, for all $x \in E$, $\sup_{A \subset \mathbb{N}} \|P_A x\| < +\infty$. If not, for some $x = \sum_n a_n e_n$, we find sets $(A_k)_{k \geq 1}$ with $\|P_{A_k} x\| \geq k$. Fix $\epsilon = 1$ in e). For the corresponding A_0 , we get $\|P_{A_k \cup A_0} x\| \xrightarrow[k \to +\infty]{} \infty$, which contradicts e).

f) $\Rightarrow$ c). Assume $\sum_n a_n e_n$ converges, and let $(n_i)_{i \in \mathbb{N}}$ be an increasing sequence of integers. Then, for $p,q \in \mathbb{N}$, $p < q$, we have :

$$\left\| \sum_{i=p+1}^{q} a_{n_i} e_{n_i} \right\| \leq K \left\| \sum_{i \geq n_{p+1}} a_i e_i \right\| ,$$

which ensures that $(\sum_{i \leq p} a_{n_i} e_{n_i})_{p \in \mathbb{N}}$ is Cauchy.

f) $\Rightarrow$ a). Let π be a permutation of $\mathbb{N}$, and $\epsilon > 0$. Choose $p_0 \in \mathbb{N}$ such that $\|\sum_{i > p_0} a_i e_i\| < \frac{\epsilon}{K}$. Then, for any finite set A containing $A_0 = \{0,1,\ldots,p_0\}$ we have :

$$\begin{aligned}\|\sum_{i \in A} a_i e_i - x\| &\leq \|\sum_{i \in A_0} a_i e_i - x\| + \|\sum_{i \in A \setminus A_0} a_i e_i\| \\ &\leq \|\sum_{i > p_0} a_i e_i\| + \|\sum_{i \in A \setminus A_0} a_i e_i\| \\ &\leq \frac{\epsilon}{K} + \epsilon = \epsilon\,(1 + \frac{1}{K}) .\end{aligned}$$

But we can find N large enough such that $\{\pi(0),\ldots,\pi(N)\}$ contains $\{0,1,\ldots,p_0\}$. This proves our assertion.

c) $\Rightarrow$ b). Let $(\epsilon_n)_{n \in \mathbb{N}}$ be a sequence of ± 1. We shall show that $\sum_m^n \epsilon_i a_i e_i \xrightarrow[m,n \to +\infty]{} 0$. Put $I_1(m,n) = \{i \in [m,n] \; ; \; \epsilon_i = \pm 1\}$,

$I_2(m,n) = \{i \in [m,n] \; ; \; \epsilon_i = -1\}$.

Then :

$$\|\sum_m^n \epsilon_i a_i e_i\| \leq \|\sum_{i \in I_1(m,n)} a_i e_i\| + \|\sum_{i \in I_2(m,n)} a_i e_i\|$$

and each term tends to zero when $m,n \to +\infty$.

b) $\Rightarrow$ c). Let $\sum_k a_k e_k$ be a convergent series, and $(n_i)_{i \in \mathbb{N}}$ an increasing sequence of integers. We may write :

$$\sum_i a_{n_i} e_{n_i} = \frac{1}{2}\left(\sum_k a_k e_k + \sum_j \epsilon_j a_j e_j\right)$$

with $\epsilon_j = -1$ if $j \notin \{(n_i)_{i \in \mathbb{N}}\}$, $\epsilon_j = +1$ if $j \in \{(n_i)_{i \in \mathbb{N}}\}$.

c) $\Rightarrow$ e). Assume e) to be false. Then for some $\epsilon > 0$, for every A_0, there is $A \supset A_0$ such that $\|\sum_{n \in A} a_n e_n - x\| \geq \epsilon$, and therefore $\|\sum_{n \in \complement A} a_n e_n\| \geq \epsilon$. Thus, for every $k \geq 1$, we can find a finite set B_k, with $\min B_k \geqslant k$, and $\|\sum_{n\in B_k} a_n e_n\| \geqslant \epsilon$. Extracting a subsequence if necessary, we may assume the B_k's to be consecutive : $\min B_{k+1} > \max B_k$ for all k. But then if we call $(n_j)_{j\in\mathbb{N}}$ the set $\bigcup_{k\geqslant 1} B_k$, the sequence $\sum_j a_{n_j} e_{n_j}$ cannot converge, thus contradicting c).

Therefore, we have proved the equivalence of all items, except d).

f) $\Rightarrow$ d). We shall introduce on E a new norm, equivalent to the original one, and having interesting properties :

LEMMA 2. - *If* $x = \sum_n a_n e_n$, *put :*

$$|x| = \sup\{\|\sum_n \epsilon_n a_n e_n\| \;;\quad \epsilon_n = \pm 1 \quad \textit{for all} \quad n\in\mathbb{N}\}$$

Then :

1) $|\cdot|$ *is equivalent to* $\|\cdot\|$, *if* f) *is satisfied.*

2) $(e_n)_{n\in\mathbb{N}}$ *is a monotone basis in* E *equipped with* $|\cdot|$.

3) *If* A *and* B *are disjoint finite subsets of* $\mathbb{N}$, *and if*

$$x = \sum_{n\in A} a_n e_n \,,\quad y = \sum_{n\in B} b_n e_n \,,$$

the function $t \to |x+ty|$ *is an even, convex function of* $t\in\mathbb{R}$.

PROOF OF LEMMA 2.

1) Let $x = \sum a_n e_n$, $\epsilon_n = \pm 1$. Let $A_+ = \{n, \epsilon_n = +1\}$, $A_- = \{n, \epsilon_n = -1\}$. Then :

$$\|\sum_n \epsilon_n a_n e_n\| \leqslant \|\sum_{n\in A_+} a_n e_n\| + \|\sum_{n\in A_-} a_n e_n\| \leqslant 2K\|x\| \,.$$

Since we have also $|x| \geqslant \|x\|$, the two norms are equivalent.

2) If B_1, B_2 are disjoint subsets of $\mathbb{N}$, we have, for every sequence $(\epsilon_n)_{n\in B_1\cup B_2}$ of ± 1, if $x = \sum_{n\in B_1\cup B_2} a_n e_n$:

$$|x| \geq \| \sum_{n \in B_1} \epsilon_n a_n e_n + \sum_{n \in B_2} \epsilon_n a_n e_n \|$$

and also

$$|x| \geq \| \sum_{n \in B_1} \epsilon_n a_n e_n - \sum_{n \in B_2} \epsilon_n a_n e_n \|$$

and consequently

$$|x| \geq \| \sum_{n \in B_1} \epsilon_n a_n e_n \|$$

from which follows

$$|x| \geq |\sum_{B_1} a_n e_n| \text{ , and } (e_n)_{n \in \mathbb{N}} \text{ is monotone in the norm } |\cdot| \text{ .}$$

3) Is obvious.

Let us now prove d). Let $(b_n)_{n \in \mathbb{N}}$, with, for all n, $|b_n| \leq |a_n|$. Put $b_n = t_n a_n$, with $|t_n| \leq 1$, and $t_n = \alpha_n + i\beta_n$, with $\alpha_n, \beta_n \in \mathbb{R}$, $|\alpha_n| \leq 1$, $|\beta_n| \leq 1$. So, in order to prove that $\sum_n b_n e_n$ converges, we have to prove that $\sum_n \alpha_n a_n e_n$ and $\sum \beta_n a_n e_n$ converge. But, for example, for the first, we have :

$$|\sum_n \alpha_n a_n e_n| \leq |\sum_n \epsilon_n a_n e_n| \text{ , with } \epsilon_n = \operatorname{sgn} \alpha_n \quad \forall n \text{ ,}$$

by 3) of the previous lemma. This last series converges when $\sum a_n e_n$ does in the norm $\|\cdot\|$, since the norms are equivalent. Finally, d) $\Rightarrow$ c) is obvious, and our proposition is proved.

The simplest example of unconditional basis is the canonical basis of c_0 or ℓ^p $(1 \leq p < +\infty)$. One can also show (but it is much harder) that the Haar system is an unconditional basis of $L^p([0,1], dt)$, if $1 < p < +\infty$ (see for example B. MAUREY [35]).

An example of a Schauder basis, which is not unconditional is the so-called "summing basis" of c_0. If $(e_n)_{n \in \mathbb{N}}$ is the canonical basis, put $s_n = e_0 + \ldots + e_n$. Then $\|s_n\|_{c_0} = 1$, for all n, and

$\| \sum_i a_i s_i \|_{c_0} = \sup_{n \in \mathbb{N}} \left| \sum_{i \geq n} a_i \right|$.

Therefore, for all N , $\left\| \sum_0^N (-1)^k s_k \right\|_{c_0} = 1$, though $\left\| \sum_0^N s_k \right\|_{c_0} = N$, and so f) cannot be satisfied.

In a space with unconditional basis, theorem 5, § 1, can be strengthened, because a concrete interpretation can be given to conditions a) and b).

THEOREM 3. (R.C. JAMES) - *Let* E *be a Banach space with unconditional basis.* E *is reflexive if and only if* E *does not contain any subspace isomorphic to* c_0 *, nor any subspace isomorphic to* ℓ_1 .

REMARK. - This condition is obviously necessary, since c_0 and ℓ_1 are not reflexive, and since every subspace of a reflexive space is reflexive.

PROOF OF THE THEOREM. - Taking into account theorem 5, § 1, it is enough to show the following two propositions :

PROPOSITION 4. - *Let* E *be a space with unconditional basis. The basis is boundedly complete if and only if* E *does not contain any subspace isomorphic to* c_0 *(we shall say, more briefly, that* E *does not contain* c_0 *).*

PROPOSITION 5. - *Let* E *be a space with unconditional basis. The basis is shrinking if and only if* E *does not contain* ℓ_1 .

PROOF OF PROPOSITION 4. - We consider E equipped with the norm $|\cdot|$ given by lemma 2. Assume that there is a series $\sum_i a_i e_i$, which is not convergent, but satisfies $\sup_n \left| \sum_{i \leq n} a_i e_i \right| < +\infty$. We may assume that for all n , $\left| \sum_{i \leq n} a_i e_i \right| \leq 1$, and therefore, for every finite set A , $\left| \sum_{i \in A} a_i e_i \right| \leq 1$. Since the series does not converge, we can find a $d > 0$, and two strictly increasing sequences of integers, $(n_k)_{k \in \mathbb{N}}$, $(m_k)_{k \in \mathbb{N}}$, with for all k ,

$$n_k < m_k < n_{k+1}$$

and

$$\left| \sum_{i=n_k}^{m_k-1} a_i e_i \right| > d .$$

Put $z_k = \sum_{n_k}^{m_k-1} a_i e_i$. We shall see that $F = \overline{\text{span}}\,[z_k]$ is isomorphic to c_o .

Let A be a finite subset of $\mathbb{N}$, and $(t_k)_{k \in A}$ a finite sequence of scalars. By lemma 2,

$$\left| \sum_{k \in A} t_k z_k \right| = \left| \sum_{k \in A} |t_k| z_k \right| \leqslant \sup_{k \in A} |t_k| \left| \sum_{k \in A} z_k \right| \leqslant \sum_{k \in A} |t_k| .$$

But, in the other direction :

$$\left| \sum_{k \in A} t_k z_k \right| \geqslant \sup_{k \in A} |t_k z_k| = \sup_{k \in A} |t_k| |z_k| \geqslant d \sup_{k \in A} |t_k| .$$

Consequently, we obtain :

$$d \sup_{k \in A} |t_k| \leqslant \left| \sum_{k \in A} t_k z_k \right| \leqslant \sup_{k \in A} |t_k| .$$

So, if $(e_n)_{n \in \mathbb{N}}$ is the canonical basis of c_o , the operator T from F into c_o , defined by $T(z_k) = e_k$, is an isomorphism from F onto c_o . This proves proposition 4.

PROOF OF PROPOSITION 5. - Here again, we put on E the norm $|\cdot|$. Assume $(e_n)_{n \in \mathbb{N}}$ is not shrinking. Then we can find a linear functional ξ , a strictly increasing sequence $(n_k)_{k \in \mathbb{N}}$, a number $\epsilon > 0$, such that for all $k \in \mathbb{N}$, $\sup\{|\xi(x)| \, ; \, x = \sum_{i \geqslant n_k} \alpha_i e_i \, , \, |x| = 1\} > \epsilon$.

We may assume $\|\xi\| = 1$. Then, there is a sequence of points $x_k \in E$, with $|x_k| = 1$, $x_k = \sum_{i \geqslant n_k} \alpha_i^{(k)} e_i$, $\xi(x_k)$ is real and $\xi(x_k) > \epsilon$ for all $k \in \mathbb{N}$. We may assume that x_k is finitely supported on the e_i's . (i.e. the sum $\sum_{i \geqslant n_k}$ is finite) : we know that, since it is a convergent series, there is a integer $n_k' > n_k$ such that $\left| \sum_{i > n_k'} \alpha_i^{(k)} e_i \right| < \frac{\epsilon}{2}$; if we put $x'_k = \sum_{n_k}^{n'_k} \alpha_i^{(k)} e_i$, $\xi(x'_k) \geq \xi(x_k) - \xi(\sum_{i>n'_k} \alpha_i^{(k)} e_i) \geq \epsilon - \frac{\epsilon}{2} = \frac{\epsilon}{2}$.

By extracting a subsequence from the n'_ks , we may also assume that x_k is supported between e_{n_k} and $e_{n_{k+1}-1}$. So we obtain :

There is a linear functional ξ , with $\|\xi\| = 1$, an $\epsilon > 0$, a strictly increasing sequence $(n_k)_{k\in\mathbb{N}}$ of integers, a sequence of points $(x_k)_{k\in\mathbb{N}}$ in E , with, for all $k \in \mathbb{N}$:

$$|x_k| = 1 , \qquad x_k = \sum_{i=n_k}^{n_{k+1}-1} \alpha_i^{(k)} e_i , \quad \xi(x_k) > \epsilon .$$

But then, if A is a finite subset of $\mathbb{N}$ and $(t_k)_{k\in A}$ are scalars :

$$\left|\sum_{k\in A} t_k x_k\right| = \left|\sum_{k\in A} |t_k| x_k\right| \geq \xi(\sum_{k\in A} |t_k| x_k) = \sum_{k\in A} |t_k| \xi(x_k) \geq \epsilon \sum_{k\in A} |t_k| .$$

But since conversely

$$\left|\sum_{k\in A} t_k x_k\right| \leq \sum_{k\in A} |t_k| ,$$

we obtain :

$$\epsilon \sum_{k\in A} |t_k| \leq \left|\sum_{k\in A} t_k x_k\right| \leq \sum_{k\in A} |t_k| .$$

So, if $(e_n)_{n\in\mathbb{N}}$ is the canonical basis of ℓ_1 , the operator T , from $\overline{\text{span}}\{(x_k)_{k\in\mathbb{N}}\}$ into ℓ_1 , defined by $T x_k = e_k$, is an isomorphism from $\overline{\text{span}}\{(x_k)_{k\in\mathbb{N}}\}$ onto ℓ_1 , which proves the proposition.

EXERCISES ON CHAPTER II.

EXERCISE 1. - Let $(e_n)_{n\in\mathbb{N}}$ be a Schauder basis in E , and $(f_n)_{n\in\mathbb{N}}$ the coordinate functionals. Show that for every $\xi \in E^\star$, one has :

$$\lim_{n\to+\infty} \sum_1^n \xi(e_k) f_k = \xi \qquad , \text{ for } \sigma(E^\star, E) .$$

EXERCISE 2. - Let $(e_n)_{n\in\mathbb{N}}$ be a Schauder basis in E , and $(f_n)_{n\in\mathbb{N}}$ the coordinate functionals. Show the equivalence of the following two properties :

a) $(f_n)_{n\in\mathbb{N}}$ is a Schauder basis of $E^\star$.

b) $(e_n)_{n\in\mathbb{N}}$ is a shrinking basis of E .

EXERCISE 3. - Let $(e_n)_{n\in\mathbb{N}}$ be a normalized Schauder basis in E , and $(f_n)_{n\in\mathbb{N}}$ the coordinate functionals. Let $(y_k)_{k\in\mathbb{N}}$ a normalized sequence in E such that, for every $m\in\mathbb{N}$, $f_m(y_k)\xrightarrow[k\to+\infty]{}0$.

Show that there is a subsequence $(y_{k_i})_{i\in\mathbb{N}}$ which is equivalent to a basic sequence made of blocks on the e_n's .

EXERCISE 4. - Let E be a Banach space with unconditional basis $(e_n)_{n\in\mathbb{N}}$. Let $(f_k)_{k\in\mathbb{N}}$ be the coordinate functionals. Assume that there is a sequence $(z_n)_{n\in\mathbb{N}}$ such that $\|z_n\|\leqslant 1$ for all $n\in\mathbb{N}$, $f_k(z_n)\xrightarrow[n\to+\infty]{}0$ for all $k\in\mathbb{N}$, but $(z_n)_{n\in\mathbb{N}}$ does not converge weakly to 0 . Show that $(z_n)_{n\in\mathbb{N}}$ has a subsequence equivalent to the ℓ_1-basis.

[Show that there is a sequence (u_n) of consecutive blocks on the e_n's (that is $u_n=\sum_{m_{n-1}+1}^{m_n}\alpha_i e_i$), with $u_n - z'_n\xrightarrow[n\to+\infty]{}0$, for a subsequence $(z'_n)_{n\in\mathbb{N}}$ of $(z_n)_{n\in\mathbb{N}}$, and that there is $f\in E^\star$, and $\delta>0$, such that $f(u_n)\geqslant\delta$ for all n]

EXERCISE 5. - Let $h_n(t)$ be the Haar system on $[0,1]$ $(n\geqslant 1)$. Put $g_0(x)=1$, $g_n(x)=\int_0^x h_n(t)dt$, $n\geqslant 1$. Show that the sequence $(g_n)_{n\geqslant 0}$ is a basis of $\mathscr{C}([0,1])$ (called the Schauder basis of $\mathscr{C}([0,1])$).

EXERCISE 6. - Let $(e_n)_{n\geqslant 1}$ be a monotone shrinking basis of a Banach space E . Let $(P_n)_{n\geqslant 1}$ be the projections associated to the basis, and $(f_k)_{k\geqslant 1}$ the coordinate functionals.

1) Show that if $\xi\in E^{\star\star}$, ${}^{tt}P_n(\xi)=\sum_{i=1}^{n}\xi(f_i)e_i$, and

$$\|\xi\|=\lim_{n\to+\infty}\|{}^{tt}P_n(\xi)\|=\sup_n\|{}^{tt}P_n(\xi)\| .$$

2) Conversely, if $(a_n)_{n\geq 1}$ is a sequence of scalars such that $\sup_n \left\| \sum_1^n a_i e_i \right\| < +\infty$, show that any limit ξ in $E^{\star\star}$, for $\sigma(E^{\star\star}, E^\star)$, of a subsequence $\sum_1^{n_j} a_i e_i$ satisfies $\xi(f_i) = a_i$. Deduce that $\sum_1^n a_i e_i$ converges to ξ in this topology.

3) Show that $E^{\star\star}$ can be isometrically identified with the space of sequences $(a_n)_{n\geq 1}$ of scalars such that $\sup_n \left\| \sum_1^n a_i e_i \right\| < +\infty$, the correspondence being $\xi \leftrightarrow (\xi(f_j))_{j\geq 1}$.

EXERCISE 7. (The James' space J). - Let J be the vector space of all sequences $x = (x(k))_{k\geq 1}$ of real numbers, such that $x(k) \xrightarrow[k\to+\infty]{} 0$ and

$$\|x\|_J = \sup\{(x(p_1) - x(p_2))^2 + \ldots + (x(p_{2n-1}) - x(p_{2n}))^2 ;$$

$$n \geq 1 ,\ p_1 < \ldots < p_{2n}\} < +\infty .$$

1) Show that $\|x\|_J$ is a norm on J, and that J is a Banach space.

2) Show that the canonical basis $(e_n)_{n\geq 1}$ of $\mathbb{K}^{(\mathbb{N}^\star)}$ is a monotone basis of J.

3) Considering $s_n = e_1 + \ldots + e_n$, $n \geq 1$, show that J is not reflexive.

4) Show that $(e_n)_{n\geq 1}$ is a shrinking basis of J.

(If not, there is a sequence of consecutive normalized blocks u_k on the e_n's, a $\delta > 0$, and a $f \in J^\star$ with $f(u_k) \geq \delta$ for all k; but $\sum_{k=1}^{\infty} \frac{u_k}{k}$ belongs to J).

5) Using exercise 6, show that $J^{\star\star}$ is the space of sequences $(a_n)_{n\geq 1}$ such that $\sup_n \left\| \sum_1^n a_i e_i \right\|_J < +\infty$, that is such that $\|(a_n)\|_J < +\infty$.

6) Show that if $\|(a_n)\|_J < +\infty$, the limit $\lim_{n\to+\infty} a_n$ exists.

7) Let ξ_0 be the limit, in $\sigma(J^{\star\star}, J^\star)$, of the sequence $(s_n)_{n\geq 1}$. Show that $\xi_0(f_j) = 1$, for all $j \geq 1$. Show that $J^{\star\star}$ is spanned by ξ_0 and J, and that $J^{\star\star}$ is isomorphic to J.

REFERENCES ON CHAPTER II.

For the various notions of convergence and the links between them, the reader may consult M.M. DAY's book [13]. In this book, beside unconditional and absolute convergence, several other types of convergence are studied.

Proposition 1, § 1, occurs in every book dealing with this topic, for example LINDENSTRAUSS-TZAFRIRI [34, vol. 1], or I. SINGER [47, vol. 1]. Proposition 3 is in BANACH's book [5].

Theorems 5 (§ 1) and 2 (§ 2) are due to R.C. JAMES [24] ; the version which we give is not complete. The reader is referred to M.M. DAY's book [13] for the complete statements and proofs, and for some extensions.

The various characterizations of unconditional convergence appear in LINDENSTRAUSS-TZAFRIRI [34].

Exercises 1 and 2 are part of R.C. JAMES' result. See [13]. Exercise 4 is in SINGER [47], exercise 6 in LINDENSTRAUSS-TZAFRIRI [34, vol. 1], and James'space is given in R.C. JAMES [24], and also in LINDENSTRAUSS-TZAFRIRI [34, vol. 1].

COMPLEMENTS ON CHAPTER II. - The Approximation Property.

A Banach space E is said to have the Approximation Property (A.P. in short) if, for every compact set $K \subset E$, every $\epsilon > 0$, there exists a finite rank operator T, with $\|Tx - x\| < \epsilon$ for all $x \in K$. An equivalent formulation is : for every Banach Y, every compact operator T, from Y into E, there exists a sequence of finite rank operators, from Y into E, converging to T in norm.

If E has a Schauder basis, E has A.P. (consider the projections on the first n coordinates). ENFLO's example [19] is a separable space without A.P. See [34], vol. I, for a detailed study of these notions.

CHAPTER III

COMPLEMENTED SUBSPACES IN BANACH SPACES

In the previous chapter, we have introduced and studied a notion of basis for Banach spaces, which may be considered as an extension of the notion of Hilbertian basis in a Hilbert space. Another result, which we have mentioned in chapter I, and which is typical of Hilbert spaces, is the fact that every closed linear subspace is the range of a linear projection (of norm 1). This is not true in every Banach space, as we shall see on an example. We shall then investigate this question in some common Banach spaces : $\ell^p(1 \leq p \leq +\infty)$, c_0 , $L^p(1 \leq p \leq +\infty)$, $\mathscr{C}(K)$; this will be done in the following chapters.

DEFINITION. - *Let* E *be a Banach space. A closed linear subspace* F *is said to be* complemented *if there is a continuous linear projection* P *from* E *onto* F .

If $k \in \mathbb{R}$, $k \geq 1$, *we say that* F *is* *k-complemented if there is a projection* P *with* $\|P\| \leq k$.

We shall first build an example of uncomplemented subspace in $\ell^p(1 \leq p \leq +\infty , p \neq 2)$, we shall see in the sequel several other examples.

It is, for this construction, more convenient to work with involutions rather than projections. Both are linked, as the following proposition shows. Let us first recall that an involution satisfies $U^2 = I$; the invariant subspace of an involution U is $\{x, Ux = x\}$.

PROPOSITION 1. - *Let* E *be a normed space,* H *a closed linear subspace of* E .

1) *If* P *is a projection from* E *onto* H , *then* $U = 2P - I$ *is an involution having* H *as invariant subspace. Conversely, if* U *is an involution having* H *as invariant subspace,* $P = \frac{1}{2}(U + I)$ *is a projection*

from E *onto* H .

2) *If* $P = \frac{1}{2}(U + I)$ *is a projection from* E *onto* H , *all the projections from* E *onto* H *are of the form* :

$$\widetilde{P} = \frac{1}{2}(U + I + V)$$

where V *is a linear operator from* E *into itself, satisfying*

$$(1) \quad UV = -VU = V .$$

PROOF. - 1) is obvious. For 2), if $\widetilde{P}$ is a projection onto H , we must have $P\widetilde{P} = \widetilde{P}$ and $\widetilde{P}P = P$. So, if we set $V = 2\widetilde{P} - (I + U)$, then, from $P = \frac{1}{2}(U + I)$ and $P\widetilde{P} = \widetilde{P}$ follows :

$$\frac{1}{4}(I + U)(I + U + V) = \frac{1}{2}(I + U) + \frac{1}{4}(V + UV) = \frac{1}{2}(I + U + V) \text{ and so } UV = V .$$

The condition $\widetilde{P}P = P$ implies, the same way, $-VU = V$.

Conversely, if $\widetilde{P} = \frac{1}{2}(I + U + V)$, where V satisfies (1) , one checks immediately that $P\widetilde{P} = \widetilde{P}$ and $\widetilde{P}P = P$, and therefore :

$$\widetilde{P}\widetilde{P} = \widetilde{P}P\widetilde{P} = P\widetilde{P} = \widetilde{P} ,$$

and $\widetilde{P}$ is a projection. This proves the proposition.

If H is a closed linear subspace of E , we define :

$$p(H) = \inf\{\|P\| ; \; P \text{ is a projection from } E \text{ onto } H\}$$
$$(= +\infty \text{ if there is no projection})$$

and :

$$u(H) = \inf\{\|U\| ; \; U \text{ is an involution with invariant subspace } H\}$$
$$(= +\infty \text{ if there is no such involution}).$$

Then the following relations follow immediately from proposition 1 :

PROPOSITION 1 bis.

$$(2) \quad \frac{1}{2}(u(H) - 1) \leqslant p(H) \leqslant \frac{1}{2}(u(H) + 1) .$$

LEMMA 2. - *Let* $\partial \in \mathbb{N}$, *and* $n = 2^{\partial}$. *Let* $p \in \mathbb{R}$, $1 \leqslant p \leqslant +\infty$, $p \neq 2$. *There is a subspace* $H^{(n)}$ *of* $\ell_p^{(n)}$ *with* $u(H^{(n)}) \geqslant n^{|\frac{1}{p} - \frac{1}{2}|}$ *(and, consequently,* $p(H^{(n)}) \geqslant \frac{1}{2}(n^{|\frac{1}{p} - \frac{1}{2}|} - 1)$ *).*

PROOF. - We may restrict ourselves to the case $1 \leqslant p < 2$, the other case being obtained by transposition.

Put $U_1 = \begin{pmatrix} 1 & 1 \\ 1 & -1 \end{pmatrix}$, and, if $U_{\partial-1}$ is defined, put

$$U_{\partial} = \begin{pmatrix} U_{\partial-1} & U_{\partial-1} \\ U_{\partial-1} & -U_{\partial-1} \end{pmatrix} \text{ and } U = \frac{1}{\sqrt{n}} U_{\partial} .$$

This is a $n \times n$ matrix, which, as one checks immediately, defines an involution of $\ell_p^{(n)}$

Let us call $H^{(n)}$ the invariant subspace of this involution. From proposition 1, 2), follows that every involution having $H^{(n)}$ as invariant subspace is of the form $U + V$, where V satisfies (1).

From (1) follows that $\operatorname{tr} V = \operatorname{tr} UV = -\operatorname{tr} VU = -\operatorname{tr} UV$, and so $\operatorname{tr} V = 0$ (let us recall that the trace is the sum of the diagonal coefficients in the matrix). Let us put

$$U = (u_{i,k})_{\substack{1 \leqslant i \leqslant n \\ 1 \leqslant k \leqslant n}} , \qquad V = (v_{i,k})_{\substack{1 \leqslant i \leqslant n \\ 1 \leqslant k \leqslant n}} .$$

Since $\sum_{k=1}^{n} v_{kk} = 0$, one of the v_{kk}'s must be $\geqslant 0$. Since $U^2 = I$, we must have (since U is symmetric) $\sum_{i=1}^{n} u_{ik}^2 = 1$, for all k . But $V = UV$,from which follows :

$$1 \leqslant 1 + v_{kk} = \sum_{i=1}^{n} u_{ik}(u_{ik} + v_{ik}) ,$$

since $\sum_{i=1}^{n} u_{ik}u_{ik} = \sum_{i} u_{ki}u_{ik} = \sum_{i} u_{ik}^2 = 1$, and

$$\sum_{i} u_{ik}v_{ik} = \sum_{i} u_{ki}v_{ik} = v_{kk} .$$

But, by construction, each u_{ik} is $\pm \frac{1}{\sqrt{n}}$. So we obtain, if $\frac{1}{q} = 1 - \frac{1}{p}$,

$$\left(\sum_{i=1}^{n} |u_{ik}|^q\right)^{1/q} = n^{(1-\frac{q}{2})^{1/q}} = n^{\frac{1}{2}-\frac{1}{p}},$$

and therefore, by Hölder's inequality :

$$1 \leqslant n^{\frac{1}{2}-\frac{1}{p}} \cdot \left(\sum_{i=1}^{n} |u_{ik} + v_{ik}|^p\right)^{1/p} = n^{\frac{1}{2}-\frac{1}{p}} \|(U+V)e_k\|_{\ell_p}$$

(where, as usually, $(e_k)_{k \in \mathbb{N}}$ is the canonical basis of ℓ_p).

So we obtain, for every V,

$$\|U + V\| \geqslant n^{\frac{1}{p}-\frac{1}{2}}$$

and consequently

$$u(H^{(n)}) \geqslant n^{\frac{1}{p}-\frac{1}{2}}, \text{ and}$$

$$p(H^{(n)}) \geqslant \tfrac{1}{2}(n^{\frac{1}{p}-\frac{1}{2}} - 1) .$$ This proves the lemma.

This lemma provides us with subspaces of ℓ_p having arbitrarily large projection constants $p(H)$. From this fact, we can build a subspace which is not complemented, by putting together the previous subspaces : we take the finite-dimensional subspaces which are worse and worse complemented, and put them one after the other. The procedure which we use for this purpose (and which we shall use again for other purposes) is the notion of "products of Banach spaces", which we shall now define.

PRODUCTS OF BANACH SPACES.

Let $(E_n)_{n \in \mathbb{N}}$ be a sequence of Banach spaces, and p, with $1 \leqslant p < +\infty$. We may consider the product $\prod_n E_n$, which is

$$\prod E_n = \{(x_n)_{n \in \mathbb{N}} \; ; \; x_n \in E_n \text{ for all } n\}$$

and, in this product, we consider the sequences $(x_n)_{n \in \mathbb{N}}$ such that :

$$\|(x_n)_n\| = \left(\sum_n \|x_n\|_{E_n}^p\right)^{1/p} < +\infty .$$

If $p = +\infty$, we define similarly :

$$\|(x_n)_n\| = \sup_n \|x_n\|_{E_n} .$$

We call $(\prod_n E_n)_p$ this subspace of $\prod E_n$, equipped with this norm. We can also consider $(\prod_n E_n)_{c_o}$, which is :

$$(\prod_n E_n)_{c_o} = \{(x_n)_{n\in\mathbb{N}} \ ; \ x_n \in E_n \ \text{ for all } n , \ \|x_n\|_{E_n} \xrightarrow[n\to+\infty]{} 0\} ,$$

equipped with the norm $\sup_n \|x_n\|_{E_n}$.

This product is also sometimes called "ℓ_p-sum of the Banach spaces E_n". The reason for that is the following : if, for example, E is a Hilbert space with $E = E_1 \oplus E_2$, then $E = (E_1 \times E_2)_2$, with the previous notations.

If all the E_n's are isometric to ℓ_p $(1 \leqslant p \leqslant +\infty)$, then the space $(\prod_n E_n)_p$ is also isometric to ℓ_p : to an element $x = (x_n)_{n\in\mathbb{N}}$, $x_n \in \ell_p$ for all n, we associate $y \in \ell_p$, defined by :

$$y(1) = x_1(1),\ y(2) = x_1(2),\ y(3) = x_2(1),\ y(4) = x_1(3),\ y(5) = x_2(2) ,$$

and so on. We obtain this way a linear isometric bijection between $(\prod \ell_p)_p$ and ℓ_p. Also, for the same reason, $(\prod c_o)_{c_o}$ is isometric to c_o.

If now the E_n's are finite-dimensional ℓ_p-spaces, $E_n = \ell_p^{(k_n)}$ $(k_n = \dim E_n)$, the space $(\prod_n E_n)_p$ is also, clearly, isometric to ℓ_p : one may apply the previous procedure, or, more simply, associate to $(x_n)_{n\in\mathbb{N}}$ the $y \in \ell_p$ obtained by putting first the k_1 components of x_1, then the k_2 components of x_2, and so on (see also first part, chap. III, exercise 2).

Let us come back to the construction of an uncomplemented subspace of ℓ_p. Put $F = (\prod_\partial \ell_p^{(2^\partial)})_p$. From what we have just seen, follows that F is isometric to ℓ_p. Put also $H = (\prod_\partial H^{(2^\partial)})_p$: this a closed linear subspace of F, and we shall see that it is not complemented in F.

Let us call Q_∂ ($\partial \in \mathbb{N}$) the projection from F onto $\ell_p^{(2^\partial)}$ which sends the other $\ell_p^{(2^{\partial'})}$ ($\partial' \neq \partial$) onto 0. More precisely, if $x = (x_{\partial'})_{\partial' \in \mathbb{N}}$, with $x_{\partial'} \in \ell_p^{(2^{\partial'})}$, then $Q^\partial(x) = x_\partial$. Then $Q_\partial(H) = H^{(2^\partial)}$, and so, if P is a projection from F onto H, $Q_\partial \circ P$ is a projection from F onto $H^{(2^\partial)}$. But $\|Q_\partial\| = 1$ for all ∂, and, if we call J_∂ the canonical injection from $\ell_p^{(2^\partial)}$ into F, we obtain

$$\|Q_\partial \circ P \circ J_\partial\| \leqslant \|Q_\partial \circ P\| \leqslant \|P\| .$$

But $Q_\partial \circ P \circ J_\partial$ is a projection from $\ell_p^{(2^\partial)}$ onto $H^{(2^\partial)}$, and so, by lemma 2, $\|Q_\partial \circ P \circ J_\partial\| \geqslant \frac{1}{2}(2^{\partial|\frac{1}{p}-\frac{1}{2}|} - 1)$ for all ∂ ; so P cannot exist, and H is not complemented F. So ℓ_p $(1 \leqslant p \leqslant +\infty,\ p \neq 2)$ contains uncomplemented subspaces. This achieves the construction of our example.

We shall meet other examples of Banach spaces with uncomplemented subspaces. Though the construction we have made does not appear completely immediate, this situation is in fact general : one can show that every Banach space which is not isomorphic to an Hilbert space contains uncomplemented subspaces (see LINDENSTRAUSS-TZAFRIRI [34]).

However, in any normed space, the finite-dimensional subspaces and finite-codimensional subspaces are always complemented (a subspace F is of finite codimension if it has a finite-dimensional topological supplement : $E = F \oplus G$, with dim G finite).

Let us show that finite-dimensional subspaces are complemented (recall that a finite-dimensional subspace is always closed). Let F be a subspace of E, of dimension n, and let $(e_1,\dots,e_n)$ be an algebraic basis of F, with $\|e_i\| = 1$ for $i = 1,\dots,n$. Let $f_1,\dots,f_n$ be the coordinate functionals.

Since F is of finite dimension, there is a $C > 0$ such that, for every sequence $(\alpha_1,\dots,\alpha_n)$ of scalars :

$$\frac{1}{C} \sup_{1 \leqslant i \leqslant n} |\alpha_i| \leqslant \left\| \sum_1^n \alpha_i e_i \right\| \leqslant C \sup_{1 \leqslant i \leqslant n} |\alpha_i| .$$

So, for $i = 1,\dots,n$, we obtain :

$$\|f_i\| = \sup\{|f_i(x)| \; ; \; x = \sum_1^n \alpha_j e_j \; , \; \left\| \sum_1^n \alpha_j e_j \right\| \leqslant 1\}$$
$$\leqslant \sup\{|f_i(x)| \; ; \; x = \sum_1^n \alpha_j e_j \; , \; \sup_j |\alpha_j| \leqslant C\} \leqslant C .$$

By Hahn-Banach Theorem, we may extend the f_i's to linear functionals on E , still denoted f_i , will $\|f_i\| \leqslant C$, for $i = 1,\dots,n$. If $x \in E$, we set :

$$Px = \sum_{i=1}^n f_i(x) e_i .$$

This a linear projection onto F , and we obtain :

$$\|Px\| = \left\| \sum_{i=1}^n f_i(x) e_i \right\| \leqslant C \sup_j \left| f_j(x) \right| \leqslant C^2 \|x\| ,$$

which proves that P is continuous and $\|P\| \leqslant C^2$.

Our assertion for finite-codimensional subspaces is a consequence of the previous one. Assume $E = F \oplus G$, with $\dim G < +\infty$. If P is a projection onto G , then $I - P$ is a projection onto F .

But this projection which exists on every finite-dimensional or finite-codimensional subspace needs not be of norm 1, contrarily to what happened in Hilbert spaces. For example, in $L^p([0,1], dt)$ $1 \leqslant p < +\infty$, $p \neq 2$, no closed hyperplane is the range of a norm 1 projection (T. ANDO [1]). The existence, in a given Banach space, of 1-complemented hyperplanes gives some informations about the geometry of the space (see B. BEAUZAMY - B. MAUREY [9]).

EXERCISES ON CHAPTER III.

EXERCISE 1. - The aim of this exercise is to show the following result : if a separable Banach space E contains a subspace F isometric to c_0 , there is a projection P from E onto F, with $\|P\| \leqslant 2$.

Let $(e_n)_{n\geqslant 1}$ be the canonical basis of c_o, $(f_n)_{n\geqslant 1}$ the canonical basis of $\ell_1 = c_o^\star$. Let $\widetilde{f}_n \in E^\star$, with $\|\widetilde{f}_n\| = 1$, and the restriction of $\widetilde{f}_n$ to c_o coïncides with f_n.

a) Let d be the distance on $E^\star$, the restriction of which to $\mathscr{B}_{E^\star}$ defines the topology $\sigma(E^\star, E)$ (see first part, chap. III, § 2, prop. 4, p. 57). Check that this distance is translation invariant : $d(y, z) = d(y - z, 0)$, $\forall\, y,z \in E^\star$.

b) Put $M = \mathscr{B}_{E^\star} \cap F^\perp$. Show that any accumulation point of the sequence $(\widetilde{f}_n)_{n\geqslant 1}$ belongs to M.

c) Deduce that $d(\widetilde{f}_n, M) \xrightarrow[n \to +\infty]{} 0$.

d) Let (g_n) be a sequence of elements in M such that $d(\widetilde{f}_n, g_n) \to 0$, that is $\widetilde{f}_n - g_n \xrightarrow[n \to +\infty]{} 0$ $\sigma(E^\star, E)$. Put, for every $x \in E$:

$$Px = (\widetilde{f}_1(x) - g_1(x) , \widetilde{f}_2(x) - g_2(x), \ldots)$$

Show that $Px \in c_o$ and that P is a projection of norm at most 2.

REFERENCES ON CHAPTER III.

The example of non-complemented subspace in ℓ_p is taken from Köthe [31].

The result given in exercise 1 is due to Sobczyk ; the proof suggested is that of Veech, and is taken from LINDENSTRAUSS-TZAFRIRI [34, vol. I].

CHAPTER IV

THE BANACH SPACES ℓ_p $(1 \leqslant p \leqslant +\infty)$ AND c_o

I - SUBSPACES OF ℓ_p $(1 \leqslant p < +\infty)$ AND c_o .

The spaces ℓ_p $(1 \leqslant p < +\infty)$ have already occured in the previous pages. Let us recall that ℓ_p $(1 \leqslant p < +\infty)$ is the space of the sequences of scalars $x = (x(k))_{k \in \mathbb{N}}$, such that $\|x\|_p = (\sum_{k \in \mathbb{N}} |x(k)|^p)^{1/p} < +\infty$. The space c_o is the set of the sequences $x = (x(k))_{k \in \mathbb{N}}$ such that $(x(k)) \xrightarrow[k \to +\infty]{} 0$, equipped with the norm $\|x\|_{c_o} = \sup_{k \in \mathbb{N}} |x(k)|$. Obviously, the canonical basis of $\mathbb{K}^{(\mathbb{N})}$, $(e_n)_{n \in \mathbb{N}}$, is an unconditional basis for ℓ_p $(1 \leqslant p < +\infty)$ or c_o . More precisely, $(e_n)_{n \in \mathbb{N}}$ is monotone unconditional : if A and B are finite subsets of $\mathbb{N}$ with $A \subset B$, we have, for every finite sequence of scalars (a_n) :

$$\|\sum_{n \in A} a_n e_n\|_p \leqslant \|\sum_{n \in B} a_n e_n\|_p ,$$

and the same in c_o .

These spaces are separable (since they have a basis) : the finite sequences with rational coefficients (or with coefficients having rational real and imaginary parts, if the space is complex) form a dense set in ℓ_p $(1 \leqslant p < +\infty)$ or c_o . The space ℓ_∞ is the set of bounded sequences of scalars, $x = (x(k))_{k \in \mathbb{N}}$, equipped with the norm $\|x\|_\infty = \sup_{k \in \mathbb{N}} |x(k)|$. This space is not separable : two sequences made only with $+1$ or -1 are at distance 2 from each other, and the set of such sequences is not countable. Therefore, the space ℓ_∞ cannot have a Schauder basis.

In ℓ_p or c_o , let us call *blocks on the canonical basis* $(e_n)_{n \in \mathbb{N}}$ elements of the type :

$$u_j = \sum_{q_j+1}^{q_{j+1}} \lambda_n e_n ,$$

where $(q_j)_{j \in \mathbb{N}}$ is a strictly increasing sequence of integers, and $(\lambda_n)_{n \in \mathbb{N}}$ are scalars.

DEFINITION. - *If* $(x_n)_{n \in \mathbb{N}}$ *and* $(y_n)_{n \in \mathbb{N}}$ *are basic sequences in a Banach space* E , *we shall say that they are* equivalent *if, for every sequence* $(a_n)_{n \in \mathbb{N}}$ *of scalars, the series* $\sum_n a_n x_n$ *converges if and only if the series* $\sum_n a_n y_n$ *converges.*

By the closed graph theorem, this happens if and only if there exist two constants m , M such that for every finite sequence (a_n) of scalars :

$$m\| \sum a_n x_n\| \leqslant \| \sum a_n y_n\| \leqslant M\| \sum a_n x_n\| .$$

If two basic sequences $(x_n)_{n \in \mathbb{N}}$ and $(y_n)_{n \in \mathbb{N}}$ are equivalent, the closed vector subspaces they generate are of course isomorphic. But the converse is not true : the canonical basis $(e_n)_{n \in \mathbb{N}}$ and the summing basis $s_n = \sum_{k \leqslant n} e_k$ are both bases of c_0 , but they are not equivalent.

An interesting property of the canonical basis $(e_n)_{n \in \mathbb{N}}$, in c_0 or ℓ_p $(1 \leqslant p < +\infty)$, is the following.

PROPOSITION 1. - *If* E *is* c_0 *or* ℓ_p $(1 \leqslant p < +\infty)$, *and* $(e_n)_{n \in \mathbb{N}}$ *is the canonical basis of* E , *then :*

a) *Any sequence of normalized blocks* $(u_j)_{j \in \mathbb{N}}$ *on the basis* $(e_n)_{n \in \mathbb{N}}$ *is a basis, equivalent to* $(e_n)_{n \in \mathbb{N}}$, *and* $\overline{\operatorname{span}}\{(u_j)_{j \in \mathbb{N}}\}$ *is isometric to* E .

b) $\overline{\operatorname{span}}\{(u_j)_{j \in \mathbb{N}}\}$ *is 1-complemented in* E .

PROOF. - We give it for $E = \ell_p$ and - since it is similar - leave it to the reader for $E = c_0$.

If $u_j = \sum_{q_j+1}^{q_{j+1}} \lambda_n e_n$, then, for all $j \in \mathbb{N}$:

$$1 = \|u_j\|_p = \left(\sum_{q_j+1}^{q_{j+1}} |\lambda_n|^p \right)^{1/p} .$$

Therefore, for every finite sequence (a_n) of scalars :

$$\| \sum_j a_j u_j \|_p = \left[\sum_j |a_j|^p \left(\sum_{n=q_j+1}^{q_{j+1}} |\lambda_n|^p \right) \right]^{1/p} = \left(\sum_j |a_j|^p \right)^{1/p} .$$

This proves that the sequence $(u_j)_{j \in \mathbb{N}}$ is equivalent to the canonical basis and that the application T , from $\overline{\text{span}}\{(u_j)_{j \in \mathbb{N}}\}$ onto E is an isometry. So a) is proved.

To show b), let us choose, for all $j \in \mathbb{N}$, a linear functional $u_j^\star$, on the vector space $\text{span}\{e_{q_j+1}, \ldots, e_{q_{j+1}}\}$, with

$$\|u_j^\star\| = 1 , \quad u_j^\star(u_j) = 1 .$$

Now if $x = \Sigma a_n e_n \in \ell_p$, put :

$$Px = \sum_{j \in \mathbb{N}} u_j^\star \left(\sum_{n=q_j+1}^{q_{j+1}} a_n e_n \right) u_j .$$

Then we obtain :

$$\|Px\|_p^p = \sum_j \left| u_j^\star \left(\sum_{n=q_j+1}^{q_{j+1}} a_n e_n \right) \right|^p \leqslant \sum_j \left\| \sum_{n=q_j+1}^{q_{j+1}} a_n e_n \right\|_p^p = \|x\|_p^p ,$$

and therefore P is a norm -1 projection from E onto $\overline{\text{span}}\{(u_j)_{j \in \mathbb{N}}\}$.

The property given in proposition 1, a) actually characterizes the spaces ℓ_p $(1 \leqslant p < +\infty)$ or c_o : one can show that if a sequence $(x_n)_{n \in \mathbb{N}}$ of points in a Banach space is equivalent to every sequence of blocks built on it, then $(x_n)_{n \in \mathbb{N}}$ is equivalent to the canonical basis of ℓ_p (for some p , $1 \leqslant p < +\infty$) or c_o (this result was proved by M. Zippin ; it can be found in LINDENSTRAUSS-TZAFRIRI [34]).

We shall now turn to subspaces of ℓ_p $(1 \leqslant p < +\infty)$ and c_o .

PROPOSITION 2. - *Let* E *be* c_o *or* $\ell_p (1 \leqslant p < +\infty)$. *Every subspace* F *of* E, *of infinite dimension, contains a subspace isomorphic to* E *and complemented in* E *(and therefore complemented in* F *).*

To establish this proposition, we shall need two lemmas, which have their own interest :

LEMMA 3.

a) *Let* $(x_n)_{n\in\mathbb{N}}$ *be a normalized Schauder basis, with basis constant* M *, in a Banach space* B *. Let* $(y_n)_{n\in\mathbb{N}}$ *be a sequence of points in* B *, with* $\sum_n \|x_n - y_n\| < \frac{1}{2M}$ *. Then* $(y_n)_{n\in\mathbb{N}}$ *is a basis, and is equivalent to* $(x_n)_{n\in\mathbb{N}}$ *.*

b) *Let* $(x_n)_{n\in\mathbb{N}}$ *be a normalized basic sequence, with basis constant* M *. Assume there is a projection* P *from* B *onto* $\overline{\text{span}}\{(x_n)_{n\in\mathbb{N}}\}$ *. Let* $(y_n)_{n\in\mathbb{N}}$ *be a sequence of points in* B *, with* $\sum_n \|x_n - y_n\| < \frac{1}{8M\|P\|}$ *. Then* $(y_n)_{n\in\mathbb{N}}$ *is a basic sequence, and* $Y = \overline{\text{span}}\{(y_n)_{n\in\mathbb{N}}\}$ *is complemented in* B *.*

PROOF OF LEMMA 3. - We know that if $(x_n)_{n\in\mathbb{N}}$ is a normalized basic sequence with basis constant M , we have, if $\sum_n a_n x_n$ converges :

$$(1) \quad \operatorname{Max}_n |a_n| \leqslant 2M \Big\| \sum_n a_n x_n \Big\| .$$

Define an operator T , from B onto $\overline{\text{span}}\{(y_n)_{n\in\mathbb{N}}\}$, by :

$$x = \sum_n a_n x_n \longrightarrow Tx = \sum_n a_n y_n .$$

This last series converges since :

$$\Big\| \sum_n a_n y_n \Big\| \leqslant \Big\| \sum_n a_n x_n \Big\| + \Big\| \sum_n a_n (x_n - y_n) \Big\| \leqslant \|x\| + \operatorname{Max} |a_k| \sum_n \|x_n - y_n\|$$

$$\leqslant \|x\| \left(1 + \frac{2M}{2M}\right) = 2\|x\| , \quad \text{and } T \text{ is continuous.}$$

Now we have :

$$\|x - Tx\| \leqslant \sum_n |a_n| \, \|x_n - y_n\| \leqslant \max_n |a_n| \sum_n \|x_n - y_n\|$$

$$(2) \quad \|x - Tx\| \leqslant 2M\|x\| \sum_n \|x_n - y_n\| .$$

So, if we put $\delta = \frac{1}{2M} \sum_n \|x_n - y_n\|$, then $\delta < 1$, and $\|x - Tx\| \leqslant \delta \|x\|$. So $\|I - T\| \leqslant \delta < 1$, and T is invertible (and $\|T\| \leqslant 1 + \delta < 2$) . From the first part, chapter I, follows that T is an isomorphism, and a) is proved.

b) If $(y_n)_{n \in \mathbb{N}}$ satisfies $\sum_n \|x_n - y_n\| < \frac{1}{8M\|P\|} \leqslant \frac{1}{8M}$, $(y_n)_{n \in \mathbb{N}}$ is a basic sequence, according to a). Let $y \in Y$. Then $y = \sum_n a_n y_n$, and :

$$\|TPy - y\| = \left\|TP\left(\sum_n a_n(x_n - y_n)\right)\right\| \leqslant \|T\| \cdot \|P\| \cdot \operatorname{Max}_n |a_n| \sum_n \|x_n - y_n\|$$

$$\leqslant \left(2M\|T\| \cdot \|P\| \sum_n \|x_n - y_n\|\right)\|x\| .$$

If $\sum_n \|x_n - y_n\| < \frac{1}{8M}$, we obtain from (2) :

$$\|x - Tx\| \leqslant \frac{1}{4} \|x\| ,$$

and so $\|x\| \leqslant \frac{4}{3} \|Tx\| = \frac{4}{3} \|y\|$

which implies

$$\|TPy - y\| \leqslant 2M\|T\| \cdot \|P\| \sum_n \|x_n - y_n\| \cdot \frac{4}{3} \|y\| \leqslant \frac{2}{3} \|y\| .$$

So, if S denotes the restriction of TP to the space Y , S is invertible on Y .

Finally, let us show that $S^{-1}TP$ is a projection from B onto Y . For $z \in B$, $TPz \in Y$, so $TP(S^{-1}TPz) = TPz$, so $S^{-1}TPS^{-1}TPz = S^{-1}TPz$. This proves our claim.

LEMMA 4. - *Let* E *be a Banach space, with Schauder basis* $(e_n)_{n \in \mathbb{N}}$ *; let* $\epsilon > 0$ *. Every closed infinite-dimensional subspace* F *of* E *contains a closed subspace* G *, which has a normalized Schauder basis* $(y_n)_{n \in \mathbb{N}}$ *equivalent to a normalized basis* $(u_n)_{n \in \mathbb{N}}$ *made of blocks on* $(e_n)_{n \in \mathbb{N}}$ *, and such that*

$$\sum_n \|y_n - u_n\| < \epsilon .$$

REMARK. - Consequently, if $(e_n)_{n \in \mathbb{N}}$ is unconditional, G has an unconditional basis, because every sequence made of blocks on an unconditional basis is itself unconditional, and every sequence equivalent to an unconditional basic sequence is also unconditional.

PROOF OF LEMMA 4. - Let us first show that, for every $p \in \mathbb{N}$, we can find in F an element y, which is not 0, and which is of the form :

$$(1) \quad y = \sum_{n>p} a_n e_n .$$

Assume this is false. Then, for some $p \in \mathbb{N}$, the projection P_p defined by $P_p(\sum_n a_n e_n) = \sum_{n \leqslant p} a_n e_n$, restricted to F, would be injective (since for every $y \in F$, $y \neq 0$ implies $P_p y \neq 0$). Its image is a finite-dimensional linear space (contained in $\text{span}\{(e_j)_{j \leqslant p}\}$), and therefore is closed. But we know that P_p is continuous. Therefore it would be an isomorphism on its image, and F would be finite-dimensional, which contradicts our assumption.

Now, choose $y_o \in F$, with $\|y_o\| = 1$. Then y_o can be written :

$$y_o = \sum_n a_n^{(o)} e_n .$$

Since the partial sums $\sum_{n \leqslant k} a_n^{(0)} e_n$ converge to y_o, we can find a p_o such that $\|\sum_{n>p_o} a_n^{(0)} e_n\| < \frac{\epsilon}{2}$. By (1), we can find in F a point y_1, with $\|y_1\| = 1$ and $y_1 = \sum_{n>p_o} a_n^{(1)} e_n$. We then choose p_1 large enough, in order to obtain $\|\sum_{n>p_1} a_n^{(1)} e_n\| < \frac{\epsilon}{2^2}$. By induction, we build this way a sequence y_j in F, with $\|y_j\| = 1$ for all j, and a sequence of integers $(p_j)_{j \in \mathbb{N}}$, such that

$$y_j = \sum_{n>p_{j-1}} a_n^{(j)} e_n \quad , \quad \|\sum_{n>p_j} a_n^{(j)} e_n\| < \frac{\epsilon}{2^{j+1}} .$$

Put now $v_j = \sum_{n=p_{j-1}+1}^{p_j} a_n^{(j)} e_n$, and $u_j = \frac{v_j}{\|v_j\|}$. Then $\|y_j - v_j\| < \frac{\epsilon}{2^{j+1}}$, so $|1 - \|v_j\|| < \frac{\epsilon}{2^{j+1}}$, and $\|y_j - u_j\| \leqslant \|y_j - v_j\| + |1 - \|v_j\|| < \frac{\epsilon}{2^j}$, and finally $\sum_{j \geqslant o} \|y_j - u_j\| < \epsilon$.

The u_j's are blocks on the e_n's, and consequently, if M is the basis constant of the sequence $(e_n)_{n \in \mathbb{N}}$, we obtain, if $p \leqslant q$, for any finite sequence of scalars $(a_n)_{n \in \mathbb{N}}$:

$$\left\| \sum_1^p a_j u_j \right\| \leq M \left\| \sum_1^q a_j u_j \right\| ,$$

and therefore the basis constant of the sequence (u_j) is at most M. So, if we took the precaution of choosing $\epsilon < \frac{1}{2M}$, we obtain by lemma 3,a), that $(y_j)_{j \in \mathbb{N}}$ is equivalent to $(u_j)_{j \in \mathbb{N}}$ and this proves lemma 4, if we take $G = \overline{\text{span}}\{(y_j)_{j \in \mathbb{N}}\}$.

Let's now turn to the proof of proposition 2. E is now c_o or $\ell_p (1 \leq p < +\infty)$. Let F be a closed infinite-dimensional subspace of E. For every $\epsilon > 0$, we can, by lemma 4, find an infinite dimensional subspace G in F, which has a normalized basis $(y_j)_{j \in \mathbb{N}}$, equivalent to a sequence of normalized blocks $(u_j)_{j \in \mathbb{N}}$ on the canonical basis $(e_n)_{n \in \mathbb{N}}$ in E, and with (if $\epsilon < \frac{1}{2}$) $\sum_{j \in \mathbb{N}} \|u_j - y_j\| < \epsilon$. By proposition 1, $\overline{\text{span}}\{(u_j)_{j \in \mathbb{N}}\}$ is isometric to E and 1-complemented in E. The basis constant of $(u_j)_{j \in \mathbb{N}}$ is also 1, so, if moreover $\epsilon < \frac{1}{8}$, the subspace $G = \overline{\text{span}}\{(y_j)_{j \in \mathbb{N}}\}$ is complemented in E (lemma 3, b)) and isomorphic to E (since $(y_j)_{j \in \mathbb{N}}$ and $(u_j)_{j \in \mathbb{N}}$ are equivalent). This ends the proof of proposition 2.

It should be observed (though we shall not prove it) that there exist subspaces of c_o or ℓ_p $(1 \leq p < +\infty)$ which are not isomorphic to the whole space. In fact, one can even find in c_o or ℓ_p subspaces which do not have Schauder bases (see LINDENSTRAUSS-TRAFRIRI [34], vol. 1). But this cannot happen for complemented subspaces of c_o or ℓ_p, as we shall see, using the previous proposition and "Pełczyński's decomposition method" :

PROPOSITION 5. - *If* E *is* c_o *or* ℓ_p $(1 \leq p < +\infty)$, *every infinite-dimensional complemented subspace of* E *is isomorphic to* E.

PROOF. - We have defined, in chapter III, the ℓ_p or c_o-products of Banach spaces, which we denoted $(\prod E_n)_p$ or $(\prod E_n)_{c_o}$. We shall take in this proof a different notation (the existence of which we already mentioned), and we shall call them

$$(E_1 \oplus E_2 \oplus \ldots \oplus E_n \oplus \ldots)_p \quad \text{or} \quad (E_1 \oplus \ldots \oplus E_n \oplus \ldots)_{c_o} .$$

A special case of these sums is the following : if we take only a finite number of spaces, $E_1,\dots,E_N$, then $(E_1 \oplus \dots \oplus E_N)_p$ can be identified to $\prod_{i=1\dots N} E_i$, equipped with the product topology. For example, if Y is a complemented subspace of E and Z a topological complement of Y (if P is the projection from E onto Y , then $Z = \operatorname{Ker} P$), then E is isomorphic to $Y \oplus Z$.

Let us come back to the proof of proposition 5. We shall denote $\approx$ the isomorphism between two spaces.

Let Y be an infinite-dimensional complemented subspace in E , Z its topological complement. Then

$$E \approx Y \oplus Z .$$

By proposition 2, Y can be written

$$Y = X_1 \oplus U , \quad \text{with} \quad X_1 \approx E .$$

So we obtain :

$$E \oplus Y \approx E \oplus (X_1 \oplus U) \approx (E \oplus X_1) \oplus U \approx (E \oplus E) \oplus U \approx E \oplus U \approx Y$$

(we have used the fact that $E \oplus E = E$, true for ℓ_p or c_o). Using the remarks we already made in chapter III, we can write :

$$\begin{aligned} E \oplus Y &\approx (E \oplus E \oplus \dots \oplus E \oplus \dots)_E \oplus Y \\ &\approx ((Y \oplus Z) \oplus \dots \oplus (Y \oplus Z) \oplus \dots)_E \oplus Y \\ &\approx (Z \oplus \dots \oplus Z \oplus \dots)_E \oplus (Y \oplus \dots \oplus Y \oplus \dots)_E \oplus Y \\ &\approx (Z \oplus \dots \oplus Z \oplus \dots)_E \oplus (Y \oplus \dots \oplus Y \oplus \dots)_E \\ &\approx ((Y \oplus Z) \oplus \dots \oplus (Y \oplus Z) \oplus \dots)_E \approx E , \end{aligned}$$

and finally $E \approx Y$, which proves the proposition.

II - SOME PARTICULAR PROPERTIES OF ℓ_1 AND ℓ_∞ .

The following theorem is due to Banach and Mazur :

THEOREM 1. - *Every separable Banach space is isometric to a quotient of* ℓ_1 .

PROOF. - Let E be a separable Banach space, and $(x_n)_{n\in\mathbb{N}}$ a dense sequence in the unit ball of E. We define a linear operator T, from ℓ_1 into E by :

$$(a_n)_{n\in\mathbb{N}} \in \ell_1 \longrightarrow \sum_{n\in\mathbb{N}} a_n x_n \in E .$$

This operator is obviously continuous ; we shall see that it is surjective. Let $z \in E$ with $\|z\| = 1$, and $\epsilon > 0$ be given. Choose n_1 such that

$$\|z - x_{n_1}\| < \epsilon ,$$

and then, if $n_1 < \ldots < n_j$ have been chosen,

$\|x - x_{n_1} - \ldots - \frac{\epsilon}{2^{j-1}} x_{n_j}\| < \frac{\epsilon}{2^j}$, so there is a $n_{j+1} > n_j$ such that

$$(1) \quad \|x - x_{n_1} - \ldots - \frac{\epsilon}{2^{j-1}} x_{n_j} - \frac{\epsilon}{2^j} x_{n_{j+1}}\| < \frac{\epsilon}{2^{j+1}}$$

and, by induction, we obtain (1) for all $j \geqslant 1$.

If, as usually, $(e_n)_{n\in\mathbb{N}}$ is the canonical basis of ℓ_1, we put

$$y = e_{n_1} + \frac{\epsilon}{2} e_{n_2} + \ldots + \frac{\epsilon}{2^{j-1}} e_{n_j} + \ldots$$

Then :

$$Ty = x_{n_1} + \frac{\epsilon}{2} x_{n_2} + \ldots + \frac{\epsilon}{2^{j-1}} x_{n_j} + \ldots ,$$

therefore $Ty = z$, and T is surjective. Moreover $\|y\|_1 = 1 + \epsilon$, which proves that

$$\|z\|_E = \inf\{\|y\|_1 \ ; \ Ty = z\} ,$$

and E is isometric to the quotient $\ell_1/_{\text{Ker } T}$, and the theorem is proved.

Theorem 1 has a dual statement, also due to Banach-Mazur.

THEOREM 2. - *Every separable Banach space is isometric to a subspace of* ℓ_∞.

PROOF. - Let us first recall that every Banach space is isometric to a subspace of a space $\mathscr{C}(K)$, space of continuous functions on a compact K : we know that if E is a Banach space, $\mathscr{B}_{E^\star}$ is $\sigma(E^\star, E)$ compact, and, since

$$\|x\| = \sup_{\|f\|_{E^\star} \leqslant 1} |f(x)| ,$$

E is isometric to a subspace of $\mathscr{C}(\mathscr{B}_{E^\star})$.

Now, we have seen that if E is separable, $\mathscr{B}_{E^\star}$ is separable for $\sigma(E^\star , E)$ (first part, chap. III, § 2, prop. 4), and so we need only to prove theorem 2 in the case $E = \mathscr{C}(K)$, K compact and separable.

Let $(x_n)_{n\in\mathbb{N}}$ be a dense sequence in K ; to each $f \in \mathscr{C}(K)$, we associate the sequence $(f(x_n))_{n\in\mathbb{N}}$: obviously, this is an element of ℓ_∞ . The application thus defined is an isometry, since

$\sup_{x\in K}|f(x)| = \sup_{n\in\mathbb{N}} |f(x_n)|$. This proves the theorem.

The applications built in theorems 1 and 2 are not unique (they depend on the choice of the dense sequence), but one can prove (cf. LINDENSTRAUSS-TZAFRIRI [34], vol. 1) that if T_1 and T_2 are two surjective mappings from ℓ_1 onto E , there exists an automorphism τ of ℓ_1 (that is a linear and bicontinuous bijection) such that $T_1 = T_2 \circ \tau$; if T_1' and T_2' are two injections from E into ℓ_∞ , there exists an automorphism τ' of ℓ_∞ such that $T_1' = \tau' \circ T_2'$.

To end this paragraph, let us prove a result about weak convergence of sequences in ℓ_1 . This result brings some light on the fact, proved in first part, chapter II, § 5, prop. 4, that E is dense in $E^{\star\star}$ for $\sigma(E^{\star\star} , E^\star)$.

PROPOSITION 3. - *In* ℓ_1 , *every sequence which converges for* $\sigma(\ell_1 , \ell_\infty)$ *is norm-convergent.*

PROOF. - Let $(x_n)_{n\in\mathbb{N}}$ be a sequence of points in ℓ_1 . Obviously, we need to prove the result only in the case when $x_n \xrightarrow[n\to+\infty]{} 0$ for $\sigma(\ell_1 , \ell_\infty)$. We put $x_n = (x_n(m))_{m\in\mathbb{N}}$.

To say that $x_n \xrightarrow[n\to+\infty]{} 0$ for $\sigma(\ell_1 , \ell_\infty)$ means that for all $c = (c(m))_{m\in\mathbb{N}} \in \ell_\infty$, we have :

$$(1) \qquad \langle x_n , c \rangle_{\ell_1 , \ell_\infty} = \sum_{m\in\mathbb{N}} x_n(m)c(m) \xrightarrow[n\to+\infty]{} 0 .$$

Choosing first $c = e_k$, for $k \in \mathbb{N}$, we obtain from (1) :

$$x_n(k) \xrightarrow[n \to +\infty]{} 0 \text{ , for all } k \in \mathbb{N} .$$

Assume now that $(x_n)_{n \in \mathbb{N}}$ is not norm-convergent to 0. Then, for some $\epsilon > 0$, $\limsup_{n \to +\infty} \|x_n\|_1 > \epsilon > 0$.

Let us now choose n_1 with $\|x_{n_1}\|_1 > \epsilon$, that is $\sum_{k \in \mathbb{N}} |x_{n_1}(k)| > \epsilon$. Now, choose r_1 such that

$$\sum_{k > r_1} |x_{n_1}(k)| < \frac{\epsilon}{2} \quad \text{(and so} \quad \sum_{k \leq r_1} |x_{n_1}(k)| > \frac{\epsilon}{2} \text{)} .$$

Assume $n_1,\ldots,n_{p-1}$, $r_1,\ldots,r_{p-1}$ have been chosen. We choose $n_p > n_{p-1}$ such that :

$$\sum_{k \in \mathbb{N}} |x_{n_p}(k)| > \epsilon$$

and

$$\sum_{k=1}^{r_{p-1}} |x_{n_p}(k)| < \frac{\epsilon}{2^p} .$$

We choose also $r_p > r_{p-1}$ such that

$$\sum_{k > r_p} |x_{n_p}(k)| < \frac{\epsilon}{2^p}$$

and so finally :

$$\sum_{r_{p-1} < k \leq r_p} |x_{n_p}(k)| > \epsilon - \frac{\epsilon}{2^{p-1}} > \frac{\epsilon}{2} .$$

We have therefore obtained the following lemma :

LEMMA 4. - *Let* E *be a Banach space with normalized Schauder basis* $(e_n)_{n \in \mathbb{N}}$. *Let* $(x_n)_{n \in \mathbb{N}}$ *a sequence of points in* E, *which does not tend to zero, and such that, for each* $k \in \mathbb{N}$, *the coordinates* $x_n(k) \xrightarrow[n \to +\infty]{} 0$. *Then a subsequence of the* x_n*'s is equivalent to a sequence of consecutive blocks on the* e_n*'s.*

This type of argument (which we proved only for the case $E = \ell_1$, but the general statement is completely similar) is called "gliding hump", and is very commonly used.

Let us come back now to the proof of our proposition. We choose

$c(m) = \operatorname{sign}(x_{n_{p+1}}(m))$ if $r_p < m \leqslant r_{p+1}$, (or 0 if $x_{n_{p+1}}(m) = 0$) if the space is real, or, if it is complex :

$$c(m) = \frac{\overline{x_{n_{p+1}}(m)}}{|x_{n_{p+1}}(m)|} \quad , \qquad r_p < m \leqslant r_{p+1}$$

(0 if $x_{n_{p+1}}(m) = 0$) . So $|c(m)| \leqslant 1$ for all m, and, by (1), $\sum_{m \in \mathbb{N}} c(m) x_{n_p}(m) \xrightarrow[p \to +\infty]{} 0$. But we have :

$$\left|\sum_{m \in \mathbb{N}} c(m) x_{n_p}(m)\right| \geqslant \left|\sum_{m=n_{p-1}+1}^{n_p} c(m) x_{n_p}(m)\right| - \sum_{m \leqslant n_{p-1}} \left|x_{n_p}(m)\right| - \sum_{m > n_p} \left|x_{n_p}(m)\right|$$

$$\geqslant \frac{\epsilon}{2} - \frac{2\epsilon}{2^p} \geqslant \frac{\epsilon}{4} \qquad \text{if} \quad p \geqslant 3 .$$

This contradiction proves ours statement.

This proposition means that, in ℓ_1, the weak convergence and the strong convergence are the same for *sequences* (this property is called the Schur property). But this does not mean that ℓ_1 is reflexive (we know it is not, since $c_0 \neq \ell_\infty$), but that points of the bidual $(\ell_1)^{\star\star}$ are limits (for $\sigma(\ell_1^{\star\star}, \ell_1^{\star})$) of uncountable families of points of ℓ_1.

Actually, the case of ℓ_1 is characteristic of this type of situation. It has been proved by E. Odell and H.P. Rosenthal (see LINDENSTRAUSS-TZAFRIRI [34], vol. 1) that if E is separable, E does not contain any subspace isomorphic to ℓ_1 if and only if every element of $E^{\star\star}$ is limit, for $\sigma(E^{\star\star}, E^{\star})$, of a *sequence* of points of E : see the complements at the end of this chapter.

Using theorem 1 and proposition 3, we can find other examples of Banach spaces having uncomplemented subspaces (see chapter III). According to theorem 1, ℓ_2 is isometric to a quotient of ℓ_1 : there exists a closed subspace X of ℓ_1 such that ℓ_2 is isometric to the quotient ℓ_1 / X.

If X was complemented in ℓ_1, it would have a topological complement Y, and then this Y would be isomorphic to ℓ_2. But this is not possible : the canonical basis of ℓ_2 is weakly convergent to zero, but not strongly. This type of argument applies as well to many other quotients of ℓ_1 : namely, to every separable space which does not have Schur property.

EXERCISES ON CHAPTER IV.

EXERCISE 1. - Let φ be a bijection from $\mathbb{N}$ onto itself, $(\epsilon_n)_{n \in \mathbb{N}}$ a sequence of ± 1. Let U be the application from ℓ_p onto itself defined by

$$(a_n)_{n \in \mathbb{N}} \in \ell_p \longrightarrow (\epsilon_n \, a_{\varphi(n)})_{n \in \mathbb{N}} \, .$$

Show that U is an isometry from ℓ_p onto itself.

EXERCISE 2. - Let p, $1 \leqslant p < +\infty$, $a = (a_n)_{n \in \mathbb{N}}$, $b = (b(n))_{n \in \mathbb{N}} \in \ell_p$. Show that if for all scalars α, β :

$$(1) \qquad \|\alpha a + \beta b\|_p^p = |\alpha|^p \|a\|_p^p + |\beta|^p \|b\|_p^p \, ,$$

then $a_n \times b_n = 0$ for all n.

[Hint : first consider the case $p = 1$. Then take $p > 2$, and then differentiate (1) twice, with respect to α or β. Then consider $1 < p < 2$ by duality].

EXERCISE 3. - Let U be a surjective isometry of ℓ_p $(1 \leqslant p < +\infty)$. Show that if $a, b \in \ell_p$ are such that $a_n \cdot b_n = 0$ for all n, then $(Ua)_n \cdot (Ub)_n = 0$ for all n.

EXERCISE 4. - Using the previous two exercises, show that any surjective isometry from ℓ_p $(1 \leqslant p < +\infty)$ onto itself is of the form given at exercise 1. What about ℓ_∞ ?

Operators from ℓ_r *into* ℓ_p $(1 \leqslant p < r < +\infty)$, or *from* c_o *into* ℓ_p.

We say that an operator T, between two Banach spaces E and F, is compact if $\overline{T(\mathscr{B}_E)}$ is compact in F. The aim of the following exercises is to show that, if $1 \leqslant p < r < +\infty$, any operator from ℓ_r into ℓ_p, or from c_o into ℓ_p, is compact.

EXERCISE 5. - Let E, F be two Banach spaces, $(T_n)_{n \in \mathbb{N}}$ compact operators from E into F, converging (in operator norm) to an operator T. Show that T is compact.

EXERCISE 6. - Show that if $\operatorname{Im} T$ is finite-dimensional, T is compact.

EXERCISE 7. - Let P_n (resp. Q_n) the projection of ℓ_p (resp. ℓ_r) onto $\operatorname{span}\{(e_k)_{k \leqslant n}\}$. Show that if $T : \ell_r \longrightarrow \ell_p$ is not compact there is a $\delta > 0$ such that

$$\operatorname*{Inf}_{n,m \in \mathbb{N}} \|(I - Q_n)T(I - P_m)\| \geqslant \delta > 0 .$$

EXERCISE 8. - We use the same notations, and assume again that T is not compact. Take $\epsilon < \delta$. Using the previous exercise and the fact that the canonical basis of ℓ_r $(r > 1)$ is shrinking (see chap. II), show that there exist a strictly increasing sequence of integers $(n_i)_{i \in \mathbb{N}}$, elements $(u_i)_{i \in \mathbb{N}} \in \ell_r$ or c_o, and $(v_i)_{i \in \mathbb{N}} \in \ell_p$, with :

$$Tu_i = v_i \qquad \forall\, i , \quad \|u_i\| = 1 ,$$

$$u_i = \sum_{j=n_i+1}^{n_{i+1}} \alpha_j e_j \quad , \quad v_i = \sum_{j=1}^{\infty} \beta_j^{(i)} e_i \ , \quad \text{and}$$

$$\Big\| \sum_{j \leqslant n_i} \beta_j^{(i)} e_j \Big\|_r \leqslant \frac{\epsilon}{2^{i+1}} \ , \quad \Big\| \sum_{n_i < j \leqslant n_{i+1}} \beta_j^{(i)} e_j \Big\|_r \geqslant \delta - \frac{\epsilon}{2} \ ,$$

$$\Big\| \sum_{j > n_{i+1}} \beta_j^{(i)} e_j \Big\|_r \leqslant \frac{\epsilon}{2^{i+2}} \ .$$

EXERCISE 9. - Deduce from the previous exercise that ℓ_r is isomorphic to ℓ_p. Show that this is not possible : therefore T is compact.

REFERENCES ON CHAPTER IV.

The first paragraph comes from LINDENSTRAUSS-TZAFRIRI [34] ; theorems 1, 2 and proposition 3 of § II are in Banach's book [5].

The description of surjective isometries of ℓ_p (exercises 1, 2, 3, 4) is in Banach's book, the fact that every operator from ℓ_r into ℓ_p is compact is in LINDENSTRAUSS-TZAFRIRI [34].

COMPLEMENTS ON CHAPTER IV.

As we mentioned, proposition 3, § II means that elements of ℓ_1^{**} are limits, for $\sigma(\ell_1^{**}, \ell_\infty)$, of filters of elements of ℓ_1, and not of sequences. Therefore, one can ask the following question : what are the non-reflexive separable spaces E in which every element of E^{**} is the limit, for $\sigma(E^{**}, E^*)$ of a sequence of elements of E ? The following theorem, due to E. ODELL and H.P. ROSENTHAL, shows that ℓ_1 is characteristic of the converse situation :

THEOREM. - *Let* E *be a separable Banach space. The following are equivalent :*

1) E *does not contain any subspace isomorphic to* ℓ_1.

2) *Every element in* E^{**} *is the limit, for* $\sigma(E^{**}, E^*)$, *of a sequence of elements in* E.

3) *The cardinality of* E^{**} *is equal to the cardinality of* E *(i.e. the cardinality of the continuum).*

4) *Every bounded sequence in* E^{**} *has a subsequence which is* $\sigma(E^{**}, E^*)$ *convergent.*

This theorem is closely related to an important result due to H.P. ROSENTHAL :

THEOREM - *Let* $(x_n)_{n\in\mathbb{N}}$ *be a bounded sequence in a Banach space* E. *Then* $(x_n)_{n\in\mathbb{N}}$ *has a subsequence* $(x_{n_i})_{i\in\mathbb{N}}$ *satisfying one of the mutually exclusive terms of the following alternative :*

1) $(x_{n_i})_{i\in\mathbb{N}}$ *is equivalent to the canonical basis of* ℓ_1.

2) $(x_{n_i})_{i \in \mathbb{N}}$ *is a weak Cauchy sequence, that is, for every* $\xi \in E^\star$, $\lim_{i \to +\infty} \xi(x_{n_i})$ *exists.*

We refer the reader to [34] for the proofs of these theorems, as well as for further developments.

CHAPTER V

EXTREME POINTS OF COMPACT CONVEX SETS AND THE BANACH SPACES $\mathscr{C}(K)$

In this chapter, we shall prove some of the known results concerning the classification of the Banach spaces of continuous functions on a compact set K . An essential tool for this study is the notion of "extreme point" of a compact convex set, which, besides, has many other applications. In the first paragraph, we establish the basic facts concerning the extreme points, and, in the second paragraph, we develop these results in the direction which interests us here. For a more detailed study of extreme points, and other possibilities of application, we refer the reader to G. CHOQUET's book [12], vol. II.

§ 1. EXTREME POINTS OF COMPACT CONVEX SETS.

Let K be a compact convex set in a topological vector space E . We say that a point $a \in K$ is an *extreme point* of K if the equality $a = \frac{x+y}{2}$, with $x,y \in K$, implies $x = y = a$. This is equivalent (as one checks immediately) to the fact that $K \setminus \{a\}$ is convex.

We call $\mathscr{E}(K)$ the set of extreme points in K . In this chapter E will be a Hausdorff Locally Convex Topological Vector space (in short H.L.C.T.V.S.), that is a topological vector space such that 0 has a fundamental system of convex neighbourhoods (we already met this notion in the first part). The reader who is not familiar with this notion may restrict himself to a normed space E endowed with the norm topology, or with a weak topology : these are the only examples of HLCTVS which we shall meet in the sequel. But to adopt a more general frame does not, for once, make things more complicated.

THEOREM 1. (BAUER's Maximum Principle). - *Let* K *be a compact convex subset of a HLCTVS* E *, and* f *an convex, upper-semi-continuous function from* K *into* $\mathbf{R}$ *. Then there is at least one point in* $\mathscr{E}(K)$ *where* f *reaches its maximum.*

PROOF. - Recall that a function f is upper-semi-continuous if, for every $\lambda \in \mathbb{R}$, $\{x, f(x) \geqslant \lambda\}$ is closed. Since K is compact, the existence of points at which f reaches its maximum follows. The problem is to find one in $\mathscr{E}(K)$.

Let us consider the familly $\mathscr{F}$ of the closed subsets F of K, which have the following property : if an open segment $]a,b[$, contained in K, meets F, it is completely contained in F. This family $\mathscr{F}$ contains at least K itself. It should be observed that these subsets F need not be convex. For example, if K is, in the plane, the closed unit disk, any closed subset of the unit cercle belongs to $\mathscr{F}$, since no open interval contained in K can meet such an F.

The family $\mathscr{F}$ is stable under intersection : if $(F_i)_{i \in I} \in \mathscr{F}$, then $\bigcap_{i \in I} F_i \in \mathscr{F}$. Let $\mathscr{F}^\star$ be the set of non-empty elements of $\mathscr{F}$. On $\mathscr{F}^\star$, we put the order given by inclusion : $F_1 > F_2$ if $F_1 \supset F_2$. With this order, this set is inductive : if $(F_i)_{i \in I}$ is totally ordered, then $\bigcap_{i \in I} F_i$ is a minorant (non-empty since all are compact). Therefore, by Zorn's axiom, $\mathscr{F}^\star$ contains minimal elements : we shall show that such an element is reduced to one point, and that this point is an extreme point.

So, let X be a minimal element of $\mathscr{F}^\star$. For any upper-semi-continuous function f, from K to $\mathbb{R}$, put $X_f = \{x \in X ; f(x) = \sup(f_{|X})\}$. This is a non-empty set, which belongs to $\mathscr{F}$: if an open interval $]a,b[$ intersects X_f, it intersects X, and so is contained in X. But f is convex and upper-semi-continuous : if f reaches its maximum at a point of $]a,b[$, it must be constant on $]a,b[$, and so $]a,b[\subset X_f$. Since X is minimal and $X \supset X_f$, then $X = X_f$, and X must be reduced to a single point : if it contained two distinct points, one could find a continuous linear functional which separates them (first part, chap. II, § 4, cor. 2), and so it would be a convex upper-semi-continuous function which would not be constant on X.

So we have seen that if f is convex and upper-semi-continuous on K, the set of points where f reaches its maximum contains a minimal element F_0 of $\mathscr{F}^\star$, and that F_0 is a single point $\{x_0\}$. It is now easy to see that x_0 is an extreme point : we cannot have $x_0 = \frac{a+b}{2}$, $a \neq x_0$, $b \neq x_0$, since the interval $\left]\frac{a+x_0}{2}, \frac{b+x_0}{2}\right[$ would meet F_0, but would not be contained in F_0. This finishes the proof of Bauer's Theorem.

THEOREM 2. (KREIN-MILMAN). - *Every convex compact subset of a HLCTVS is the closed convex hull of its extreme points.*

PROOF. - Let K be a compact convex set. Put $K' = \overline{\text{conv}}\ \mathscr{E}(K)$. Obviously $K' \subset K$. If there was a point x in K but not in K' , we could find a $\alpha > 0$ and a real linear functional f with $f(x) > \alpha$, and $f(y) < \alpha$ for every $y \in K'$. This linear functional would not reach its maximum on $\mathscr{E}(K)$, which contradicts the previous theorem.

PROPOSITION 3. - *Let* K *be a compact convex subset of a real HLCTVS* E . *Every extreme point of* K *has a base of neighbourhoods (in* K *) made of open slices, that is, sets of the form :*

$$K \cap \{y \ ; \ f(y) > \alpha\} ,$$

where f *is a real linear functional, and* $\alpha \in \mathbb{R}$.

PROOF. - Since K is compact, the topologies of E and $\sigma(E, E^\star)$ coïncide on K . Let V be an open neighbourhood, for $\sigma(E, E^\star)_{|K}$, of a point $x \in K$: by definition (see first part, chapter II), one can find $\epsilon > 0$ and real linear functionals $f_1,\dots,f_n$ such that $V \supset \{y \in K \ ; \ |f_i(x) - f_i(y)| < \epsilon \ , \ i = 1,\dots,n\}$. Therefore the sets $\{y \in K \ ; \ f_i(y) \geqslant f_i(x) + \epsilon\}$ or $\{y \in K \ ; \ f_i(y) \leqslant f_i(x) - \epsilon\}$ $(i = 1,\dots,n)$ are compact, convex, and do not contain x .

If x is an extreme point, their convex hull does not contain x either : otherwise, we could write $x = \sum_i \alpha_i x_i$, finite decomposition with $x_i \in K$, $\alpha_i \geqslant 0$, $\Sigma\, \alpha_i = 1$, and this contradicts the extremality of x . But this convex hull is also compact (if K_i $(i = 1,\dots,2n)$ are the previous sets, $\text{conv}\{K_i\}$ is the image of $A \times \prod_{i=1}^{2n} K_i$, in K , by the application $((a_1,\dots,a_{2n}),(x_1,\dots,x_{2n})) \to \sum_1^{2n} a_i x_i$, where $A \subset \mathbb{R}^{2n}$ is $\{(a_1,\dots,a_{2n}),\ a_i \geqslant 0,\ \sum_1^{2n} a_i = 1\}$) . So we can find an hyperplane strictly separating x and $\text{conv}\{K_i\}$ (first part, chapter II, § 4, cor. 5) : there is a linear functional f and a real α such that $\{y \in K,\ f(y) > \alpha\}$ contains x and is contained in V ; this proves the proposition.

We shall say that a hyperplane $H = \{z ; f(z) = \alpha\}$ (f real) is a *supporting* hyperplane for a convex K if K is contained in one of the half-spaces $\{z ; f(z) \leqslant \alpha\}$ or $\{z ; f(z) \geqslant \alpha\}$, and if $K \cap H$ is non-empty.

PROPOSITION 4. - *Let K be a compact convex subset in a real HLCTVS E, and U a compact subset of K. The following conditions are equivalent :*

a) *K is the closed convex hull of U,*

b) *U meets the intersection of K and any of its supporting hyperplanes,*

c) *U contains $\mathscr{E}(K)$.*

PROOF.

a) ⇒ b). Assume that there is a supporting hyperplane $H = \{z ; f(z) = \alpha\}$ such that $K \cap H \cap U = \phi$. Assume for example that $f(x) \geqslant \alpha$, for all $x \in K$. Then $f(x) > \alpha$ for all $x \in U$, and, since U is compact, $\beta = \inf_{x \in U} f(x) > \alpha$. So U is contained in the half-space $\{z ; f(z) \geqslant \beta\}$, and $\overline{\text{conv}}\, U$ also, and K cannot be equal to $\overline{\text{conv}}\, U$.

b) ⇒ c). Assume that some extreme point x of K does not belong to U. Then there is a neighbourhood V of x such $V \cap U$ is empty. By proposition 3, V contains an open slice : there is a linear functional f and a real α such that $H = \{z ; f(z) = \alpha\}$ strictly separates U and x. Assume for example that $f(x) > \alpha$. Then, for some $\gamma > \alpha$, $\{z ; f(z) = \alpha\}$ is a supporting hyperplane, which cannot meet U.

c) ⇒ a). Is a consequence of Krein-Milman's Theorem.

§ 2. THE BANACH SPACES $\mathscr{C}(K)$.

If K is a compact set, we call $\mathscr{C}(K)$ the Banach space of real-valued continuous functions on K, with the norm of uniform convergence :

$$\|f\|_{\mathscr{C}(K)} = \sup_{x \in K} |f(x)| .$$

The dual of $\mathscr{C}(K)$, $\mathscr{M}(K)$, is the space of Radon measures on K.

Please note that in this paragraph , K needs not be convex.

The spaces $\mathscr{C}(K)$, and specially $\mathscr{C}([0,1])$, play an important role in the classification of Banach spaces, as already shows the following proposition, due to Banach and Mazur :

PROPOSITION 1. - *Every separable Banach space is isometric to a subspace of* $\mathscr{C}([0,1])$.

PROOF. - If E is separable, the unit ball $\mathscr{B}_{E^\star}$ of its dual is compact and metrizable for $\sigma(E^\star, E)$ (first part, chap. III, § II, proposition 3). We need a lemma :

LEMMA 2. - *Every metrizable compact set is a continuous image of the Cantor set* $\mathbf{C}$.

PROOF OF LEMMA 2. - We first recall briefly the construction of the Cantor set in $[0,1]$. At the first stage, withdraw $]\frac{1}{3}, \frac{2}{3}[$ from $[0,1]$. At the second, withdraw $]\frac{1}{9}, \frac{2}{9}[$ and $]\frac{7}{9}, \frac{8}{9}[$: so 2^2 closed intervals remain. At the $n^{\underline{th}}$ stage one has 2^n intervals of same length ; each is divided into three parts, and the middle one is removed at the $(n+1)^{\underline{th}}$ stage, so 2^{n+1} closed intervals remain.

Now, let $(n_k)_{k \geqslant 1}$ be any sequence of strictly positive natural numbers (not necessarily increasing). Then any sequence $(m_k)_{k \geqslant 1}$ with $m_k \leqslant 2^{n_k}$ for every $k \geqslant 1$ determinates a point of the Cantor set : at the $n_1{}^{\underline{th}}$ stage, there are 2^{n_1} intervals, we choose the $m_1{}^{\underline{th}}$. This interval will in its turn be devided into 2^{n_2} intervals (at the stage $2^{n_1+n_2}$), we shall choose the $m_2{}^{\underline{th}}$, and so on. Since $n_k \geqslant 1$ for all k , the diameters of these intervals tend to zero, and there is only one point in their intersection.

Let K be a metrizable compact set, and d be the distance which defines the topology. First we cover K by a finite number (say 2^{n_1}) of closed balls of radius $\frac{1}{2}$. Then the intersection with K of each of these closed balls will be covered by a finite number (say 2^{n_2}) of balls of radius $\frac{1}{4}$, and so on : the sequence $(n_k)_{n\geq 1}$ is constructed inductively.

We shall now construct a surjective continuous mapping from $\mathbf{C}$ onto K . Take any point $x \in \mathbf{C}$: it determinates biunivocally a sequence $(m_k)_{k\geq 1}$, with $m_k \leq 2^{n_k}$ for all $k \geq 1$ (m_1 is the number of the interval of length $(\frac{1}{3})^{n_1}$ to which x belongs, m_2 is the number, among the 2^{n_2} intervals obtained in the previous one at the $2^{n_1+n_2}$ th stage, of the interval of length $(\frac{1}{3})^{n_1+n_2}$ which contains x , and so on). Then this sequence $(m_k)_{k\geq 1}$ in its turn determinates a point of K : among the 2^{n_1} balls of radius $\frac{1}{2}$, take the ball of number m_1 ; among the 2^{n_2} balls of radius $\frac{1}{4}$ covering it, take the ball of number m_2 , and so on : these balls form a sequence of decreasing compacts, so the intersection is non-empty, and consists in a single point, since the radii tend to zero. Since for each k , the balls of radius $\frac{1}{2^k}$ cover K , each point of K belongs to such a chain, and the mapping is surjective. It is also continuous : take any $y_0 \in K$, and any open neighbourhood of it : it contains some closed ball of radius $\frac{1}{2^k}$, for $k \geq 1$. The converse image of this ball is, by definition of the mapping, an interval of $[0,1]$, which is a neighbourhood, in $\mathbf{C}$, of the point x_0 , converse image of y_0 . This ends the proof of our lemma.

If $t \in \mathbf{C}$, let f_t the element of $\mathscr{B}_{E^\star}$ which corresponds to t , by the application given by the lemma. Let $x \in E$. We define a continuous function on $[0,1]$ by :

$$y(t) = f_t(x) \qquad \text{if} \quad t \in \mathbf{C} ,$$

extended linearly on $[0,1] \setminus \mathbf{C}$.

This function $y(t)$ is continuous : if $t_n \in \mathbf{C}$ and $t_n \to t_0$, then $(f_{t_n})_n \xrightarrow[n\to+\infty]{} f_{t_0}$ for $\sigma(E^\star, E)$, and so $y(t_n) \to y(t_0)$. On $[0,1] \setminus \mathbf{C}$, $y(t)$ is affine, therefore continuous.

Now, let us compute $\|y(t)\|_{\mathscr{C}([0,1])} = \sup_{0\leq t\leq 1} |y(t)|$. Choose $f \in \mathscr{B}_{E^\star}$, with $f(x) = \|x\|$. By construction, there is a $t_0 \in \mathbf{C}$ such that $f = f_{t_0}$.

So we have $|y(t_0)| = |f_{t_0}(x)| = \|x\|$, and since

$$|y(t)| = |f_t(x)| \leqslant \|f_t\| \cdot \|x\| \leqslant \|x\| , \quad \text{for all} \quad t \in \mathbf{C} ,$$

we obtain

$$\|y(t)\|_{\mathscr{C}([0,1])} = \|x\| .$$

Let U be the application $x \longrightarrow y(t)$: this is a linear isometry from E into $\mathscr{C}([0,1])$, and our proposition is proved.

This result shows that $\mathscr{C}([0,1])$ is a "universal space" for separable Banach spaces : it is separable (since polynomials with rational coefficients are dense in it), and contains all separable spaces. So there is no need to wonder what are the subspaces of $\mathscr{C}([0,1])$, but one can wonder what are the complemented subspaces. The answer to this question is not perfectly known in general (see also exercise 7, below), but a description can be given in some special cases. Among them, there is the case of 1-complemented subspaces of $\mathscr{C}(K)$. We shall mention this result without giving its proof (the reader is referred to LINDENSTRAUSS-TZAFRIRI [34]).

Let σ be an involutive homeomorphism (i.e. $\sigma^2 = Id$) of K into itself. Call $\mathscr{C}_\sigma(K)$ the set of functions of $\mathscr{C}(K)$ which satisfy $f(\sigma x) = -f(x)$ for all $x \in K$ (for example if K is the unit cercle and σ the symmetry of center 0 , $\mathscr{C}_\sigma(K)$ is the set of functions such that $f(-x) = -f(x)$).

PROPOSITION. (J. LINDENSTRAUSS - D. WULBERT). - *A Banach space* E *is isometric to a 1-complemented subspace of a space* $\mathscr{C}(K)$ *if and only if it is isometric to a space* $\mathscr{C}_\sigma(H)$, *for some compact* H *and some involutive homeomorphisme* σ *of* H .

Concerning the complemented subspaces of $\mathscr{C}(K)$, the following two results are known, but do not provide a complete description :

- If $X \oplus Y$ is isomorphic to a $\mathscr{C}(K)$, either X or Y is isomorphic to $\mathscr{C}(K)$.

- Every complemented subspace of $\mathscr{C}([0,1])$ with non-separable dual is isomorphic to $\mathscr{C}([0,1])$ (H.P. ROSENTHAL [42]).

It is interesting to compare two $\mathcal{C}(K)$-spaces . The first result in this direction is due to Banach and Stone :

THEOREM 3. (BANACH - STONE). - *Let* K *and* H *be two compact sets. The spaces* $\mathcal{C}(K)$ *and* $\mathcal{C}(H)$ *are isometric if and only if* K *and* H *are homeomorphic.*

PROOF.

a) If φ is an homeomorphism between K and H , the application

$$f \in \mathcal{C}(K) \longrightarrow f \circ \varphi \in \mathcal{C}(H)$$

is a surjective isometry between $\mathcal{C}(K)$ and $\mathcal{C}(H)$.

b) To show the converse implication, the basic tool will be the extreme points of the unit balls of $\mathcal{C}(H)$ and $\mathcal{C}(K)$. We shall first describe them :

LEMMA 4. - *The extreme points of the unit ball of* $\mathcal{C}(K)$ *are the continuous functions* f *such that* $|f(x)| = 1$ *for all* $x \in K$.

PROOF OF LEMMA 4. - Obviously such an f is an extreme point of $\mathcal{B}_{\mathcal{C}(K)}$. Conversely, if for some x_0 , $|f(x_0)| < 1$, one may find a decomposition $f = \frac{f_1 + f_2}{2}$, with $|f_1(x)| \leq 1$, $|f_2(x)| \leq 1$ for all x , but $f_1(x_0) \neq f_2(x_0)$, and f is not extreme.

REMARK. - Since the functions are real-valued, the functions f with $|f| = 1$ are constant, with values $+1$ or -1 , on each connected component of K . If K is connected, then there are only two extreme points of $\mathcal{B}_{\mathcal{C}(K)}$: the function constantly equal to $+1$, and the function equal to -1 . So $\mathcal{B}_{\mathcal{C}(K)}$ is certainly not the closed convex hull of its extreme points (closed for the norm, or for $\sigma(\mathcal{C}(K), \mathcal{M}(K))$). But this does not contradict Krein-Milman's theorem, since $\mathcal{B}_{\mathcal{C}(K)}$ is compact for none of these topologies.

LEMMA 5. - *The extreme points of the unit ball* $\mathcal{B}^\star$ *of* $\mathcal{M}(K)$ *are the Dirac measures* $\pm \delta_x$, $x \in K$.

PROOF. - Put $A = \{ \pm\delta_x , x \in K\}$.

1) Let us show first that $\mathrm{Ext}\, \mathscr{B}^\star \subset A$.

Of course, $A \subset \mathscr{B}^\star$, and consequently $\overline{\mathrm{conv}}\, A \subset \mathscr{B}^\star$ (the closure is taken for $\sigma(\mathscr{M}(K), \mathscr{C}(K))$). Assume that this inclusion is strict. Then there is a $\xi \in \mathscr{B}^\star$, $\xi \notin \overline{\mathrm{conv}}\, A$. By Hahn-Banach theorem, there is a function $f \in \mathscr{C}(K)$, with $\|f\| = 1$, and a number $a > 0$ such that $\xi(f) > a$, and $\eta(f) > a$ for all $\eta \in \overline{\mathrm{conv}}\, A$. This implies $|f(x)| < a$, $\forall\, x \in K$. Since $\|\xi\| \leqslant 1$, and $\|\xi\| \cdot \|f\| > a$, then $a < 1$, and so $|f(x)| < a < 1$ for all $x \in K$. This contradicts $\|f\|_{\mathscr{C}(K)} = \sup_{x \in K} |f(x)| = 1$, and so $\overline{\mathrm{conv}}\, A = \mathscr{B}^\star$. Using proposition 4 of the previous paragraph, we obtain $\mathrm{Ext}\, \mathscr{B}^\star \subset A$: this proposition can be applied, since $\mathscr{B}^\star$ is compact for $\sigma(\mathscr{M}(K), \mathscr{C}(K))$.

2) Conversely, let us show that any element of A is an extreme point. Assume for example that $\delta_x = \alpha\mu_1 + \beta\mu_2$, $\mu_1 , \mu_2 \in \mathscr{M}(K)$, $\|\mu_1\| = \|\mu_2\| = 1$, $\alpha, \beta \geqslant 0$, $\alpha + \beta = 1$ (the argument applies also to $-\delta_x$). We call 1_K the constant function, equals to 1 on K . Then :

$$\delta_x(1_K) = 1 = \alpha\mu_1(1_K) + \beta\mu_2(1_K) ,$$

and therefore, since $|\mu_1(1_K)| \leqslant 1$ and $|\mu_2(1_K)| \leqslant 1$,

$$\mu_1(1_K) = \mu_2(1_K) = 1 .$$

Take now $f \in \mathscr{C}(K)$, with $\|f\| \leqslant 1$, $f \geqslant 0$ on K and $f(x) = 0$. Then $\|1_K - f\|_{\mathscr{C}(K)} = 1$, and, again :

$$1 = \delta_x(1_K - f) = \alpha\mu_1(1_K - f) + \beta\mu_2(1_K - f)$$

and therefore

$$\mu_1(1_K - f) = \mu_2(1_K - f) = 1 ,$$

and

$$\mu_1(f) = \mu_2(f) = 0 .$$

If $f \in \mathscr{C}(K)$, with $\|f\| \leqslant 1$, $f(x) = 0$, we decompose f into : $f = f_+ - f_-$, where f_+ , f_- are positive, disjointly supported, and

$$\|f_+\| \leq 1 , \quad \|f_-\| \leq 1 , \quad f_+(x) = f_-(x) = 0 .$$

So we obtain $\mu_1(f_+) = \mu_1(f_-) = \mu_2(f_+) = \mu_2(f_-) = 0$,
and

$$\mu_1(f) = \mu_2(f) = 0 .$$

So, for all $f \in \mathscr{C}(K)$, the condition $f(x) = 0$ implies $\mu_1(f) = \mu_2(f) = 0$. Therefore, we can find two numbers α' , β' , such that $\mu_1 = \alpha'\delta_x$ and $\mu_2 = \beta'\delta_x$ (the linear functionals μ_1 and δ_x on $\mathscr{C}(K)$ have the same Kernel, so they are proportional). Since $\|\mu_1\| \leq 1$ and $\|\mu_2\| \leq 1$, we have $|\alpha'| \leq 1$ and $|\beta'| \leq 1$, and since $\delta_x = \alpha\alpha'\delta_x + \beta\beta'\delta_x$, then $\alpha\alpha' + \beta\beta' = 1$. This implies $\alpha' = \beta' = 1$, and so $\mu_1 = \mu_2 = \delta_x$, and δ_x is an extreme point. The lemma is proved.

Let us come back to the proof of Banach-Stone's theorem. Let T be an isometry from $\mathscr{C}(K)$ onto $\mathscr{C}(H)$. Since the function 1_K is an extreme point of $\mathscr{B}_{\mathscr{C}(K)}$, its image $T1_K$ must be an extreme point of $\mathscr{B}_{\mathscr{C}(H)}$. By lemma 4, this implies

$$|T(1_K)(y)| = 1 \qquad \text{for all} \quad y \in H .$$

For all $f \in \mathscr{C}(K)$, let us define

$$T_o(f) = \frac{T(f)}{T(1_K)} .$$

This is also an isometry from $\mathscr{C}(K)$ onto $\mathscr{C}(H)$, which sends 1_K on 1_H ; From this follows that T_o sends positive functions onto positive functions : if $f(x) \geq 0$ for all $x \in K$, then $\left\|1_K - \frac{f}{\|f\|}\right\| \leq 1$, so $\left\| 1_H - \frac{T_o(f)}{\|T_o(f)\|} \right\| \leq 1$, and $T_o(f)(y) \geq 0$, for all $y \in H$.

The transpose of T_o , called ${}^tT_o$, is a surjective isometry between $\mathscr{M}(H)$ and $\mathscr{M}(K)$, and the image of a positive measure is a positive measure : if $\mu \in \mathscr{M}(H)$ is a positive measure, then $\mu(f) \geq 0$ for all $f \geq 0$ in $\mathscr{C}(H)$, so $({}^tT_o\,\mu)(g) = \mu(T_o\,g) \geq 0$, for all $g \geq 0$ in $\mathscr{C}(K)$.

Since ${}^tT_o$ sends extreme points of $\mathscr{B}_{\mathscr{M}(H)}$ on extreme points of $\mathscr{B}_{\mathscr{M}(K)}$, it sends the Dirac masses δ_y at the points of H onto the Dirac masses δ_x at the points of K. Therefore, for every $y \in H$, there is an $x \in K$ such that $\delta_x = {}^tT_o\, \delta_y$. Let φ be application thus defined : φ is a bijection from H onto K, continuous, since the convergence $y_i \to y$ in H is equivalent to the convergence $\delta_{y_i} \longrightarrow \delta_y$ in $\mathscr{M}(H)$, for $\sigma(\mathscr{M}(H), \mathscr{C}(H))$, and therefore φ is bicontinuous, whichs ends the proof.

The conclusion of proposition 3 remains true if we assume only that there exists an isomorphism T between $\mathscr{C}(K)$ and $\mathscr{C}(H)$ with $\|T\| \cdot \|T^{-1}\| < 2$, but it is false general. In fact, one can prove that, for every metric uncountable compact K, $\mathscr{C}(K)$ is isomorphic to $\mathscr{C}([0,1])$ (MILUTIN [36]).

EXERCISES ON CHAPTER V.

EXERCISE 1. - Show that the unit ball of c_o does not have extreme points. What are the extreme points of the unit ball of ℓ_∞ ?

EXERCISE 2. - Let $(e_n)_{n\in\mathbb{N}}$ be the canonical basis of c_o, and A the closed convex hull of 0 and of the set $\{ \frac{e_n}{n+1} ; n \in \mathbb{N} \}$. Show that A is convex and compact, but is not equal to the convex hull of its extreme points.

EXERCISE 3. - The aim of this exercise is to show that the following two assertions are equivalent, for a Banach space E :

1) For every sequence $(x_n)_{n\in\mathbb{N}}$ in E, tending to 0 for $\sigma(E, E^\star)$, and every sequence $(f_n)_{n\in\mathbb{N}}$ in $E^\star$, tending to 0 for $\sigma(E^\star, E^{\star\star})$, $f_n(x_n) \xrightarrow[n \to +\infty]{} 0$.

2) For every Banach space Y, every weakly compact operator T from E into Y, every weakly compact set K in E, $T(K)$ is norm-compact in Y.

(See first part chapter III, exercise 7, for the definition of a weakly compact operator).

a) Let $x_n \in K$, with $x_n \xrightarrow[n \to +\infty]{} 0$ for $\sigma(E, E^\star)$. Let $g_n \in Y^\star$, with $\|g_n\| = 1$ and $g_n(Tx_n) = \|Tx_n\|$. Considering ${}^tTg_n$, and using 1), show that $Tx_n \xrightarrow[n \to +\infty]{} 0$ in norm (use the fact that ${}^tT$ is weakly compact (see first part chapter III, exercise 7). Deduce that 1) ⇒ 2).

b) Let $x_n \in E$, $x_n \xrightarrow[n \to +\infty]{} 0$ for $\sigma(E, E^\star)$, $f_n \in E^\star$, $f_n \xrightarrow[n \to +\infty]{} 0$ for $\sigma(E^\star, E^{\star\star})$. Consider the operator T, from E into c_0, defined by $Tx = (f_n(x))_{n \in \mathbb{N}}$. Show that T is weakly compact. Show that $Tx_n \xrightarrow[n \to +\infty]{} 0$, and thus that $f_n(x_n) \to 0$. So 2) ⇒ 1).

A Banach space E with 1) and 2) is said to have the Dunford-Pettis Property.

EXERCISE 4. - Show that if E has the Dunford-Pettis Property, for every Banach spaces Y_1, Y_2 and every weakly compact operators $T_1 : Y_1 \to E$ and $T_2 : E \to Y_2$, the composition $T_2 \circ T_1$ is compact.

EXERCISE 5. - Show that if E has the Dunford-Pettis Property, if F is a reflexive complemented subspace of E, F is finite-dimensional (use the previous exercise).

EXERCISE 6. - Show that if $E^\star$ has the Dunford-Pettis Property, so does E.

EXERCICE 7. - The aim of this exercise is to show that $\mathscr{C}([0,1])$ (and, more generally, $\mathscr{C}(K)$, K compact) has the Dunford-Pettis Property.

For this, we admit the following fact :

(☆) If $(\mu_n)_{n \geqslant 0}$ is a sequence of measures on $[0,1]$, with $\mu_n \to 0$ for $\sigma(\mathscr{M}([0,1]), \mathscr{M}^\star([0,1]))$, there is a positive $\mu \in \mathscr{M}([0,1])$, such that :

$\forall \epsilon > 0$, $\exists \delta(\epsilon) > 0$, such that for every $A \subset [0,1]$ with $\mu(A) < \delta$, then $|\mu_n(A)| < \epsilon$ for all n.

This fact will be proved in the next chapter (§ 2, prop. 5) for probability measures. The general case follows by decomposing each μ_n in $\mu_n^+ - \mu_n^-$, where μ_n^+ and μ_n^- are positive measures.

a) Let $f_n \xrightarrow[n \to +\infty]{} 0$ for $\sigma(\mathscr{C}([0,1]), \mathscr{M}([0,1]))$. Show that $\sup_n \|f_n\| < +\infty$ and that $f_n(t) \xrightarrow[n \to +\infty]{} 0$ for all $t \in [0,1]$.

b) Let $\mu_n \in \mathscr{M}([0,1])$, $\mu_n \xrightarrow[n \to +\infty]{} 0$ for $\sigma(\mathscr{M}([0,1]), \mathscr{M}^\star([0,1]))$.

Show that $\sup_n \|\mu_n\| < +\infty$. Show that, for every $\epsilon > 0$, there is a set $A \subset [0,1]$, with $\mu(A) < \delta(\epsilon)$, and a number $n(\epsilon)$ such that, for $n \geqslant n(\epsilon)$, $|f_n(t)| < \epsilon$ on $[0,1] \setminus A$ (use (☆) and Egoroff's Theorem). Deduce that $\int f_n d\mu_n \xrightarrow[n \to +\infty]{} 0$.

Observe that, by proposition 1, § 2, and exercise 5, every separable, reflexive, infinite-dimensional Banach space is isometric to an uncomplemented subspace of $\mathscr{C}([0,1])$.

REFERENCES ON CHAPTER V.

The proof of Bauer's maximum principle comes from G. CHOQUET's book [12]. Proposition 2, § I, follows BOURBAKI [11].

The proof of Banach-Mazur's Theorem (§ II) is the original proof (S. BANACH [5]) ; that of Banach-Stone's Theorem, using extreme points, can be found, for example, in LINDENSTRAUSS-TZAFRIRI [34].

Exercise 2 comes from BOURBAKI [11], Exercises 3 to 7 from LINDENSTRAUSS-TZAFRIRI [34].

CHAPTER VI

THE BANACH SPACES $L_p(\Omega, \mathscr{A}, \mu)$, $1 \leq p < +\infty$

We have already studied the case $\Omega = \mathbb{N}$, $\mathscr{A} = \mathscr{P}(\mathbb{N})$, and μ defined by $\mu(A) = |A|$, if $A \subset \mathbb{N}$ ($|A|$ is the cardinality of A , i.e. the number of points in A) : we obtained the ℓ_p-spaces .

We shall now consider the case of a purely non-atomic measure μ : for all $\omega \in \Omega$, $\mu(\{\omega\}) = 0$. One can show that if the space $L_p(\Omega, \mathscr{A}, \mu)$ is separable and if μ is purely non-atomic, then $L_p(\Omega, \mathscr{A}, \mu)$ is isometric to $L_p([0,1], dt)$. Therefore, we shall restrict ourselves to the study of this last space, which we call L_p in short.

Let us recall that L_p is the space of (real or complex) classes of measurable functions f such that $\left(\int_0^1 |f(t)|^p dt\right)^{1/p} < +\infty$, if $p < +\infty$ (two elements in the class are equal almost everywhere). We put

$$\|f\|_p = \left(\int_0^1 |f(t)|^p dt\right)^{1/p} ;$$

endowed with this norm, L_p is a Banach space. If $p = +\infty$, the norm is $\|f\|_\infty = \operatorname*{supess}_{0 \leq t \leq 1} |f(t)|$.

The dual of L_p $(1 \leq p < +\infty)$ can be identified with L_q , $\frac{1}{p} + \frac{1}{q} = 1$. An element g of L_q defines a linear functional ξ_g on L_p by means of the formula :

$$\xi_g(f) = \int f(t)g(t)dt .$$

The continuity of ξ_g comes from Hölder's inequality :

$$\left|\int f(t)g(t)dt\right| \leq \|f\|_p \cdot \|g\|_q .$$

Therefore, for $1 < p < +\infty$, L_p is reflexive. L_1 is not reflexive : the dual of L_1 is L_∞, but the dual of L_∞ is strictly larger than L_1. Also we shall see that L_1 has subspaces isomorphic to ℓ_1, and we know that this last space is not reflexive.

By definition of the duality, a sequence $(f_n)_{n \geq 1}$ of elements of L_p converges to an element f for $\sigma(L_p, L_q)$ if, for all $g \in L_q$,

$$\int f_n(t)g(t)dt \xrightarrow[n \to +\infty]{} \int f(t)g(t)dt .$$

One sees easily (see exercise 4, chap. II, first part) that it is enough for that to have :

$$\sup_n \|f_n\|_p < +\infty \qquad \text{(the sequence is bounded)}$$

and, for all $u \in [0,1]$, $\int_0^u f_n(t)dt \longrightarrow \int_0^u f(t)dt$.

In chapter II, we have introduced the Haar System. If we normalize each function in L_p, we obtain the Normalized Haar System :

$$h_0(t) = 1 \quad \text{for all} \quad t \in [0,1] ,$$

for all $n \geq 0$, if $2^n \leq m < 2^{n+1}$, put $k = m - 2^n$, then :

$$h_m(t) = 2^{\frac{n}{p}} \left[1_{\left[\frac{2k}{2^{n+1}}, \frac{2k+1}{2^{n+1}}\right[}(t) - 1_{\left[\frac{2k+1}{2^{n+1}}, \frac{2k+2}{2^{n+1}}\right[}(t) \right]$$

(where $1_A(t)$ is the characteristic function of A : $1_A(t) = 1$ if $t \in A$, 0 if not).

Let us now define an increasing sequence $(\mathcal{B}_m)_{m \in \mathbb{N}}$ of σ-fields, contained in the Borel σ-field of $[0,1]$, by :

$$\mathcal{B}_0 = \{\phi , [0,1]\} ,$$

and $\mathscr{B}_m$ is the σ-field generated by the random variables $h_o, h_1, \ldots, h_m$. So $\mathscr{B}_m$ contains all the dyadic intervals of length at most 2^{-n} , and contains also the intervals

$$\left[\frac{2j}{2^{n+1}}, \frac{2j+1}{2^{n+1}}\right], \left[\frac{2j+1}{2^{n+1}}, \frac{2j+2}{2^{n+1}}\right] \quad \text{for} \quad j \leq k \quad (k = m - 2^n).$$

The set $A_{m+1} = \{h_{m+1} \neq 0\}$ is an atom of $\mathscr{B}_m$, and the difference between $\mathscr{B}_m$ and $\mathscr{B}_{m+1}$ is exactly that this atom has been cut into two equal parts : the set $\{h_{m+1} < 0\}$ and the set $\{h_{m+1} > 0\}$. On these two sets, the values of h_{m+1} are opposite. So, clearly, $|h_{m+1}| = 2^{n/p} 1_{A_{m+1}}$ is $\mathscr{B}_m$-mesurable.

Let us call $E^{\mathscr{B}_m}$ the conditional expectation on the σ-field $\mathscr{B}_m$ (here, since $\mathscr{B}_m$ is made of a finite number of disjoint atoms, the definition of $E^{\mathscr{B}_m}$ is specially simple : if $C_1 \ldots C_K$ are these atoms, $E^{\mathscr{B}_m} f = \sum_{i=1}^{K} \frac{1}{P(C_i)} \left(\int_{C_i} f \, dP \right) 1_{C_i}$, where P is the Lebesgue measure on $[0,1]$).

We have, for all $m \geq 0$:

$$E^{\mathscr{B}_m}(h_{m+1}) = \frac{1}{P(A_{m+1})} \left(\int_{A_{m+1}} h_{m+1} \, dP \right) 1_{A_{m+1}} = 0 \, .$$

But the conditional expectation on a sub-σ-field $\mathscr{B}$ is always a norm-one projection from $L^p([0,1], dt)$ onto $L^p([0,1], \mathscr{B}; dt)$ (one checks immediately on the definition that $\|E^{\mathscr{B}} f\|_p \leq \|f\|_p$) so, for every sequence of scalars $a_1, \ldots, a_m$ and every $m' \leq m$, we can write :

$$\left\| \sum_{i=o}^{m'} a_i h_i \right\|_p = \left\| E^{\mathscr{B}_{m'}} \sum_{1}^{m} a_i h_i \right\|_p \leq \left\| \sum_{1}^{m} a_i h_i \right\|_p ,$$

which proves that the sequence $(h_m)_{m \in \mathbb{N}}$ is a monotone basic sequence. It is a basis, because the reunion $\bigcup_{m \in \mathbb{N}} \mathscr{B}_m$ of the σ-fields generated by $h_o, \ldots, h_m$ contains all the open intervals with dyadic extremities (that is, of the form $\frac{r}{2^s}$, $r, s \in \mathbb{N}$) , and, for every given $\epsilon > 0$, one can approximate any function f in L_p by a function g which is constant on dyadic intervals and satisfies $\|f - g\|_p < \epsilon$; therefore $\overline{\text{span}}\{(h_m)_{m \geq 0}\} = L_p$, and we have obtained :

PROPOSITION. - *The Haar System is a Schauder basis in* L_p $(1 \leq p < +\infty)$.

One can show that it is in fact an unconditional basis in L_p $(1 < p < +\infty)$, but this is much harder to prove (see for example B. MAUREY [35]). It is not an unconditional basis in L_1 , and, in fact, L_1 does not have any unconditional basis.

§ 1. SUBSPACES OF L_p $(1 \leq p < +\infty)$.

To start with, we shall exhibit a remarkable subspace of L_p $(1 \leq p < +\infty)$. It is spanned by the so-called "Rademacher functions". These functions are defined by :

$$\begin{cases} r_o(t) = 1 & \text{for all} \quad t \in [0,1] , \\ r_n(t) = \text{sign}(\sin 2^n \pi t) & n = 1,2,\dots \end{cases}$$

Obviously, $|r_n| = 1$ and $\int_0^1 r_n(t) r_m(t) dt = 0$ if $n \neq m$, so these functions form an orthonormal system in $L_2([0,1], dt)$.

So if we call R_2 the closed subspace spanned by the r_n's in $L_2([0,1], dt)$, the Rademacher functions form a Hilbertian basis of R_2 . They do not span the whole space $L_2([0,1], dt)$: as immediately checked, the product $r_1 . r_2$ (or any product $r_{n_1} . r_{n_2} . \dots . r_{n_k}$) is orthogonal to each of the r_n's , and therefore to R_2 . If one considers all these products, one obtains the so called "Walsh System" which is now a basis for $L_2([0,1], dt)$.

The remarkable fact is that the Rademacher functions span ℓ_2 also in L_p , $1 \leq p < +\infty$:

PROPOSITION 1. (Khintchine's Inequalities). - *For every* p $(1 \leq p < +\infty)$, *there are strictly positive numbers* A_p, B_p, *such that, for every* $n \in \mathbb{N}^\star$, *every finite sequence of scalars* $a_1,\dots,a_n$, *one has :*

$$A_p\Big(\sum_{i=1}^n |a_i|^2\Big)^{1/2} \leq \Big(\int_0^1 \Big|\sum_{i=1}^n a_i\, r_i(t)\Big|^p dt\Big)^{1/p} \leq B_p\Big(\sum_{i=1}^n |a_i|^2\Big)^{1/2} .$$

PROOF. - Assume first that the a_i's are real and that p is an even integer: $p = 2k$, $k \geq 1$. Put, if α_i's are integers :

$$A_{\alpha_1,\dots,\alpha_n} = \frac{(\alpha_1 + \dots + \alpha_n)!}{\alpha_1! \dots \alpha_n!} .$$

Then :

$$\int\Big(\sum_{i=1}^n a_i\, r_i(t)\Big)^{2k} dt$$

$$= \sum_{\alpha_1+\dots+\alpha_n=2k} A_{\alpha_1,\dots,\alpha_n}\, a_1^{\alpha_1} \dots a_n^{\alpha_n} \int_0^1 (r_1^{\alpha_1} \dots r_n^{\alpha_n})dt .$$

The integral $\int_0^1 (r_1^{\alpha_1} \dots r_n^{\alpha_n})dt$ is equal to zero, unless $\alpha_1,\dots,\alpha_n$ are all even numbers (since the r_i's equal ± 1 , one has $r_i^\alpha = r_i$ if α is odd, $= 1$ if α is even).

But we have :

$$\sum_{\beta_1+\dots+\beta_n=k} A_{\beta_1,\dots,\beta_n}\, a_1^{2\beta_1} \dots a_n^{2\beta_n} = (a_1^2 + \dots + a_n^2)^k$$

and so :

$$\int\Big(\sum_1^n a_i\, r_i(t)\Big)^{2k} dt = \sum_{\beta_1+\dots+\beta_n=k} A_{2\beta_1,\dots,2\beta_n}\, a_1^{2\beta_1} \dots a_n^{2\beta_n}$$

$$\leq B_{2k}^{2k}\,(a_1^2 + \dots + a_n^2)^k ,$$

where we put $B_{2k}^{2k} = \sup_{\beta_1+\dots+\beta_n=k} \left[\dfrac{A_{2\beta_1,\dots,2\beta_n}}{A_{\beta_1,\dots,\beta_n}}\right]$.

But :

$$\frac{A_{2\beta_1,\dots,2\beta_n}}{A_{\beta_1,\dots,\beta_n}} = \frac{(k+1)(k+2)\dots(2k)}{\prod_{j=1}^{n}(\beta_j+1)\dots(2\beta_j)} .$$

For the non-zero β_j's , we have :

$(\beta_j+1)\ \dots\ (2\beta_j) \geqslant 2^{\beta_j}$, and since $\beta_1 + \dots + \beta_n = k$,

we obtain :

$$\frac{A_{2\beta_1,\dots,2\beta_n}}{A_{\beta_1,\dots,\beta_n}} \leqslant \frac{(k+1)\dots(2k)}{2^k} .$$

Therefore $B_{2k}^{2k} \leqslant k^k$, and $B_{2k} \leqslant \sqrt{k}$. We have obtained, for $k = 1,2,\dots$

$$\left(\int_0^1 \left(\sum_{i=1}^{n} a_i\ r_i(t)\right)^{2k} dt\right)^{1/2k} \leqslant \sqrt{k}\ \left(\sum_{i=1}^{n} a_i^2\right)^{1/2} .$$

Now, we know that (since we are on a probability space), the function $p \to \|f\|_p$ is increasing (for fixed $f \in L_p$). So if $p \geqslant 1$ and if k is the smallest integer such that $2k \geqslant p$, we have again :

$$\left(\int_0^1 \left|\sum_{i=1}^{n} a_i\ r_i(t)\right|^p dt\right)^{1/p} \leqslant \sqrt{k}\ \left(\sum_{i=1}^{n} a_i^2\right)^{1/2}$$

and this proves the right hand side inequality.

For the left one, let us observe that, if $p \geqslant 2$:

$$\left(\int_0^1 \left|\sum_{i=1}^{n} a_i\ r_i(t)\right|^p dt\right)^{1/p} \geqslant \left(\int_0^1 \left|\sum_{i=1}^{n} a_i\ r_i(t)\right|^2 dt\right)^{1/2} = \left(\sum_{i=1}^{n} a_i^2\right)^{1/2} .$$

Let us assume now $1 \leqslant p \leqslant 2$. By Hölder's inequality,

$$\left(\int f^2 dt\right)^{1/2} = \left(\int |f|^{1/2} \cdot |f|^{3/2} dt\right)^{1/2} \leqslant \left[\left(\int |f| dt\right)^{1/2} \cdot \left(\int |f|^3 dt\right)^{1/2}\right]^{1/2}$$

$$\leq (\int |f|dt)^{1/4} \cdot (\int |f|^3 dt)^{1/4}$$

and, if we take $f(t) = \sum_1^n a_i \, r_i(t)$, and put $\gamma = (\sum_1^n a_i^2)^{1/2}$, we obtain :

$$\gamma \leq (\int |f|dt)^{1/4} \cdot (\int |f|^3 dt)^{1/4} \leq \|f\|_1^{1/4} \cdot (\sqrt{2}\,\gamma)^{3/4} ,$$

using the right hand side inequality, and so $\|f\|_p \geq \|f\|_1 \geq 2^{-3/8}\,\gamma$,

and our proposition is proved for real scalars. If now the a_j's are complex, we write $a_j = \alpha_j + i\,\beta_j$, with α_j , β_j real, and use the fact that there are constants A'_p , B'_p such that :

$$A'_p(\left|\sum_{j=1}^n \alpha_j r_j(t)\right|^p + \left|\sum_{j=1}^n \beta_j r_j(t)\right|^p)^{1/p} \leq \left|\sum_{j=1}^n a_j r_j(t)\right|$$

$$\leq B'_p(\left|\sum_{j=1}^n \alpha_j r_j(t)\right|^p + \left|\sum_{j=1}^n \beta_j r_j(t)\right|^p)^{1/p} .$$

This proves the proposition.

In L_∞ , the subspace spanned by the Rademacher functions is quite different :

PROPOSITION 2.

- *In* $L_\infty^{\mathbb{R}}([0,1], dt)$ *(real-valued classes of essentially bounded functions), the subspace spanned by the Rademacher functions is isometric to* ℓ_1 .

- *In* $L_\infty^{\mathbb{C}}$ *(complex-valued classes), this subspace is isomorphic to* ℓ_1 .

PROOF. - Let $a_1,\ldots,a_n$ be a finite sequence of reals, all different from 0 . There is a $t_o \in [0,1]$ such that $r_1(t_o) = \text{sign}\, a_1,\ldots,r_n(t_o) = \text{sign}\, a_n$. Therefore :

$$\sup_{t\in[0,1]} \left|\sum_1^n a_k \, r_k(t)\right| \geq \left|\sum_1^n a_k \, r_k(t_o)\right| = \sum_1^n |a_k|$$

and since

$$\sup_{t\in[0,1]} \left| \sum_1^n a_k r_k(t) \right| \leqslant \sum_1^n |a_k| ,$$

our first assertion is proved. For the second, if the a_k's are complex, put $a_k = \alpha_k + i\beta_k$, $\alpha_k \in \mathbb{R}$, $\beta_k \in \mathbb{R}$. Then :

$$\sup_{t\in[0,1]} \left| \sum_1^n a_k r_k(t) \right| \geqslant \max \left\{ \sup_t \left| \sum_1^n \alpha_k r_k(t) \right| , \sup_t \left| \sum_1^n \beta_k r_k(t) \right| \right\}$$

$$\geqslant \max\left(\sum_1^n |\alpha_k| , \sum_1^n |\beta_k| \right)$$

$$\geqslant \frac{1}{2} \sum_1^n (|\alpha_k| + |\beta_k|) \geqslant \frac{1}{2} \sum_1^n |a_k| .$$

One sees immediately, in this second case, that $\mathrm{span}\{(r_k)_{k\in\mathbb{N}}\}$ is not isometric to ℓ_1 : it is enough, for example, to take $a_1 = 1$, $a_2 = i$.

Let us now come back to the Rademacher functions in L_p , $p < +\infty$.

Let us call R_p the closed subspace spanned in L_p , $1 \leqslant p < +\infty$ by the Rademacher functions. The Khintchine's Inequalities say that R_p is isomorphic to ℓ_2 , and, more precisely, that the sequence (r_n) , $n \geqslant 1$, is equivalent to the canonical basis of ℓ_2 . They also say that the sequence $(r_n)_{n\geqslant 1}$ is basic and unconditional.

PROPOSITION 3. - *If* $1 < p < +\infty$, R_p *is complemented in* L_p .

PROOF. - Assume first $p \geqslant 2$. Then L_p is algebraically contained in L_2 , and, if $f \in L_p$, $\|f\|_2 \leqslant \|f\|_p$.

For all $f \in L_p$, let us define :

$$a_n = \int f(\tau) r_n(\tau) d\tau \qquad n \geqslant 1 ,$$

and $Pf = \sum_1^\infty a_n r_n$. This series converges in L_p , because $\sum_n |a_n|^2 < +\infty$, since $f \in L_2$.

We have :

$$\|Pf\|_p = \left\| \sum_1^\infty a_n r_n \right\|_p \leqslant B_p \left(\sum_1^\infty |a_n|^2 \right)^{1/2} = B_p \left\| \sum_1^\infty a_n r_n \right\|_2 \leqslant B_p \|f\|_2 \leqslant B_p \|f\|_p ,$$

and this proves that P is a linear operator from L_p into itself. If $f \in R_p$, then f can be written

$$f = \sum_{n=1}^{\infty} a_n r_n ,$$

and this series converges in L_p (this is so because the r_n's form a basic sequence). Therefore $Pf = f$, and P is a continuous linear projection from L_p onto R_p .

Now, if $1 < p \leq 2$, Pf can be defined by the same formula when $f \in L_p \cap L_2$, which is a dense subset of L_p . We have :

$$\|Pf\|_p = \left\| \sum_1^{\infty} a_n r_n \right\|_p \leq B_p \left(\sum_1^{\infty} |a_n|^2 \right)^{1/2}$$

$$\leq B_p \sup \left\{ \left| \sum_1^{\infty} \lambda_n a_n \right| ; \quad (\lambda_n) \in \ell_2 , \sum_{n \geq 1} |\lambda_n|^2 \leq 1 \right\}$$

$$\leq B_p \sup \left\{ \left| \int f(\tau) \left(\sum_1^{\infty} \lambda_n \, r_n(\tau) \right) d\tau \right| ; \Sigma |\lambda_n|^2 \leq 1 \right\} .$$

But, if $\frac{1}{p} + \frac{1}{q} = 1$, by Hölder's inequality :

$$\left| \int f(\tau) \left(\sum_1^{\infty} \lambda_n \, r_n(\tau) \right) d\tau \right| \leq \|f\|_p \left\| \sum_1^{\infty} \lambda_n r_n \right\|_q \leq \|f\|_p \cdot B_q \cdot \left(\sum_1^{\infty} |\lambda_n|^2 \right)^{1/2} ,$$

since $q < +\infty$ (that is, since $p > 1$). Finally, we obtain

$$\|Pf\|_p \leq B_p \cdot B_q \|f\|_p .$$

Therefore, P can be extended by continuity to the whole space L_p , and we obtain a projection from L_p to R_p , with norm at most $B_p \cdot B_q$. This proves our proposition.

One can show that R_1 is not complemented in L_1 .

We shall now describe another type of subspace of L_p $(1 \leq p < +\infty)$, spanned by functions with disjoint supports.

Let $(A_n)_{n \geq 1}$ be a sequence of measurable subsets of $[0,1]$, with $P(A_n) > 0$ for all n , and $P(A_i \cap A_j) = 0$ if $i \neq j$.

Let $(f_n)_{n\geqslant 1}$ be a sequence of norm-one functions in L_p, each f_n being supported by A_n, that is satisfying :

$$f_n = 1_{A_n} \cdot f_n \qquad \text{almost everywhere, for } n = 1,2,\ldots$$

Then we have :

PROPOSITION 4. - $F = \overline{\text{span}}\{(f_n), n \geqslant 1\}$ *is isometric to* ℓ_p *and 1-complemented in* L_p, $(1 \leqslant p < +\infty)$.

PROOF. - The first claim is obvious : if (a_k) is a finite sequence of scalars, we have :

$$\left(\int \left| \Sigma\, a_k\, f_k(t)\right|^p dt\right)^{1/p} = \left(\int_{\cup A_i} \left| \Sigma\, a_k\, f_k(t)\right|^p dt\right)^{1/p}$$

$$= \left(\sum_i \int_{A_i} \left| \sum_k a_k\, f_k(t)\right|^p dt\right)^{1/p} = \left(\Sigma\, |a_i|^p\right)^{1/p},$$

and this proves that $(f_n)_{n\geqslant 1}$ is equivalent to the canonical basis of ℓ_p and that $\overline{\text{span}}\{(f_n)_{n\geqslant 1}\}$ is isometric to ℓ_p.

For the second, we choose for all $n \geqslant 1$ a function g_n in L_q $(\frac{1}{p} + \frac{1}{q} = 1)$ with $\|g_n\|_q = 1$ and

$$\int \overline{f_n(t)}\, g_n(t)\, dt = 1 .$$

Now, if $f \in L_p$, we define

$$a_n = \int_{A_n} g_n(t)\, \overline{f(t)}\, dt$$

and

$$Pf = \sum_1^\infty a_n f_n .$$

This series converges if $(a_n)_{n\geqslant 1} \in \ell_p$, and satisfies $\|Pf\|_p = (\Sigma\, |a_k|^p)^{1/p}$. But this is the case, since :

$$(\Sigma |a_k|^p)^{1/p} = (\sum_k |\int_{A_k} g_k(t)\overline{f(t)}dt|^p)^{1/p} = (\sum_k |\int 1_{A_k}(t) g_k(t)\overline{f(t)}dt|^p)^{1/p}$$

$$\leqslant \left[\sum_k (\|g_k\|_q \cdot \|1_{A_k} f\|_p)^p \right]^{1/p} \leqslant \|1_{\cup A_k} f\|_p \leqslant \|f\|_p \ .$$

Therefore P is a continuous projection from L_p onto $\overline{\text{span}}\{(f_n)_{n \geqslant 1}\}$. Since $\|Pf\|_p \leqslant \|f\|_p$, P is of norm one.

We have described two examples of complemented subspaces of L_p : ℓ_2 , obtained with the Rademacher functions, and ℓ_p , obtained with any sequence of disjointly supported functions. We shall use these two types to obtain a classification of the subspaces of L_p . The results which follow now are due to KADEC and PELCZYNSKI [30].

For every $\epsilon > 0$ and p , $1 \leqslant p < +\infty$, we put :

$$A(\epsilon,p) = \{f \in L_p([0,1], dt) \ , \ P\{t, |f(t)| \geqslant \epsilon \|f\|_p\} \geqslant \epsilon \} .$$

LEMMA 5. - *For every* r , $1 \leqslant r < p$, *every* $f \in A(\epsilon,p)$, *one has*

$$\epsilon^{1+\frac{1}{r}} \|f\|_p \leqslant \|f\|_r \leqslant \|f\|_p \ .$$

PROOF. - We can write :

$$\|f\|_r = (\int |f(t)|^r dt)^{1/r} \geqslant (\int_{\{|f| \geqslant \epsilon \|f\|_p\}} |f(t)|^r dt)^{1/r}$$

$$\geqslant \epsilon \|f\|_p \cdot \left[P\{t \ ; \ |f(t)| \geqslant \epsilon \|f\|_p\} \right]^{1/r} \geqslant \epsilon^{1+\frac{1}{r}} \|f\|_p \ .$$

The sets $A(\epsilon,p)$ have some easy properties, which we shall gather in the following lemma :

LEMMA 6.

a) *If* $\epsilon_2 < \epsilon_1$, $A(\epsilon_2,p) \supset A(\epsilon_1,p)$.

b) $\bigcup_{0<\epsilon<1} A(\epsilon,p) = L_p$.

c) *If* f *does not belong to* $A(\epsilon,p)$ *, there is a measurable subset* A *of* $[0,1]$ *with* $P(A) < \epsilon$ *and*

$$\int_A |f(t)|^p \, dt \geq (1 - \epsilon^p)\|f\|_p^p \ .$$

PROOF.

a) and b) are obvious. Let us show c). Assume $f \notin A(\epsilon,p)$. Take $A = \{t, \ |f(t)| \geq \epsilon \|f\|_p\}$. Then $P(A) < \epsilon$, and

$$\int_{\complement A} |f(t)|^p \, dt \leq \epsilon^p \|f\|_p^p \ .$$

Therefore

$$\int_A |f(t)|^p \, dt \geq \|f\|_p^p - \epsilon^p \|f\|_p^p \geq (1 - \epsilon^p)\|f\|_p^p \ ,$$

and the lemma is proved.

LEMMA 7. - *Let* p , $1 \leq p < +\infty$, *and let* $(f_n)_{n \geq 1}$ *be a sequence of norm-one functions in* L_p *, which is not completely contained in any of the sets* $A(\epsilon,p)$, $0 < \epsilon < 1$.

Then there is a subsequence $(f'_n)_{n \geq 1}$ *which is equivalent to the canonical basis of* ℓ_p *(and therefore* $\overline{\text{span}}\{(f'_n)_{n \geq 1}\}$ *is isomorphic to* ℓ_p *), and* $\overline{\text{span}}\{(f'_n)_{n \geq 1}\}$ *is complemented in* L_p .

PROOF. - Let us first observe that, for every given $\epsilon > 0$, there are infinitely many f_n's which do not belong to $A(\epsilon,p)$. Indeed, assume that for some $\epsilon > 0$, only a finite number of functions, $f_{n_1}, \ldots, f_{n_k}$ do not belong to $A(\epsilon,p)$. Since $L_p = \bigcup_{0<\epsilon<1} A(\epsilon,p)$, there is $\epsilon_1 > 0$ such that $f_{n_1} \in A(\epsilon_1,p), \ldots, \epsilon_k > 0$ such that $f_{n_k} \in A(\epsilon_k,p)$, and so, if $\epsilon' = \min(\epsilon,\epsilon_1,\ldots,\epsilon_k)$, then $f_n \in A(\epsilon',p)$ for all $n \geq 1$, which contradicts the assumption.

From this remark follows that, for every $\epsilon > 0$ and every $n_o \geq 1$, there is $n \geq n_o$ such that $f_n \notin A(\epsilon,p)$.

Let now $\eta > 0$, $\eta < 1$. Put $\epsilon_1 = \frac{\eta}{4^2}$. Since $(f_n)_{n \geq 1}$ is not contained in $A(\epsilon_1,p)$, we can find an index n_1 such that $f_{n_1} \notin A(\epsilon_1,p)$. By lemma 6, c), there is a measurable subset A_1 of $[0,1]$ such that $P(A_1) < \epsilon_1$ and

$$\int_{A_1} |f_{n_1}(t)|^p \, dt \geq 1 - \epsilon_1^p .$$

Choose now $\epsilon_2 < \frac{\eta}{4^3}$, and small enough to have :

$$\int_A |f_{n_1}(t)|^p \, dt < \epsilon_1^p ,$$

for every measurable set A , with $P(A) < \epsilon_2$.

There is an index $n_2 > n_1$ such that f_{n_2} is not in $A(\epsilon_2,p)$. Therefore, there is a set A_2 , with $P(A_2) < \epsilon_2$, and

$$\int_{A_2} |f_{n_2}(t)|^p \, dt \geq 1 - \epsilon_2^p .$$

Assume we have found measurable subsets $A_1,\dots,A_k$ in $[0,1]$, integers $n_1 < n_2 < \dots < n_k$, positive numbers $\epsilon_1 > \epsilon_2 > \dots > \epsilon_k$, with, for $i = 1,\dots,k$:

(1) $\quad P(A_i) < \epsilon_i < \frac{\eta}{4^{1+i}}$.

(2) $\quad \int_{A_i} |f_{n_i}(t)|^p \, dt \geq 1 - \epsilon_i^p$.

(3) $\quad \int_{A_i} (|f_{n_1}(t)|^p + \dots + |f_{n_{i-1}}(t)|^p) dt < \epsilon_{i-1}^p$.

We then choose $\epsilon_{k+1} < \frac{\eta}{4^{k+2}}$, and small enough to obtain, for all A with $P(A) < \epsilon_{k+1}$,

$$\int_A (|f_{n_1}(t)|^p + \dots + |f_{n_k}(t)|^p) dt < \epsilon_k^p .$$

We then take $n_{k+1} > n_k$ such that $f_{n_{k+1}} \notin A(\epsilon_{k+1},p)$ and A_{k+1} will be given by lemma 6, c), for $f_{n_{k+1}}$.

Let us define now $A'_k = A_k \setminus \bigcup_{j>k} A_j$, for $k = 1,2,\dots$ These sets are mutually disjoint. We have :

$$1 \geqslant \int_{A'_k} |f_{n_k}(t)|^p \, dt \geqslant \int_{A_k} |f_{n_k}(t)|^p \, dt - \sum_{j>k} \int_{A_j} |f_{n_j}(t)|^p \, dt$$

$$\geqslant 1 - \epsilon_k^p - (\epsilon_k^p + \epsilon_{k+1}^p + \dots) \geqslant 1 - (\frac{\eta}{4^{k+1}})^p - \left[(\frac{\eta}{4^{k+1}})^p + (\frac{\eta}{4^{k+2}})^p + \dots \right]$$

$$\geqslant 1 - \frac{\eta^p}{4^{kp}} (\frac{1}{4^p} + \frac{1}{4^p - 1}) \geqslant 1 - \frac{\eta^p}{4^{kp}} .$$

If we put $f'_k = \dfrac{f_{n_k} \cdot 1_{A'_k}}{\|f_{n_k} \cdot 1_{A'_k}\|_p}$, we obtain disjointly supported normalized functions : the sequence $(f'_k)_{k \geqslant 1}$ is 1-equivalent to the canonical basis of ℓ_p , and $\overline{\mathrm{span}}\{(f'_k)_{k \geqslant 1}\}$ is 1-complemented in L_p , by proposition 4.

Finally, we have :

$$\|f_{n_k} - f'_k\|_p \leqslant \|f_{n_k} - f_{n_k} \cdot 1_{A'_k}\|_p + \|f_{n_k} \cdot 1_{A'_k} - f'_k\|_p$$

and

$$\|f_{n_k} - f_{n_k} \cdot 1_{A'_k}\|_p = (\int_{\complement A'_k} |f_{n_k}(t)|^p \, dt)^{1/p} \leqslant \frac{\eta}{4^k} ,$$

$$\|f_{n_k} \cdot 1_{A'_k} - f'_k\|_p^p = \left\| f_{n_k} \cdot 1_{A'_k} - \frac{f_{n_k} \cdot 1_{A'_k}}{\|f_{n_k} \cdot 1_{A'_k}\|_p} \right\|_p^p$$

$$= (1 - \|f_{n_k} \cdot 1_{A'_k}\|_p)^p \leqslant 1 - \|f_{n_k} \cdot 1_{A'_k}\|_p^p \leqslant \frac{\eta^p}{4^{kp}} .$$

Therefore

$$\|f_{n_k} 1_{A'_k} - f'_k\|_p \leqslant \frac{\eta}{4^k}$$

and

$$\|f_{n_k} - f'_k\|_p \leqslant 2 \cdot \frac{\eta}{4^k} , \qquad \sum_{k=1}^{\infty} \|f_{n_k} - f'_k\|_p \leqslant \frac{8}{3} \eta .$$

If we choose η small enough to have $\frac{8}{3}\eta < \frac{1}{8}$, that is $\eta < \frac{3}{64}$, the lemma follows now from chapter IV, § I, lemma 3, b), p. 110.

THEOREM 8. (Kadec-Pełczyński). - *If* $2 \leq p < +\infty$, *every closed infinite-dimensional subspace* X *of* L_p *is isomorphic to* ℓ_2 , *or contains a closed subspace* Y , *isomorphic to* ℓ_p *and complemented in* L_p .

If X *is isomorphic to* ℓ_2 , *it is complemented in* L_p .

PROOF. - Let X be a closed infinite-dimensional subspace of L_p , $2 < p < +\infty$. Put $S = \{x \in X , \|x\|_p = 1\}$, (the unit sphere of X). Two cases can occur :

- either there is an $\epsilon > 0$ such that $S \subset A(\epsilon,p)$.

In this case, by lemma 5, there is a constant $M > 0$ such that $M\|f\|_2 \leq \|f\|_p \leq \|f\|_2$, for all $f \in X$, and X is isomorphic to L_2 , and therefore to ℓ_2 .

- or, for all $\epsilon > 0$, there is a $f_\epsilon \in S$, not belonging to $A(\epsilon,p)$. Taking successively $\epsilon = \frac{1}{2^n}$, $n \geq 1$, we obtain a sequence (f_n) , with $\|f_n\|_p = 1$, and $f_n \notin A(\frac{1}{2^n}, p)$, for all n . By lemma 7, the sequence $(f_n)_{n\geq 1}$ contains a subsequence $(f'_n)_{n\geq 1}$ which is equivalent to the canonical basis of ℓ_p , and $\overline{\text{span}}\{(f'_n)_{n\geq 1}\}$ is complemented in L_p . This proves the first part of the theorem.

For the second part, we assume that X is isomorphic to ℓ_2 . We shall show that X is contained in a set $A(\epsilon,p)$.

LEMMA 9. - *Let* X *be a subspace of* L_p , $2 < p < +\infty$, isomorphic to ℓ_2 . *Then, for some* $\epsilon > 0$, X *is contained in* $A(\epsilon,p)$.

PROOF. - This is clear from lemma 7 : if the conclusion failed, we could find in X a sequence equivalent to the canonical basis of ℓ_p . This is obviously impossible if X is isomorphic to a Hilbert space, since all its subspaces are then also isomorphic to Hilbert space, and the lemma is proved.

By lemma 5, there is a constant $C \geqslant 1$ such that, for all $f \in X$,

$$\|f\|_2 \leqslant \|f\|_p \leqslant C\|f\|_2 ,$$

and therefore, the topologies induced on X by L_p and L_2 coïncide (X can be considered as a subspace of L_2, since $X \subset L_p \subset L_2$). Consequently, X is a closed subspace of L_2, and must be the range of a projection P, from L_2 onto X. For every f in L_p, we have :

$$\frac{1}{C}\|Pf\|_p \leqslant \|Pf\|_2 \leqslant \|f\|_2 \leqslant \|f\|_p ,$$

and P is a continuous projection from L_p onto X. This proves the theorem.

By duality, we obtain a description of the complemented subspaces of L_p, $1 < p \leqslant 2$:

THEOREM 10. - *Every infinite-dimensional complemented subspace of* L_p $(1 < p < +\infty)$ *which is not isomorphic to* ℓ_2 *contains a subspace isomorphic to* ℓ_p *and complemented in* L_p.

PROOF. - For $2 \leqslant p < +\infty$, the proof has already been given : we assume $1 < p < 2$. Let X be a complemented, infinite-dimensional subspace of L_p, and let P be the projection from L_p onto X.

By first part, chap. II, § 7, prop. 2 (p. 42), we know that the dual $X^\star$ of X is isomorphic to a complemented subspace of L_q $\left(\frac{1}{p}+\frac{1}{q}=1\right)$. If X is not isomorphic to a Hilbert space, $X^\star$ is not, either. By theorem 8, $X^\star$ contains a subspace Z, isomorphic to ℓ_q and complemented in L_q. $Z^\star$ is therefore isomorphic to a subspace of X, and is complemented in L_p. But $Z^\star$ is isomorphic to ℓ_p : this proves the theorem.

The description of the subspaces of L_p, $1 < p < 2$, is not so well-known as in the case $p > 2$. (Theorem 10 applies only to *complemented* subspaces). Let us only mention the following results : If X is a closed subspace of L_p, $1 < p < 2$, either there is a $r > p$ such that X is isomorphic to a subspace of L_r, or X contains a subspace isomorphic to ℓ_p, and complemented in L_p (H.P. ROSENTHAL [41]). But the analogue of theorem 8 is false : one can show that if $1 \leqslant p \leqslant r \leqslant 2$, the space L_r is isometric to a subspace of L_p .

In the third part, we shall meet some other properties of the L_p spaces, linked with a metric property of the norm, which is called uniform convexity. We postpone this study, and turn now to some specific properties of L_1 .

§ 2. THE SPACE L_1 .

In order not to change our notations, we shall restrict ourselves to the space $L_1([0,1], dt)$ (denoted L_1 , as before), but what follows is valid for any space $L_1(\Omega, \mathscr{A}, P)$, where $\mathscr{A}$ is a σ-field and P a probability measure.

A very important tool for the study of L_1 will be the notion of equi-integrability :

1°) *Equi-integrable sets of functions.*

Let F be a subset of L_1 . We shall say that the functions in F are equi-integrable (or, more briefly, that F is equi-integrable) if :

$$\sup_{f \in F} \int_{\{|f| > a\}} |f(t)|dt \xrightarrow[a \to +\infty]{} 0 .$$

PROPOSITION 1. - *If some integrable function* g *dominates the moduli of all the functions in* F , F *is equi-integrable. In particular, every finite set of functions is equi-integrable.*

PROOF. - Let us assume that, for all $f \in F$, $|f| \leqslant g$ a.e. . Since $\{|f| > a\} \subset \{g > a\}$, we have :

$$\int_{\{|f| > a\}} |f(t)|dt \leqslant \int_{\{g > a\}} g(t)dt .$$

But $\int_{\{g \leqslant a\}} g(t)dt \xrightarrow[a \to +\infty]{} \int g(t)dt$, since $g \in L^1$, and $\int_{\{g > a\}} g(t)dt \xrightarrow[a \to +\infty]{} 0$.

Using proposition 1, it is easy to build examples of equi-integrable sets of functions. Conversely, the sequence $(f_n)_{n\geqslant 1}$ defined by $f_n = n \, . \, 1_{[0,\frac{1}{n}]}$ is not equi-integrable.

PROPOSITION 2. - *A subset* F *of* L_1 *is equi-integrable if and only if the following two conditions are simultaneously satisfied :*

a) (equi-continuity) : *For every* $\epsilon > 0$, *there is a* $\delta > 0$ *such that, for every measurable subset* $A \subset [0,1]$ *with* $P(A) < \delta$, *for every element* $f \in F$, *one has*

$$\int_A |f(t)|dt < \epsilon \, .$$

b) (boundedness) : F *is bounded in* L_1 : *there is a constant* $M > 0$ *such that, for all* $f \in F$, $\|f\|_1 \leqslant M$.

PROOF.

1°) Let us first assume that F is equi-integrable. Then, for all $f \in F$, for all measurable A :

$$\int_A |f(t)|dt = \int_{A\cap\{|f|\leqslant a\}} |f(t)|dt + \int_{A\cap\{|f|>a\}} |f(t)|dt \leqslant aP(A) + \int_{\{|f|>a\}} |f(t)|dt \, .$$

Fix $\epsilon > 0$. We can (by definition) find a large enough to have, for all $f \in F$:

$$\int_{\{|f|>a\}} |f(t)|dt \leqslant \frac{\epsilon}{2} \, .$$

We then choose δ small enough to obtain $a\delta < \frac{\epsilon}{2}$. So, if A satisfies $P(A) < \delta$, then $\int_A |f(t)|dt < \epsilon$, for all $f \in F$.

The same way, but taking $A = [0,1]$, we obtain, for all $a > 0$,

$$\int |f(t)|dt \leqslant a + \int_{\{|f|>a\}} |f(t)|dt \, .$$

Choose a large enough to obtain

$$\int_{|f|>a} |f(t)|dt \leq 1 ,$$

for all $f \in F$: it follows that $\int |f(t)|dt \leq a + 1$, and F is bounded in L_1 .

2°) Assume now conditions a) and b) to be satisfied. We know that, for every function f in L_1 , we have, for all $a > 0$:

$$P\{|f| > a\} \leq \frac{\|f\|_1}{a} .$$

Therefore, for all $f \in F$, for all $a > 0$,

$$P\{|f| > a\} \leq \frac{M}{a} \xrightarrow[a \to +\infty]{} 0 .$$

Fix $\epsilon > 0$. If F is equi-continuous (condition a), there is a $\delta > 0$ such that $P(A) < \delta$ implies $\int_A |f| < \epsilon$, for all $f \in F$. Choose a large enough, so that $\frac{M}{a} < \delta$. We then have $P\{|f| > a\} \leq \delta$, and so $\int_{\{|f|>a\}} |f(t)|dt < \epsilon$, which proves the equi-integrability of F .

We shall now investigate the links, for a sequence of functions, between equi-integrability and convergence for the topology $\sigma(L_1 , L_\infty)$.

PROPOSITION 3. - *Every sequence* $(f_n)_{n\geq 1}$ *of integrable functions such that, for all measurable subsets* A *of* $[0,1]$ *, the limit* $\lim_{n\to+\infty} \int_A f_n(t)dt$ *exists and is finite, is equi-integrable.*

Moreover, the sequence $(f_n)_{n\geq 1}$ *converges, for* $\sigma(L_1 , L_\infty)$ *, to an integrable function* f *, and, for all* A *measurable in* $[0,1]$ *,*

$$\int_A f_n(t)dt \to \int_A f(t)dt .$$

PROOF. - We shall first fix some notations.

We call $\mathscr{A}$ the Borel σ-field of $[0,1]$ (or, more generally a σ-field on Ω) , and call $\tilde{\mathscr{A}}$ the quotient σ-field by the subsets of measure zero : that is, we identify two measurable sets which differ only by a set of measure zero. If Δ is the symmetric difference between two sets (that is $A \Delta A' = A \cup A' \setminus A \cap A'$), then we can define a distance on $\tilde{\mathscr{A}}$ by

$$d(A, A') = P(A \Delta A') ,$$

and, equipped with this distance, $\tilde{\mathscr{A}}$ is a complete metric space. To see this, let $(A_n)_{n \in \mathbb{N}}$ be a Cauchy sequence. From the formula :

$$\int |1_{A_n}(t) - 1_{A_m}(t)|dt = \int 1_{A_n \Delta A_m}(t)dt = P(A_n \Delta A_m)$$

follows that the sequence $(1_{A_n})_{n \in \mathbb{N}}$ is Cauchy in L_1, and therefore converges to a function f_o. Since a subsequence of $(1_{A_n})_{n \in \mathbb{N}}$ converges almost everywhere to f_o, f_o takes only the values 0 and 1(a.e.), and so, f_o is a characteristic function : $f_o = 1_A$, and since

$$\int |1_{A_n}(t) - 1_A(t)|dt \xrightarrow[n \to +\infty]{} 0 , \quad P(A_n \Delta A) \xrightarrow[n \to +\infty]{} 0 .$$

If f is a function in L_1, we consider the application φ, from $\tilde{\mathscr{A}}$ into $\mathbb{K}$, defined by $\varphi(A) = \int_A f(t)dt$.

LEMMA 4. - *The application* φ *is uniformly continuous from* $\tilde{\mathscr{A}}$ *into* $\mathbb{K}$.

PROOF. - We write

$$\int_A f(t)dt - \int_{A'} f(t)dt = \int_{A \setminus A'} f(t)dt - \int_{A' \setminus A} f(t)dt ,$$

and thus

$$\left|\int_A f(t)dt - \int_{A'} f(t)dt\right| \leq \int_{A \Delta A'} |f(t)|dt ,$$

and this last term can be made smaller than a given $\epsilon > 0$ if $P(A \Delta A')$ is small enough. This proves the lemma.

From this lemma follows that, for any $\epsilon > 0$, any $f \in L_1$, the set

$$\{A \in \tilde{\mathscr{A}} , \left|\int_A f(t)dt\right| \leq \epsilon \}$$ is closed in $\tilde{\mathscr{A}}$ for this topology.

Let us now start the proof of proposition 3. Let $(f_n)_{n\geq 1}$ be a sequence of integrable functions, such that, for all $A \in \widetilde{\mathcal{A}}$,

$\lim_{n\to+\infty} \int_A f_n(t)dt$ exists and is finite.

For any $\epsilon > 0$, we define, for $N = 1,2,\ldots,$

$$F_N = \{A \in \widetilde{\mathcal{A}} \ ; \ \forall\, m,n \geq N \ , \ \left|\int_A (f_m(t) - f_n(t))dt\right| \leq \epsilon \ \} .$$

Each F_N , by the previous remark, is closed in $\widetilde{\mathcal{A}}$. By assumption, for every $A \in \widetilde{\mathcal{A}}$, $\lim_{n\to+\infty} \int_A f_n(t)dt$ exists, and we can find a N such that, if $m,n \geq N$, $\left|\int_A (f_m(t) - f_n(t))dt\right| < \epsilon$. This says that $\bigcup_{N\geq 1} F_N = \widetilde{\mathcal{A}}$.

By Baire's Property (first part, chapter I), we can find an integer N_0 such that F_{N_0} has non-empty interior : this means that there is $A_0 \in \widetilde{\mathcal{A}}$, and $r > 0$ such that, for all $A \in \widetilde{\mathcal{A}}$ with $P(A \,\Delta\, A_0) < r$, then $A \in F_{N_0}$, that is

$$(1) \qquad \left|\int_A (f_m(t) - f_n(t))dt\right| \leq \epsilon \qquad \text{if } m,n \geq N_0 .$$

Now, take any $B \in \widetilde{\mathcal{A}}$. We have :

$$\int_B (f_m(t) - f_n(t))dt = \int_{A_0 \cup B} (f_m(t) - f_n(t))dt - \int_{A_0 \setminus B} (f_m(t) - f_n(t))dt ,$$

and

$$\begin{cases} (A_0 \cup B) \,\Delta\, A_0 \subset B \\ (A_0 \setminus B) \,\Delta\, A_0 \subset B . \end{cases}$$

Therefore, if $P(B) < r$, $A_0 \cup B$ and $A_0 \setminus B$ are in F_{N_0} , which means :

$$(2) \qquad \left|\int_B (f_m(t) - f_n(t))dt\right| \leq 2\epsilon .$$

If the f_n's are real-valued, we can apply the previous reasoning to the sets $B \cap \{f_m \geq f_n\}$ and $B \cap \{f_m < f_n\}$, instead of B , and, if $P(B) < r$, we obtain, instead of (2) :

$$\begin{cases} \displaystyle\int_{B \cap \{f_m \geqslant f_n\}} (f_m(t) - f_n(t))dt \leqslant 2\epsilon \\ \displaystyle\int_{B \cap \{f_m < f_n\}} (f_n(t) - f_m(t))dt \leqslant 2\epsilon \end{cases}$$

that is

$$\int_B |f_m(t) - f_n(t)|dt = \int_{B \cap \{f_m \geqslant f_n\}} (f_m(t) - f_n(t))dt + \int_{B \cap \{f_m < f_n\}} (f_n(t) - f_m(t))dt \leqslant 4\epsilon.$$

If the f_n's are complex valued, we obtain first from (2) :

$$\left|\int \mathcal{R}e(f_m(t) - f_n(t))dt\right| \leqslant 2\epsilon \quad \text{and} \quad \left|\int Im(f_m(t) - f_n(t))dt\right| \leqslant 2\epsilon \ ,$$

from which follows, as previously,

$$\int \left|\mathcal{R}e(f_m(t) - f_n(t))\right|dt \leqslant 4\epsilon\ , \quad \int |Im(f_m(t) - f_n(t))|dt \leqslant 4\epsilon\ ,$$

and

$$\int |f_m(t) - f_n(t)|dt \leqslant 8\epsilon\ ,$$

and we shall keep this last estimate for further computations.

We can find r' small enough to obtain

$$\int_B |f_m(t)|dt < \epsilon$$

if $P(B) \leqslant r'$, for $m = 1,2,\ldots,N_o$.

If $m \geqslant N_o$, and if $P(B) \leqslant \min(r, r')$, we have :

$$\int_B |f_m(t)|dt \leqslant \int_B |f_m(t) - f_{N_o}(t)|dt + \int_B |f_{N_o}(t)|dt \leqslant 9\epsilon$$

So finally, we have obtained that, for all $m \geqslant 1$, if $P(B) \leqslant \min(r, r')$, then $\int_B |f_m(t)|dt \leqslant 9\epsilon$.

and, if we call φ_n the application $A \to \int_A f_n(t)dt$, this proves that the family of functions $(\varphi_n)_{n\geqslant 1}$ is equi-continuous (a) of proposition 2).

In order to prove the equi-integrability of $(f_n)_{n\geqslant 1}$, according to proposition 2, it remains to show that $\sup_{n\geqslant 1} \|f_n\|_1 < +\infty$. For this, we take $(A_j)_{j=1,\dots,K}$ a partition of $[0,1]$ in K sets of measure at most $\min(r, r')$. For $j = 1,\dots,K$, we have, for all $m \geqslant 1$,

$$\int_{A_j} |f_m(t)|dt \leqslant 9\epsilon ,$$

and so

$$\int |f_m(t)|dt \leqslant 9\epsilon K ,$$

which proves the first part of the proposition.

Now, if $A \in \widetilde{\mathscr{A}}$, we put $Q(A) = \lim_{n\to+\infty} \int_A f_n(t)dt$. From the fact that the φ_n's are equi-continuous, one deduces easily that Q is σ-additive, and, obviously, it is absolutely continuous with respect to the Lebesgue measure on $[0,1]$. By Radon-Nikodym Theorem, there is a function $f \in L_1$ such that $Q(A) = \int_A f(t)dt$, for all $A \in \widetilde{\mathscr{A}}$.

We shall now show that $f_n \xrightarrow[n\to+\infty]{} f$ for $\sigma(L_1 , L_\infty)$.

We know that, for all $A \in \widetilde{\mathscr{A}}$, $\int_A f_n(t)dt \to \int_A f(t)dt$. Therefore, if g is a step-function (that is, a linear combination of characteristic functions), then $\int f_n(t)g(t)dt \to \int f(t)g(t)dt$. If now h is a bounded function, there is a sequence $(g_q)_{q\geqslant 1}$ of step-functions which converges to h in L_∞ . We can write :

$$\left|\int f_n(t)h(t)dt - \int f(t)h(t)dt\right| \leqslant$$

$$\leqslant \left|\int f_n(t)g_q(t)dt - \int f(t)g_q(t)dt\right| + \int |f_n(t)||h(t) - g_q(t)|dt$$

$$+ \int |f(t)||h(t) - g_q(t)|dt$$

$$\leqslant \left|\int f_n(t)g_q(t)dt - \int f(t)g_q(t)dt\right| + (\sup_n \|f_n\|_1 + \|f\|_1)\|h - g_q\|_\infty .$$

We know that $\sup_n \|f_n\|_1 = M < +\infty$. Now, for a given $\epsilon > 0$, we choose q large enough to have $(M + \|f\|_1)\|h - g_q\|_\infty < \frac{\epsilon}{2}$, and then, q being fixed, n_0 large enough so that if $n \geqslant n_0$, $\left|\int (f_n(t) - f(t))g_q(t)dt\right| < \frac{\epsilon}{2}$. This proves that, for every bounded function h,

$$\int f_n(t)h(t)dt \xrightarrow[n \to +\infty]{} \int f(t)h(t)dt ,$$

that is, $f_n \xrightarrow[n \to +\infty]{} f$ for $\sigma(L_1, L_\infty)$, and our proposition is proved.

This proposition has several corollaries, which we shall now give (though they do not concern directly our aim), because they have their own interest.

PROPOSITION 5. (Vitali-Hahn-Saks Theorem). - *Let* $(P_n)_{n \geqslant 1}$ *be a sequence of probability measures defined on* $(\Omega, \mathcal{A})$ *(here* $\Omega = [0,1]$ *), such that the limits* $Q(A) = \lim_{n \to +\infty} P_n(A)$ *exist for every* $A \in \mathcal{A}$. *Then* Q *is a probability on* $(\Omega, \mathcal{A})$. *If* P_0 *is the probability defined by* $P_0 = \sum_{n \geqslant 1} 2^{-n} P_n$, *each* P_n *can be written* $P_n = f_n \cdot P_0$, *where* $f_n \in L^1(\Omega, \mathcal{A}, P_0)$, *the sequence* (f_n) *is equi-integrable, and converges weakly to a function* $f \in L^1(\Omega, \mathcal{A}, P_0)$, *such that* $Q = f \cdot P_0$. *Therefore, if* (A_i) *is a decreasing family of sets such that* $P_0(A_i) \xrightarrow[i \to +\infty]{} 0$, *then* $\sup_n P_n(A_i) \xrightarrow[i \to +\infty]{} 0$.

PROOF. - If $P_0(A) = 0$, for $A \in \mathcal{A}$, then $\sum_{n \geqslant 1} 2^{-n} P_n(A) = 0$, and $P_n(A) = 0$ for all n.

Therefore, each P_n is absolutely continuous with respect to P_o , and so there is a function f_n in L_1 such that $P_n = f_n \cdot P_o$. Let us use proposition 3 for the f_n's , on $(\Omega,\mathcal{A},P_o)$. By assumption, for all $A \in \mathcal{A}$, $\lim_{n\to+\infty} \int_A f_n \, dP_o = \lim_{n\to+\infty} P_n(A)$ exists. So there is a function f in $L_1(\Omega,\mathcal{A},P_o)$ such that $f_n \xrightarrow[n\to+\infty]{} f$ for $\sigma(L_1(\Omega,\mathcal{A},P_o), L_\infty(\Omega,\mathcal{A},P_o))$. We have

$$Q(A) = \lim_{n\to+\infty} P_n(A) = \lim_{n\to+\infty} \int_A f_n \, dP_o = \int_A f dP_o .$$

Also, $Q(\Omega) = \lim_{n\to+\infty} P_n(\Omega) = 1$, since the P_n's are probabilities. Moreover, $Q(A) \geqslant 0$ for all $A \in \mathcal{A}$, and Q is a probability.

Finally, if $(A_i)_{i\in\mathbb{N}}$ is a decreasing family, with $P_o(A_i) \xrightarrow[i\to+\infty]{} 0$, we have

$$\sup_n P_n(A_i) = \sup_n \int_{A_i} f_n \, dP_o < \epsilon$$

if $P_o(A_i)$ is small enough, since the f_n's are equi-integrable (proposition 2, a)).

PROPOSITION 6. (Sequential Cauchy criterion). - A sequence $(f_n)_{n\geqslant 1}$ *of integrable functions converges for the topology* $\sigma(L_1 , L_\infty)$ *if and only if it is a Cauchy sequence for this topology.*

(In other words, L_1 is weakly sequentially complete).

PROOF. - For a sequence to converge, it is clearly necessary to be Cauchy. Conversely, assume $(f_n)_{n\geqslant 1}$ to be a Cauchy sequence for $\sigma(L_1 , L_\infty)$: this means that, for every $g \in L_\infty$, $\int f_n(t)g(t)dt$ is a Cauchy sequence in $\mathbb{K}$. In particular, for every $A \in \mathcal{A}$, $\int_A f_n(t)dt$ has a finite limit, when $n \to +\infty$. Proposition 3 says then that the sequence $(f_n)_{n\geqslant 1}$ converges for $\sigma(L_1 , L_\infty)$.

We can now give a topological characterization of the equi-integrability.

PROPOSITION 7. - *A subset* F *of* L_1 *is relatively compact for* $\sigma(L_1, L_\infty)$ *if and only if it is equi-integrable.*

PROOF.

a) Let us assume first that F is not equi-integrable. Then there is a number $\eta > 0$ and a sequence $(f_n)_{n \geq 1}$ of elements of F such that, for all $n \geq 1$,

$$\int_{\{|f_n| > n\}} |f_n(t)| dt > \eta .$$

Therefore, no subsequence of $(f_n)_{n \geq 1}$ is equi-integrable, and, by proposition 3, no subsequence is $\sigma(L_1, L_\infty)$ convergent. From first part, chapter III, § 3, proposition 1, p. 60, follows that $\{(f_n)_{n \geq 1}\}$ (and therefore F) is not weakly relatively compact.

b) Conversely, let us assume F equi-integrable. Then F is bounded in L_1 (prop. 2, b)), and thus in $(L_\infty)^\star$. Let $\overline{F}$ be the closure of F for $\sigma((L_\infty)^\star, L_\infty)$: $\overline{F}$ is compact for this topology. Since the restriction of it is $\sigma(L_1, L_\infty)$, all we have to show is that $\overline{F} \subset L_1$. Take $X \in \overline{F}$; there is a set I, a filter $\mathcal{F}$ on I and, for each $\alpha \in I$, an element $f_\alpha \in F$, such that $f_\alpha \xrightarrow[\mathcal{F}]{} X$ for $\sigma((L_\infty)^\star, L_\infty)$, that is :

$$\int f_\alpha(t) g(t) dt \xrightarrow[\mathcal{F}]{} X(g) , \quad \text{for all} \quad g \in L_\infty .$$

Since F is equi-continuous, (prop. 2, a)), for every $\epsilon > 0$, there is a $\delta > 0$ such that, if $P(A) < \delta$, then $\sup\limits_{f \in F} \int_A |f(t)| dt \leq \epsilon$, and, in particular, for every α, $\int_A |f_\alpha(t)| dt \leq \epsilon$. This implies that $X(1_A) \leq \epsilon$. If we put $Q(A) = X(1_A)$, we obtain a probability on $\mathcal{A}$, which is clearly absolutely continuous with respect to Lebesgue measure. So we can find $f \in L_1$ such that $Q = f \cdot P$, that is $\int \varphi \, dQ = \int \varphi(t) f(t) dt$, for all measurable φ, and $X(1_A) = \int_A f(t) dt$, for all $A \in \mathcal{A}$. From this we obtain, for every $h \in L_\infty$, $X(h) = \int f(t) h(t) dt$ which implies that $f_\alpha \xrightarrow[\mathcal{F}]{} f$ for $\sigma((L_\infty)^\star, L_\infty)$. This ends the proof of our proposition.

Now, we shall be able to give a classification of the subspaces of L_1.

2°) *The subspaces of* L_1 .

PROPOSITION 8. - *Let* F *be a closed subspace of* L_1 . *Then :*

- *either* F *is reflexive,*

- *or* F *contains a subspace isomorphic to* ℓ_1 *and complemented in* L_1 . *(And, obviously, both terms of the alternative exclude each other).*

PROOF. - Let S be the unit sphere of F (that is $S = \{f \in F, \|f\|_1 = 1\}$). Two cases can occur : either S is an equi-integrable subset of L_1 , or not.

a) If S is equi-integrable, the unit ball $\mathcal{B}$ of F is also equi-integrable. Therefore, $\mathcal{B}$ is $\sigma(L_1 , L_\infty)$ relatively compact, and, since it is closed, it is $\sigma(L_1 , L_\infty)$ compact. But we know that the restriction to F of the topology $\sigma(L_1 , L_\infty)$ is the topology $\sigma(F, F^\star)$ (first part, chapter II, § 7, prop. 1, p. 41), this implies that F is reflexive (first part, chapter III, § 1, p. 49).

b) Let us now assume that S is not equi-integrable.

Then, there is a number $\delta > 0$ such that

$$(1) \qquad \lim_{a \to +\infty} \sup_{f \in S} \int_{\{|f|>a\}} |f(t)|dt = \delta > 0$$

(because the function $a \longrightarrow \sup_{f \in S} \int_{\{|f|>a\}} |f(t)|dt$ is a decreasing function of a).

There is an increasing sequence $(a_n)_{n \geqslant 1}$ of positive real numbers, tending to infinity, such that

$$(2) \qquad \delta(1 - \frac{1}{2n}) < \sup_{f \in S} \int_{\{|f|>a_n\}} |f(t)|dt < \delta(1 + \frac{1}{2n}) .$$

Therefore, there is a sequence $(f_n)_{n \geqslant 1}$ of elements of S such that, for all $n \geqslant 1$:

$$(3) \quad \delta(1-\frac{1}{2n}) < \int_{\{|f_n|>a_n\}} |f_n(t)|\,dt < \delta(1+\frac{1}{2n}) .$$

We put $g_n = f_n \cdot 1_{\{|f_n|>a_n\}}$, and $h_n = f_n - g_n$. For every $\epsilon > 0$, we have :

$$P\{|g_n| > \epsilon \|g_n\|_1\} \leqslant P\{|g_n| > 0\} \leqslant P\{|f_n| > a_n\} \leqslant \frac{1}{a_n} .$$

Since $\frac{1}{a_n} \xrightarrow[n\to+\infty]{} 0$, we see that the sequence $(g_n)_{n\geqslant 1}$ is not contained in any of the sets $A(\epsilon,1)$, introduced in § 1. Lemma 7, § 1, shows that we can extract from the sequence $(g_n)_{n\geqslant 1}$ a subsequence $(g'_n)_{n\geqslant 1}$ which is equivalent to the canonical basis of ℓ_1 (let us observe that it follows from (3) that $\frac{\delta}{2} \leqslant \|g'_n\|_1 \leqslant \frac{3\delta}{2}$) and such that $\overline{\text{span}}\{(g'_n)_{n\geqslant 1}\}$ is complemented in L_1 .

We shall now turn to the sequence $h_n = f_n - g_n$, and show that it is equi-integrable.

For any $n \geqslant 1$, if $p \leqslant n$, the sets $\{|h_p| \geqslant a_n\}$ are empty, and therefore :

$$\sup_p \int_{\{|h_p|>a_n\}} |h_p(t)|\,dt = \sup_{p>n} \int_{\{|h_p|>a_n\}} |h_p(t)|\,dt .$$

But

$$\int_{\{|h_p|>a_n\}} |h_p(t)|\,dt = \int_{\{|h_p|\leqslant a_p\}\cap\{|h_p|>a_n\}} |h_p(t)|\,dt = \int_{\{|h_p|>a_n\}} |h_p(t)|\,dt - \int_{\{|h_p|>a_p\}} |h_p(t)|\,dt .$$

So we obtain :

$$\sup_p \int_{\{|h_p|>a_n\}} |h_p(t)|\,dt \leqslant \delta(1+\frac{1}{2n}) - \delta(1-\frac{1}{2p}) \leqslant \frac{\delta}{n} \xrightarrow[n\to+\infty]{} 0 ,$$

which proves the equi-integrability of the sequence $(h_n)_{n\geqslant 1}$. We keep from this sequence only the subsequence $(h'_n)_{n\geqslant 1}$ with indices corresponding to $(g'_n)_{n\geqslant 1}$: we still have, of course, an equi-integrable sequence. From proposition 8 (and first part, chapter III, § 3, prop. 1, p. 60) follows that we can extract from the sequence $(h'_n)_{n\geqslant 1}$ a subsequence

$(h''_n)_{n\geq 1}$ which converges weakly to a limit h_o . We keep the subsequence $(g''_n)_{n\geq 1}$ with the same indices. The consecutive differences $h''_{2k} - h''_{2k+1}$ converge weakly to zero, and, consequently, 0 is in the closure (for the norm) of the set $\mathrm{conv}\{h''_{2k} - h''_{2k+1} \; ; \; k \geq 1\}$. Therefore, we can find an increasing sequence $(k_j)_{j\geq 1}$ of integers, and positive numbers $(a_i)_{i\in \mathbb{N}}$, with, for all j , $\sum_{i=k_j+1}^{k_{j+1}} a_i = 1$, such that, if we set :

$$u_j = \sum_{i=k_j+1}^{k_{j+1}} a_i(f''_{2i} - f''_{2i+1})$$

$$v_j = \sum_{i=k_j+1}^{k_{j+1}} a_i(g''_{2i} - g''_{2i+1})$$

$$w_j = \sum_{i=k_j+1}^{k_{j+1}} a_i(h''_{2i} - h''_{2i+1}) \; ,$$

then $u_j = v_j + w_j$, and :

$$\lim_{j\to+\infty} \|u_j - v_j\| = \lim_{j\to+\infty} \|w_j\| = 0 \; .$$

We have seen that the sequence $(g'_n)_{n\geq 1}$ (and also $(g''_n)_{n\geq 1}$) was equivalent to the canonical basis of ℓ_1 , and that $\overline{\mathrm{span}}\{(g'_n)_{n\geq 1}\}$ was complemented in L_1 . We shall see that the sequence $(v_j)_{j\geq 1}$, which is made of blocks on the previous one, has the same properties :

LEMMA 9. - *The sequence* $(v_j)_{j\geq 1}$ *is equivalent to the canonical basis of* ℓ_1 *, and* $\overline{\mathrm{span}}\{(v_j)_{j\geq 1}\}$ *is complemented in* L_1 .

PROOF. - The first fact is obvious, since the v_j's are blocks one the g'_i's , with $\sum_{k_j+1}^{k_{j+1}} a_i = 1$: there are two constants C_1 , C_2 , such that, for every finite sequence (t_j) of scalars :

$$C_1 \sum_j |t_j| \leq \left\| \sum t_j v_j \right\|_1 \leq C_2 \sum |t_j| \; ,$$

and, in particular, $C_1 \leqslant \|v_j\|_1 \leqslant C_2$, for all j.

(except for the normalisation, the proof is the same as in chapter IV, prop. 1, p. 108).

For the second claim, let us choose, for $j \geqslant 1$, an element $v_j^\star \in L_\infty = (L_1)^\star$, with $v_j^\star(v_j) = 1$, and $\|v_j^\star\|_\infty \leqslant C_2$. If $f = \sum_j t_j g'_j$, we put $Pf = \sum_{j=1}^{\infty} v_j^\star \Big(\sum_{i=k_j+1}^{k_{j+1}} t_i g'_i \Big) v_j$. We define this way a projection from $\overline{\text{span}}\{(g'_i)_{i\geqslant 1}\}$ onto $\overline{\text{span}}\{(v_j)_{j\geqslant 1}\}$, which is continuous, since :

$$\begin{aligned} \|Pf\|_1 &\leqslant C_2 \sum_{j=1}^{\infty} \Big| v_j^\star \Big(\sum_{i=k_j+1}^{k_{j+1}} t_i g'_i \Big) \Big| \\ &\leqslant C_2^2 \sum_{j=1}^{\infty} \Big\| \sum_{k_j+1}^{k_{j+1}} t_i g'_i \Big\|_1 \\ &\leqslant \frac{3\delta}{2} C_2^2 \sum_{j=1}^{\infty} |t_j| \leqslant \frac{3\delta C_2^2}{2C'_1} \Big\| \sum_j t_j g'_j \Big\|_1 \\ &\leqslant \frac{3\delta C_2^2}{2C'_1} \|f\|_1 , \end{aligned}$$

where C'_1 is the constant such that $C'_1 \sum_j |t_j| \leqslant \Big\| \sum t_j g'_j \Big\|_1$.

Therefore, we have obtained a projection from $\overline{\text{span}}\{(g'_j)_{j\geqslant 1}\}$ onto $\overline{\text{span}}\{(v_j)_{j\geqslant 1}\}$. If we compose with the projection from L_1 onto $\overline{\text{span}}\{(g'_j)_{j\geqslant 1}\}$, we obtain a projection from L_1 onto $\overline{\text{span}}\{(v_j)_{j\geqslant 1}\}$ (since $\overline{\text{span}}\{(v_j)_{j\geqslant 1}\}$ is a subspace of $\overline{\text{span}}\{(g'_j)_{j\geqslant 1}\}$). This proves the lemma.

Let us now come back to the proof of the proposition. We know that $\|u_j - v_j\|_1 \xrightarrow[j\to+\infty]{} 0$. If, once more, we extract a subsequence, and renumber, we may assume that, $\epsilon > 0$ being given, we have $\|u_j - v_j\|_1 < \frac{\epsilon}{2^j}$, for all $j \geqslant 1$. It follows that $\sum_{j\geqslant 1} \|u_j - v_j\|_1 < \epsilon$, and, if ϵ has been chosen small enough, $(u_j)_{j\geqslant 1}$ is equivalent to the canonical basis of ℓ_1 and $\overline{\text{span}}\{(u_j)_{j\geqslant 1}\}$ is complemented in L_1 (chapter IV, lemma 3, p. 110). This ends the proof of our proposition.

As we mentioned for L_p-spaces $(p > 1)$, proposition 8 can be improved : H.P. ROSENTHAL [41] as shown that if a subspace F of L_1 is reflexive, there is a $r > 1$ such that F is isomorphic to a subspace of L_r .

§ 3. BANACH VALUED FUNCTIONS.

In the previous paragraphs, we considered scalar-valued functions ; we turn now to Banach-valued ones.

Let E a Banach space, and $(\Omega,\mathcal{A},\mu)$ a measurable space.

A function f , from Ω into E , will be called a *simple function* if it can be written

$$f = \sum_i x_i 1_{A_i} ,$$

where the sum is finite, the x_i's are points in E , and the A_i's are mutually disjoint measurable subsets of Ω . A function f , from Ω into E , will be called measurable if there is a sequence of simple functions $(f_n)_{n\geq 1}$ which converges to f almost everywhere, that is :

$$\text{for almost all } \omega , \quad \|f_n(\omega) - f(\omega)\| \xrightarrow[n \to +\infty]{} 0 .$$

PROPOSITION 1.

- *If* $f : (\Omega,\mathcal{A},\mu) \to E$ *is measurable, then, for every closed set* C *(or open set) in* E , $f^{-1}(C) \in \mathcal{A}$.

- *If* E *is separable and if, for every closed set* C *(or open set) in* E , $f^{-1}(C) \in \mathcal{A}$, *then* f *is measurable.*

PROOF.

1°) It is enough, obviously, to prove our assertion when C is a closed ball B . Then, let f be measurable, $(f_n)_{n\geq 1}$ a sequence of simple functions converging to f a.e. . For each $n \geq 1$, $k \geq 1$, we put

$$F_{n,k} = \{\omega \in \Omega , \text{dist}(f_n(\omega), B) \leq \frac{1}{k} \} .$$

Then, each $F_{n,k}$ is obviously measurable, and

$$f^{-1}(B) = \bigcap_{k \geq 1} \liminf_{n \to +\infty} F_{n,k}$$

and so, $f^{-1}(B) \in \mathscr{A}$.

2°) Now, we assume E to be separable. Let $(x_n)_{n \geq 1}$ be a dense sequence in E . For each $n \geq 1$, we consider the following list of closed balls :

$$B(x_1, \frac{1}{n}), \ldots, B(x_n, \frac{1}{n}),\ B(x_1, \frac{1}{n-1}), \ldots, B(x_n, \frac{1}{n-1}), \ldots, B(x_1, 1), \ldots, B(x_n, 1).$$

If $f(\omega)$ does not belong to any of these balls, we put $f_n(\omega) = 0$; if $f(\omega)$ belongs to at least one of them, let $B(x_j, \frac{1}{k})$ $(1 \leq j \leq n ,$ $1 \leq k \leq n)$ be the first in the list to which $f(\omega)$ belongs. Then we put $f_n(\omega) = x_j$.

The function f_n thus defined takes only finitely many values (in fact : $0, x_1, \ldots, x_n$), and for each j , $f_n^{-1}(\{x_j\})$ is measurable (since $f^{-1}(B)$ is measurable for every ball B) , so f_n is a simple function. We shall see that the sequence $(f_n)_{n \geq 1}$ converges to f everywhere.

For this, let $\omega \in \Omega$, and let $\epsilon > 0$. Take m_0 with $\frac{1}{m_0} \leq \epsilon$. Since the sequence $(x_j)_{j \geq 1}$ is dense in E , there is an index n_0 such that $\|x_{n_0} - f(\omega)\| \leq \frac{1}{m_0}$. Now, for $n \geq \max(n_0 , m_0)$, there is a j such that $f_n(\omega) = x_j$: this corresponds to the fact that $f(\omega)$ belongs to a set $B(x_j , \frac{1}{k})$, and $k \geq m_0$. Therefore, we get $\|f_n(\omega) - f(\omega)\| \leq \frac{1}{k} \leq \frac{1}{m_0} \leq \epsilon$, and our assertion follows.

We denote by $L^p(\Omega, \mathscr{A}, \mu ; E)$ $(1 \leq p < +\infty)$ the space of equivalence classes of measurable functions f (two functions being equivalent if they are equal almost everywhere), such that :

$$\left(\int \|f(\omega)\|_E^p \, d\mu(\omega) \right)^{1/p} < +\infty ,$$

$$\|f\|_{L^p(\Omega\,;\,E)} = \left(\int \|f(\omega)\|_E^p \, d\mu(\omega) \right)^{1/p} .$$

For $p = +\infty$, $L^\infty\ (\Omega,\mathscr{A},\mu\ ;\ E)$ is the space of essentially bounded classes of functions : $f \in L^\infty\ (\Omega,\mathscr{A},\mu\ ;\ E)$ if there is a $M \geqslant 0$ such that $\|f(\omega)\|_E \leqslant M$ a.e. , and $\|f\|_{L^\infty(\Omega\,;\,E)}$ is the infimum of such M's .

These spaces are Banach spaces, just as in the scalar case, but the duals are not so easy to characterize (see L. SCHWARTZ [46]). One can show, however, that if E is reflexive, $L^p\ (\Omega\ ;\ E)$ is reflexive $(1 < p < +\infty)$ and its dual is $L^q\ (\Omega\ ;\ E^\star)$, $\frac{1}{p} + \frac{1}{q} = 1$. This result is related to the Radon-Nikodym Property : see the "complements" at the end of this chapter.

EXERCISES ON CHAPTER VI.

EXERCISE 1. - Let $1 \leqslant p \leqslant 2$; show that for any $f_1,\dots,f_n$ in $L_p\ (\Omega,\mathscr{A},\mu)$, one has, for almost all ω :

$$\left(\int \left| \sum_1^n r_k(t) f_k(\omega) \right|^p dt \right)^{1/p} \leqslant \left(\sum_{k=1}^n |f_k(\omega)|^p \right)^{1/p} ,$$

where (r_k) are the Rademacher functions.

Deduce that :

$$\left(\int \left\| \sum_1^n r_k(t) f_k \right\|_{L^p}^p dt \right)^{1/p} \leqslant \left(\sum_1^n \|f_k\|_{L^p}^p \right)^{1/p} .$$

[A Banach space satisfying, for some $C > 0$, all $n \geqslant 1$, all $x_1,\dots,x_n$:

$$\left(\int \left\| \sum_1^n r_k(t) x_k \right\|^p dt \right)^{1/p} \leqslant C \left(\sum_1^n \|x_k\|^p \right)^{1/p}$$

is said to be of *type* p-*Rademacher*].

EXERCISE 2. - Let $2 \leqslant q < +\infty$. Show that, for any $f_1,\dots,f_n$ in $L_q\ (\Omega,\mathscr{A},\mu)$, one has, for almost all ω :

$$\left(\int \left| \sum_1^n r_k(t) f_k(\omega) \right|^q dt \right)^{1/q} \leqslant B_q \left(\sum_1^n |f_k(\omega)|^2 \right)^{1/2} ,$$

where B_q depends only on q, and deduce that :

$$\left(\int \left\| \sum_1^n r_k(t) f_k \right\|_{L^q}^q dt \right)^{1/q} \leqslant B_q \left(\sum_1^n \|f_k\|_{L^q}^2 \right)^{1/2},$$

that is, L_q, $2 \leqslant q < +\infty$, is of type 2-Rademacher.

EXERCISE 3. - Let $1 \leqslant p \leqslant 2$. Show that there is a constant A_p such that, for all $f_1, \ldots, f_n \in L_p(\Omega, \mathcal{A}, \mu)$, one has, for almost all ω :

$$\left(\int \left| \sum_1^n r_k(t) f_k(\omega) \right|^p dt \right)^{1/p} \geqslant A_p \left(\sum_1^n |f_k(\omega)|^2 \right)^{1/2}.$$

Deduce that

$$\left(\int \left\| \sum_1^n r_k(t) f_k \right\|_p^p dt \right)^{1/p} \geqslant A_p \left(\sum_1^n \|f_k\|_p^2 \right)^{1/2}.$$

(Use the fact that, for any norm, $\int \|x(\omega)\| d\mu \geqslant \left\| \int x(\omega) d\mu \right\|$).

A Banach space for which, for some $r > 0$, there is a constant $C_r > 0$ such that for all $n \geqslant 1$, all $x_1, \ldots, x_n$

$$\left(\int \left\| \sum_1^n r_k(t) x_k \right\|^r dt \right)^{1/r} \geqslant C_r \left(\sum_1^n \|x_k\|^q \right)^{1/q}$$

is said to be of *cotype* *q-Rademacher*. This property does not depend on r : for another r', there is a $C_{r'}$, satisfying the same property. This last fact was established by J.P. Kahane. The same holds for the type-p property. A proof can be found in [34], vol. II. So L_p, $1 \leqslant p \leqslant 2$, is of cotype 2.

EXERCISE 4. - Let $2 \leqslant q < +\infty$. Show that, for all $f_1, \ldots, f_n$ in $L_q(\Omega, \mathcal{A}, \mu)$, one has, for almost all $\omega \in \Omega$:

$$\left(\int \left| \sum_1^n r_k(t) f_k(\omega) \right|^q dt \right)^{1/q} \geqslant \left(\sum_1^n |f_k(\omega)|^q \right)^{1/q}.$$

Deduce that :

$$\left(\int \left\| \sum_1^n r_k(t) f_k \right\|_q^q dt \right)^{1/q} \geqslant \left(\sum_1^n \|f_k\|_q^q \right)^{1/q},$$

so L_q is of cotype q, $2 \leqslant q < +\infty$.

REFERENCES ON CHAPTER VI.

The main results of § I are due to KADEC and PELCZYNSKI [30] ; they can also be found in LINDENSTRAUSS-TZAFRIRI [34].

For § II, propositions 1 to 7 come principally from J. NEVEU's book [38]. The classification of subspaces of L_1 is due to KADEC and PELCZYNSKI [30] ; it can be found in a lecture of G. PISIER [39].

COMPLEMENTS ON CHAPTER VI.

We shall give, without proofs, a few results which complete those of § 3. We refer the reader, for example, to J. DIESTEL [14] or J. DIESTEL and J. UHL [15] for the proofs.

First, if $f : (\Omega,\mathcal{A},\mu) \to E$, one can wonder what are the relations between the fact that f is measurable and the fact that it is scalarly measurable, i.e. for all $\xi \in E^\star$, $\xi \circ f$ is measurable. We have :

PROPOSITION. - *If* E *is separable,* f *is measurable if and only if it is scalarly measurable.*

VECTOR MEASURES. - Let $(\Omega,\mathcal{A},\mu)$ be a measurable space (μ positive, finite), E a Banach space. A function $F : \mathcal{A} \to E$ is a *vector measure* if it countably additive : if $A_n \in \mathcal{A}$, pairwise disjoint, $F(\bigcup_n A_n) = \sum_n F(A_n)$.

The *variation* of F , denoted by $|F|$, is a positive function, defined by $|F|(A) = \sup \{ \sum_{B\in\pi} \|F(B)\| \;;\; \pi$ partition of $A \in \mathcal{A}$ into a finite number of pairwise disjoint subsets$\}$. If $|F|(\Omega) < +\infty$, F is said to be of bounded variation.

The Banach space E is said to have the *Radon-Nikodym Property* (R.N.P. in short) if, for any $(\Omega,\mathcal{A},\mu)$, any vector measure F with values in E , with bounded variation, such that $|F|$ is absolutely continuous with respect to μ , there exists a function f in $L^1(\Omega,\mathcal{A},\mu\,;E)$ such that

$$F(A) = \int_A f(\omega)\,d\mu(\omega) , \quad \text{for all} \quad A \in \mathcal{A} .$$

This means that F has a density with respect to μ, and f is this density (for $E = \mathbb{R}$, the existence of this density is precisely the Radon-Nikodym Theorem).

It can be shown (see [14], [15], [46]) that every reflexive space, and every space which is a separable dual, have R.N.P.

This property has a geometrical characterization (compare with Fourth part, chapter I, § 2, lemma 5). We say that a subset $D \subset E$ is *dentable* if, for every $\epsilon > 0$, there is an $x_\epsilon \in D$ such that

$$x_\epsilon \notin \overline{\text{conv}}\{D \setminus \{y \in E\ ,\ \|x_\epsilon - y\| < \epsilon\}\}$$

(If D is convex, this means that one can withdraw from D a slice of total diameter $< \epsilon$, and what remains is still convex).

Then we have (see [14], [15]) :

PROPOSITION. (Rieffel - Maynard - Huff - Davis - Phelps). - E *has R.N.P. if and only if every bounded subset of* E *is dentable.*

For the dual of a separable space, the R.N.P. can be characterized simply :

PROPOSITION. (Stegall). - *If* E *is separable,* $E^\star$ *has R.N.P. if and only if* $E^\star$ *is separable.*

Another geometric property is the following. We say that E has Krein - Milman Property (K.M.P. in short) if every closed bounded convex set in E is the closed convex hull of its extreme points. Then R.N.P. implies K.M.P. (LINDENSTRAUSS, see [14]) ; the converse is not known.

The duality between $L^p(\Omega, \mathcal{A}, \mu\ ;\ E)$ and $L^{p'}(\Omega, \mathcal{A}, \mu\ ;\ E^\star)$ is related to the R.N.P. as follows (see L. SCHWARTZ [46] for a detailed study) :

PROPOSITION. - *The dual of* $L^p(\Omega, \mathcal{A}, \mu\ ;\ E)$ *coïncides with* $L^{p'}(\Omega, \mathcal{A}, \mu\ ;\ E^\star)$ *if and only if* $E^\star$ *has R.N.P.* .

We shall relate R.N.P. with martingale's convergence in the complements to the fourth chapter, fourth part.

PART 3

SOME METRIC PROPERTIES IN BANACH SPACES

We have already made the distinction between a "topological property" and a "metric property". The first depends only on the topology of the space, and will therefore be conserved if we replace the norm by an equivalent one. The second is a characteristic of the given metric, and may disappear in an equivalent change of norm.

As examples of properties of the first type, we met reflexivity, the existence of a subspace isomorphic to ℓ_1 or c_0 , the existence of a Schauder basis, and several others. We have not seen so many properties of the second type : one is the existence of a monotone Schauder basis (if the norm changes, the basis remains a basis, but not necessarily monotone). Of course, the Parallelogram Identity, which characterizes prehilbertian spaces, is a metric property. In this third part, we shall now describe several important metric properties which can be satisfied in a Banach space.

CHAPTER I

STRICT CONVEXITY AND SMOOTHNESS

In this chapter, we investigate two metric properties : strict convexity and smoothness. These properties are "local", in the sense that they are not uniform on the unit sphere. The uniform analogues will be studied in the next chapter.

§ 1. STRICT CONVEXITY.

Let E be a Banach space. We say that E is *strictly convex* if, whenever x and y are not colinear, then :

(1) $\|x + y\| < \|x\| + \|y\|$.

We shall see later that the ℓ_p , L_p spaces $(1 < p < +\infty)$ are strictly convex. The spaces c_0 and ℓ_1 are not strictly convex : in the second, take $x = e_1$, $y = e_2$ (the first two vectors of the canonical basis), then $\|x\|_1 = \|y\|_1 = 1$; $\|x + y\|_1 = 2$; in the first, take $x = e_1 + e_2$, $y = e_1 - e_2$: then $\|x\|_0 = \|y\|_0 = 1$, $\|x + y\|_0 = 2$.

We shall give two equivalent characterizations of strict convexity, which, very often, prove to be more convenient for computations :

PROPOSITION 1.

a) E *is strictly convex if and only if, for all* $x,y \in E$, $x \neq y$,

(2) *if* $\|x\| = \|y\| = 1$, *then* $\| \frac{x+y}{2} \| < 1$.

b) E *is strictly convex if and only if, for every* p , $1 < p < +\infty$ *for all* $x,y \in E$, $x \neq y$,

(3) $\| \frac{x+y}{2} \|^p < \frac{1}{2} (\|x\|^p + \|y\|^p)$.

PROOF.

a) If E is strictly convex, then, if $x \neq y$, $\|x\| = 1$, $\|y\| = 1$, one has $\| \frac{x+y}{2} \| < 1$. Conversely, let us assume that, for two points x, y, not colinear :

$$\|x + y\| = \|x\| + \|y\| .$$

We may of course assume $\|y\| = 1$, $\|x\| \leqslant 1$. We put, for $t \in \mathbb{R}$, $\varphi_1(t) = \|tx + y\|$, $\varphi_2(t) = t\|x\| + \|y\|$. For every $t \geqslant 0$, we have $\varphi_1(t) \leqslant \varphi_2(t)$. We also have $\varphi_1(0) = \varphi_2(0)$, $\varphi_1(1) = \varphi_2(1)$. Since φ_1 is convex and φ_2 is linear, it follows that $\varphi_1(t) = \varphi_2(t)$ for every $t \geqslant 0$. This implies $\left\| \frac{x}{\|x\|} + y \right\| = 2$, and proves that (2) cannot be satisfied.

b) Let us assume E to be strictly convex. If x and y are not colinear, by strict convexity :

$$\| \frac{x+y}{2} \|^p \leqslant \left(\frac{\|x\| + \|y\|}{2} \right)^p \leqslant \frac{1}{2}(\|x\|^p + \|y\|^p)$$

(the last inequality holds since the function $t \to t^p$, for $1 \leqslant p < +\infty$, is a convex function of t).

If x and y are colinear, say $y = tx$, then :

$$\| \frac{x+y}{2} \|^p = \| \frac{x+tx}{2} \|^p = \left| \frac{1+t}{2} \right|^p \|x\|^p .$$

But $\left| \frac{1+t}{2} \right|^p < \frac{1}{2}(1 + |t|^p)$, for all $t \neq 1$, since $p > 1$: for $t \geqslant 0$, one computes the derivatives (recall that, for $p > 1$, the function $t \to |t|^p$ is derivable, and has for derivative $p\, t^{p-1}\, \mathrm{Sgn}t$). For $t \leqslant 0$, one writes $\left| \frac{1+t}{2} \right|^p \leqslant \left| \frac{1-t}{2} \right|^p$, and our proposition is proved.

Geometrically, strict convexity means that no line segment $[x,y]$ can be contained in the unit sphere : (2) says that if x, y belong to the unit sphere, their middle $\frac{x+y}{2}$ does not. But we have no information on the quantity $1 - \| \frac{x+y}{2} \|$, except that it is strictly positive. Let us also observe that, if $0 < \lambda < 1$, we also have $\|\lambda x + (1-\lambda)y\| < 1$.

§ 2. SMOTHNESS AND GATEAUX-DIFFERENTIABILITY OF THE NORM.

Let E be a Banach space. We say that E is smooth, if, for every $x \neq 0$, there is a unique linear functional which norms the point x, that is a unique $f \in E^\star$, with $\|f\| = 1$, such that $f(x) = \|x\|$. This functional will be denoted by f_x. Obviously, it is enough to check this property on the points of norm 1 in E.

From this uniqueness follows immediately a continuity result :

PROPOSITION 1. - *If* E *is smooth, the mapping* $x \to f_x$, *defined on* $E - \{0\}$, *is continuous on* $E - \{0\}$ *endowed with the norm, into* $E^\star$ *endowed with* $\sigma(E^\star, E)$.

PROOF. - Let $x_0 \in E$, $x_0 \neq 0$, and let $(x_i)_{i \in I}$ be any family converging to x_0 for some filter $\mathscr{F}$ on I. Let f_{x_i} be the norming functionals of x_i's. Observe first that the limit of $f_{x_i}(x_0)$ exists, and is equal to one, since :

$$f_{x_i}(x_0) = f_{x_i}(x_i) + f_{x_i}(x_i - x_0) ,$$

and thus

$$\left| f_{x_i}(x_0) - 1 \right| \leqslant \|x_i - x_0\| .$$

Now, we know that, since $\mathscr{B}_{E^\star}$ is $\sigma(E^\star, E)$-compact, the family $(f_{x_i})_{i \in I}$ has at least a cluster point f_0. It follows from the previous remark that $f_0(x_0) = 1$, and therefore $f_0 = f_{x_0}$, the unique linear functional norming x_0 (since we know also that $f_0 \in \mathscr{B}_{E^\star}$). Therefore, every cluster point of $(f_{x_i})_{i \in I}$ is equal to f_{x_0}, and, since we are on a compact set, the family $(f_{x_i})_{i \in I}$ converges to f_{x_0} for the filter $\mathscr{F}$, and our proposition is proved.

The use of filters, here, is necessary since $\mathscr{B}_{E^\star}$ needs not be $\sigma(E^\star, E)$-metrizable. If E is separable, then it is enough to consider a sequence $x_n \xrightarrow[n \to +\infty]{} x_o$, and the corresponding functionals $(f_{x_n})_{n \in \mathbb{N}}$.

We shall now relate smoothness to a weak differentiability property of the norm :

We say that the norm of E is *Gâteaux-differentiable* if, for all $x \neq 0$ in E , for all $h \in E$, the limit

$$\lim_{t \to o} \frac{\|x + th\| - \|x\|}{t}$$

exists when $t \to 0$, $t \in \mathbb{R}$. We call it $G(x,h)$. We shall see that, for a fixed $x \in E$, $h \to G(x,h)$ is a real linear functional on E (that is, $G(x,\lambda h) = \lambda G(x,h)$, for $\lambda \in \mathbb{R}$). Indeed, one has :

- $G(x, -h) = -G(x,h)$

- For every h_1 , h_2 , every λ , $0 \leqslant \lambda \leqslant 1$, $G(x, \lambda h_1 + (1-\lambda)h_2) \leqslant$ $\leqslant \lambda G(x, h_1) + (1 - \lambda)G(x, h_2)$.

This last property says that G is convex, and the first, that $-G$ is also convex : therefore, G is $\mathbb{R}$-linear. We use the notation $G_x(h)$, to recall that x is fixed.

For all $x,h \in E$, we have

$$\left| \frac{\|x + th\| - \|x\|}{t} \right| \leqslant \frac{\|x + th - x\|}{|t|} = \|h\|$$

and consequently, $|G_x(h)| \leqslant \|h\|$.

Now, if we take $h = \frac{x}{\|x\|}$, we obtain

$$G_x(h) = \lim_{t \to o} \frac{(1 + \frac{t}{\|x\|})\|x\| - \|x\|}{t} = \frac{\|x\|}{\|x\|} = 1$$

and we have proved that, for every $x \in E$, $x \neq 0$, $h \longrightarrow G(x,h)$ was a norm-one real linear functional on E (considered as a vector space on $\mathbb{R}$).

If E is real, it is a norm-one linear functional. If E is complex we define $\widetilde{G}_x(h) = G_x(h) - iG_x(ih)$. This is a $\mathbb{C}$-linear functional on E , and $G_x(h) = \mathcal{R}e\, \widetilde{G}_x(h)$. The functional $\widetilde{G}_x$ is also of norm-one : for $h \in E$, set $\theta = \operatorname{Arg} \widetilde{G}_x(h)$. Then :

$$|\widetilde{G}_x(h)| = e^{-i\theta}\widetilde{G}_x(h) = \mathcal{R}e(e^{-i\theta}\widetilde{G}_x(h)) = \mathcal{R}e\, \widetilde{G}_x(e^{-i\theta}h) = G_x(e^{-i\theta}h) \leqslant \|h\| .$$

One checks also, immediately in both cases, that

$$G_x(x) = \|x\| \quad , \quad \widetilde{G}_x(x) = \|x\| ,$$

and we have proved that, for every $x \in E$, $x \neq 0$, $G_x(h)$ was a norm-one real linear functional on E real, and $\widetilde{G}_x(h)$ on E complex.

REMARK. - In any normed space E , the function $t \longrightarrow \|x + th\|$ is convex and continuous, and has at every point a right and a left derivative. Let us put :

$$G_x^+(h) = \lim_{t \to o^+} \frac{\|x + th\| - \|x\|}{t} ,$$

$$G_x^-(h) = \lim_{t \to o^-} \frac{\|x + th\| - \|x\|}{t} .$$

As previously, one checks immediately that G_x^+ is positively homogeneous, subadditive, and continuous, and that

$$G_x^-(h) = -G_x^+(-h) \quad ; \quad G_x^-(h) \leqslant G_x^+(h) .$$

The norm is Gâteaux-differentiable if and only if $G_x^+(h) = G_x^-(h)$, for all $h \in E$.

PROPOSITION 2. - E *is smooth if and only if its norm is Gâteaux-differentiable.*

PROOF. - Let us first assume that the norm is Gâteaux-differentiable. We know that $G_x(h)$ is a norm-one real linear functional. But also :

$$G_x(x) = \|x\| ,$$

and G_x norms the point x . We shall see it is the only one. Assume f_x is a norm-one linear functional norming x . Then

$$\mathcal{R}e\, f_x(x + th) \leqslant \|x + th\| = \|x\| + tG_x(h) + t\epsilon(t) \text{ , with } \epsilon(t) \xrightarrow[t \to 0]{} 0 \text{ ,}$$

by definition of $G_x(h)$.

But since $f_x(x) = \|x\|$, we obtain :

$$t\, \mathcal{R}e\, f_x(h) \leqslant t\, G_x(h) + t\epsilon(t) \text{ ,}$$

that is $\mathcal{R}e\, f_x(h) \leqslant G_x(h)$, for all h , and therefore, since both are linear :

$$\mathcal{R}e\, f_x(h) = G_x(h) \text{ , and } f_x = G_x \text{ (E real) , or } \widetilde{G}_x \text{ (E complex).}$$

Conversely, let us assume that, for some $x_0 \in E$, $x_0 \neq 0$, $G^-_{x_0} \neq G^+_{x_0}$. Since the function $t \longrightarrow \|x_0 + th\|$ is convex, we know that $G^-_{x_0}(h) \leqslant G^+_{x_0}(h)$, for all h : therefore, there is a point $h_0 \in E$ such that $G^-_{x_0}(h_0) < G^+_{x_0}(h_0)$. Choose γ with $G^-_{x_0}(h_0) < \gamma < G^+_{x_0}(h_0)$. We may also assume $\|x_0\| = 1$.

Let F be the linear subspace spanned, in E , by x_0 and h_0 . On F , we define a linear functional, ℓ , by :

$$\ell(z) = \alpha + \beta\gamma \quad \text{, if } z = \alpha x_0 + \beta h_0 \text{ .}$$

We have :

$$\frac{\|x_0 + tz\| - \|x_0\|}{t} = \frac{\|x_0(1 + \alpha t) + t\beta h_0\| - \|x_0\|}{t} =$$

$$= \left(\frac{1 + \alpha t}{t}\right) \left\|x_0 + \frac{t}{1 + \alpha t}\beta h_0\right\| - \frac{1}{t} \text{ ,}$$

(if t is small enough to satisfy $1 + \alpha t \geqslant 0$)

$$= \frac{1 + \alpha t}{t} \left(\left\|x_0 + \frac{t}{1 + \alpha t}\beta h_0\right\| - 1\right) + \alpha \text{ .}$$

From this, we obtain, letting $t \to 0^+$:

$$G^+_{x_0}(z) = \alpha + G^+_{x_0}(\beta h_0) .$$

But, for all $\beta \in \mathbb{R}$, we have $\beta\gamma < G^+_{x_0}(\beta h_0)$ (this is clear, when $\beta > 0$, and, if $\beta < 0$, we write $G^+_{x_0}(\beta h_0) = \beta G^-_{x_0}(h_0)$) . Therefore, we obtain :

$$\ell(z) = \alpha + \beta\gamma \leqslant \alpha + G^+_{x_0}(\beta h_0) = G^+_{x_0}(z) ,$$

and ℓ is dominated by $G^+_{x_0}$. By Hahn-Banach Theorem, we can extend ℓ into a linear functional $\widetilde{\ell}$, defined on E , dominated by $G^+_{x_0}$, that is $\widetilde{\ell}(z) \leqslant G^+_{x_0}(z)$, for all $z \in E$. We obtain :

$$\widetilde{\ell}(-z) \leqslant G^+_{x_0}(-z) = -G^-_{x_0}(z) ,$$

and thus

$$\widetilde{\ell}(z) \geqslant G^-_{x_0}(z) .$$

Since $|G^+_{x_0}(z)| \leqslant \|z\|$, we have $|\widetilde{\ell}(z)| \leqslant \|z\|$, and $\widetilde{\ell}$ has norm at most one. Since $\ell(x_0) = 1$, $\widetilde{\ell}$ norms x_0 . Since each γ with $G^-_{x_0}(h_0) < \gamma < G^+_{x_0}(h_0)$ gives a different extension $\widetilde{\ell}$, there is no uniqueness of the functional which norms x_0 , and E is not smooth.

REMARKS.

1°) The Gâteaux-differentiability may be defined locally : we say that the norm is Gâteaux-differentiable at a point $x_0 \neq 0$ if, for all $h \in E$,

$$\lim_{t \to 0} \frac{\|x_0 + th\| - \|x_0\|}{t} \quad \text{exists} .$$

Proposition 2 can be extended as follows : the norm is Gâteaux-differentiable at $x_0 \neq 0$ if and only if x_0 has a unique norming functional.

2°) The Gâteaux-differentiability condition is often written under the form :

$$\|x + th\| = G_x(x + th) + t\epsilon(t) \ , \quad \epsilon(t) \xrightarrow[t \to o]{} 0 \ ,$$

where $\epsilon(t)$ depends on $x \in E$ $(x \neq 0)$ and on $h \in E$.

3°) More generally, one can define the Gâteaux-differentiability of convex functions : this means the existence of the limit

$$\lim_{t \to o} \frac{\varphi(x + th) - \varphi(x)}{t}$$

for all $x, h \in E$ $(x \neq 0)$. The reader may consult the book by I. EKELAND and R. TEMAM [18], for these questions.

4°) If the norm of E is Gâteaux-differentiable, and if $1 < p < +\infty$, the limit

$$\lim_{t \to o} \frac{\|x + th\|^p - \|x\|^p}{t}$$

exists for all $x, h \in E$: if $x \neq 0$, it comes from the fact that both $t \to \|x + th\|$ and $t \to t^p$ are derivable, and, if $x = 0$, it can be checked directly, using this last fact.

5°) Geometrically, the smoothness condition means that, at each point x of the unit sphere, there is a unique supporting hyperplane : the hyperplane $\{\widetilde{G}_x = 1\}$ is tangent at x_0 to the unit ball, and this unit ball is contained in the half space $\{G_x \leqslant 1\}$.

6°) Let us come back on our notations about norming functionals. If the norm is Gâteaux-differentiable, the linear functional G_x is real, is of norm one, and therefore, if the space is real, coïncides with the norming functional f_x (which, in this case, is unique). If the space is complex, the complexification of G_x , which we called $\widetilde{G}_x$, coïncides with f_x , and $G_x = \mathcal{R}e\, f_x$. So the notations G_x , $\widetilde{G}_x$ exist only in the smooth case , whereas f_x indicates any norming functional for x (which always exists, by Hahn-Banach Theorem).

We shall now give a geometric characterization of smoothness :

PROPOSITION 3.

1°) *Let* E *be a smooth space. For every* x , $\|x\| = 1$, *let* f_x *be its norming functional. The reunion of the homothetics, in all homotheties of center* x *and positive coefficients, of the open unit ball* $\mathring{\mathscr{B}}_E$ *is the open half space* $\{\mathcal{R}e\, f_x < 1\}$.

2°) *Conversely, if for every* x *with* $\|x\| = 1$, *the reunion of these homothetics is an open half space* $\{\mathcal{R}e\, f < 1\}$ $(\|f\| = 1)$, *then* E *has a Gâteaux-differentiable norm, and* f *is the norming functional for* x .

PROOF.

1°) Let $y \in E$, with $G_x(y) < 1$. All we have to show is that some point of the segment $]x,y]$ is contained in $\mathring{\mathscr{B}}_E$. This means, that, for some $\lambda < 1$, $\lambda x + (1 - \lambda)y \in \mathring{\mathscr{B}}_E$.

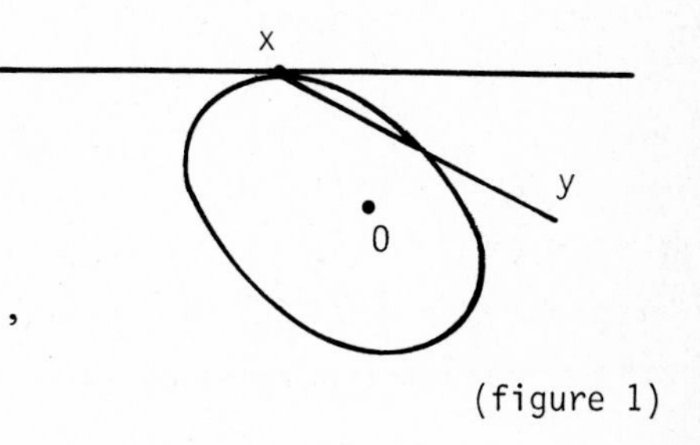

(figure 1)

But $x + (1 - \lambda)y = x + (1 - \lambda)(y - x)$. Set $t = 1 - \lambda$. Then $\lambda < 1 \Leftrightarrow t > 0$. By Gâteaux-differentiability, we have :

$$\|x + t(y - x)\| = \|x\| + t\, G_x(y - x) + t\epsilon(t) ,$$

where $\epsilon(t)$ depends on x, y , and tends to zero when $t \to 0$. Thus we obtain :

$$\|x + t(y - x)\| = 1 + t(G_x(y) - 1) + t\epsilon(t) .$$

But $G_x(y) - 1$ is strictly negative by assumption ; therefore, if $t > 0$, we have $\|x + t(y - x)\| < 1$ for t small enough.

2°) Assume E is not smooth. Then some x , $\|x\| = 1$, has two distinct norming functionals f, g . Then $\mathring{\mathscr{B}}_E \subset \{\mathcal{R}e\, f < 1\}$ and $\mathring{\mathscr{B}}_E \subset \{\mathcal{R}e\, g < 1\}$, that is $\mathring{\mathscr{B}}_E \subset \{\mathcal{R}e\, f < 1\} \cap \{\mathcal{R}e\, g < 1\}$. This last set is invariant under the considered homotheties.

REMARK. - In fact, this proposition is purely local : if we define, as we did in remark 1 above, the Gâteaux-differentiability at a point $x \neq 0$, the proposition says that the norm is Gâteaux-differentiable at x if and only if the geometric characterization holds.

We shall prove later that the following spaces are smooth : ℓ_p , L_p , for $1 < p < +\infty$. But ℓ_1 , c_0 , L_1 , L_∞ , $\mathscr{C}(K)$, ℓ_∞ , are not smooth. For example, in the case of $L_1(\Omega, \mu)$ (μ finite or not : we cover also the case of ℓ_1), choose $\Omega' \subset \Omega$, such that $\mu(\Omega \setminus \Omega') > 0$, and let f be a positive function with support in Ω' , such that $\int f d\mu = 1$. Then the function f is normed by any function $g \in L_\infty$, with $g = 1$ a.e. on Ω' and $|g| \leqslant 1$ on Ω : there is no uniqueness.

We shall now study the duality between strict convexity and smoothness.

§ 3. DUALITY BETWEEN STRICT CONVEXITY AND SMOOTHNESS.

PROPOSITION 1. - *If* $E^\star$ *is smooth,* E *is strictly convex ; if* $E^\star$ *is strictly convex,* E *is smooth.*

PROOF.

1°) If E is not strictly convex, we can find two points a, b, $a \neq b$, with $\|a\| = \|b\| = \| \frac{a+b}{2} \| = 1$. The line segment $[a,b]$ is thus disjoint from the open unit ball. By Hahn-Banach Theorem (geometric form), there is an hyperplane which contains this segment and does not meet the open unit ball. This hyperplane is of the form $\{x \in E \ ; \ f(x) = 1\}$, where $f \in E^\star$ and $\|f\| = 1$. But then a and $b \in E^{\star\star}$ are two distinct linear functionals on $E^\star$ which norm f , and therefore $E^\star$ is not smooth.

2°) If E is not smooth, one can find a point $x \neq 0$ which is normed by two distinct linear functionals f and g : $\|f\| = \|g\| = 1$, $f(x) = g(x) = \|x\|$. But $\| \frac{f+g}{2} \| \leqslant 1$, and $\frac{f+g}{2}(x) = \|x\|$, therefore $\| \frac{f+g}{2} \| = 1$, and $E^\star$ is not strictly convex.

One can give an example showing that E may be strictly convex, but $E^\star$ not smooth (see exercise 5). The duality does not exist in general between these two notions, but it does, of course, when E is reflexive :

COROLLARY 2. - *If* E *is reflexive,* E *is strictly convex (resp. smooth) if and only if* $E^\star$ *is smooth (resp. strictly convex).*

Let us now build an example of a strictly convex space which we shall use in the sequel. Let $(E_n)_{n\in\mathbb{N}}$ be a sequence of Banach spaces. Recall that the space $(\prod_{n\in\mathbb{N}} E_n)_p$ $(1 \leq p < +\infty)$, introduced in second part, chap. III, is :

$$(\prod E_n)_p = \{x = (x_n)_{n\in\mathbb{N}} \ ; \ x_n \in E_n \quad \text{for all} \quad n ,$$

$$\text{and} \qquad \sum_{n\in\mathbb{N}} \|x_n\|^p_{E_n} < +\infty \}$$

endowed with the norm

$$\|x\| = \Big(\sum_{n\in\mathbb{N}} \|x_n\|^p_{E_n} \Big)^{1/p} .$$

PROPOSITION 3. - *If* $1 < p < +\infty$, *and if the* E_n's *are all strictly convex,* $(\prod_{n\in\mathbb{N}} E_n)_p$ *is strictly convex.*

PROOF. - Let $x = (x_n)_{n\in\mathbb{N}}$, $y = (y_n)_{n\in\mathbb{N}}$ be two distinct elements of $(\prod_{n\in\mathbb{N}} E_n)_p$: there is an index n_0 for which $x_{n_0} \neq y_{n_0}$. Since E_{n_0} is strictly convex, we have, by § 1, prop. 1, b) :

$$\Big\| \frac{x_{n_0} + y_{n_0}}{2} \Big\|^p_{E_{n_0}} < \frac{1}{2} \big(\|x_{n_0}\|^p_{E_{n_0}} + \|y_{n_0}\|^p_{E_{n_0}} \big) .$$

For $n \neq n_0$, we just write

$$\Big\| \frac{x_n + y_n}{2} \Big\|^p_{E_n} \leq \frac{1}{2} \big(\|x_n\|^p_{E_n} + \|y_n\|^p_{E_n} \big)$$

and we obtain

$$\sum_{n\in\mathbb{N}} \Big\| \frac{x_n + y_n}{2} \Big\|^p_{E_n} < \frac{1}{2} \Big(\sum_n \|x_n\|^p_{E_n} + \sum_n \|y_n\|^p_{E_n} \Big) ,$$

and this proves our assertion, using again § 1, prop. 1 b).

This proposition shows in particular (taking $E_n = \mathbb{K}$ for all n) that the spaces ℓ_p, $1 < p < +\infty$, are strictly convex. If we take all the E_n's equal to a given Banach space E, then $(\prod E_n)_p$ is the space of p-summable sequences with values in E, and is denoted $\ell_p(E)$. The proposition says that if E is strictly convex, so is $\ell_p(E)$, $1 < p < +\infty$.

EXERCISES ON CHAPTER I.

EXERCISE 1. - Let E and F be two Banach spaces, F being strictly convex. Show that, if there exists a continuous injection from E into F, E admits an equivalent strictly convex norm.

EXERCISE 2. - Show that a subspace of a strictly convex space is strictly convex, and that a subspace of a smooth space is smooth. Same question for quotients.

EXERCISE 3. - Build on ℓ_2 an equivalent norm which is not strictly convex.

EXERCISE 4. - On $L^2([0,1], dt)$, we consider the norm

$$\|f\| = \left[\frac{1}{2}\left(\|f\|_2^2 + \|f\|_1^2\right)\right]^{1/2} .$$

Show that this norm is equivalent to $\|\cdot\|_2$, but is not smooth.

EXERCISE 5.

a) On ℓ_1, we consider the norm :

$$\text{if } x = (x_n)_{n\in\mathbb{N}}, \qquad \|x\| = \left(\|x\|_1^2 + \|x\|_2^2\right)^{1/2}$$

$$\left(\text{where } \|x\|_1 = \sum_{n\in\mathbb{N}} |x_n| , \quad \|x\|_2 = \left(\sum_{n\in\mathbb{N}} |x_n|^2\right)^{1/2}\right).$$

Show that this norm is equivalent to the ℓ_1-norm, and that it is strictly convex.

b) On ℓ_∞, consider the dual norm, denoted $\|\cdot\|^*$, of $\|\cdot\|$. This norm is equivalent to $\|\ \|_\infty$. Show that the constant sequence $y = (y_n)_{n\in\mathbb{N}}$, with $y_n = 1$, for all n , satisfies $\|y\|^\star = 1$.

c) Let z with $\|z\|^\star = 1$; $(z = (z_n)_{n \in \mathbb{N}})$. Show that for all $n \geqslant 1$, $\frac{1}{\sqrt{n^2+n}} \sum_1^n |z_k| \leqslant 1$. (Consider a sequence $x = (\alpha, \ldots, \alpha, 0, \ldots)$, α repeated n times).

d) Let $\mathcal{U}$ be an ultrafilter on $\mathbb{N}$. On ℓ_∞ , we define a linear functional by :

$$z = (z_n)_{n \in \mathbb{N}} \to \lim_{\mathcal{U}} z_n \quad ;$$

we call $\xi_{\mathcal{U}}$ this linear functional.

From b), c) deduce that $\|\xi_{\mathcal{U}}\| = 1$.

e) Take now $\mathcal{U}_1$, $\mathcal{U}_2$ two distinct ultrafilters on $\mathbb{N}$. Then $\xi_{\mathcal{U}_1}$ and $\xi_{\mathcal{U}_2}$ are two distinct norm one functionals, with $\xi_{\mathcal{U}_1}(y) = \xi_{\mathcal{U}_2}(y) = 1$. Therefore ℓ_∞ endowed with $\|\cdot\|^\star$ is not smooth.

REFERENCES ON CHAPTER I.

The definitions and propositions presented in this chapter can be found in most of the books dealing with metric properties of Banach spaces (see for example M.M. DAY [13] or G. KÖTHE [31]), except prop. 3, § 2, which is inspired by a paper of B. MAUREY and the author [9]. Also, as far as we know, exercise 5 is new.

CHAPTER II

UNIFORM CONVEXITY AND UNIFORM SMOOTHNESS

§ 1. UNIFORM CONVEXITY.

We have seen, in the previous chapter, that a Banach space E was strictly convex if, for all $x,y \in E$, with $x \neq y$, $\|x\| = \|y\| = 1$, then $\left\| \frac{x+y}{2} \right\| < 1$. But we observed that the difference $1 - \left\| \frac{x+y}{2} \right\|$ needed not be uniformly bounded from below. This justifies the following definition :

DEFINITION. - *A Banach space* E *is said to be* uniformly convex *if, for every* $\epsilon > 0$, *there is a number* $\delta > 0$ *such that, for all* x, y *in* E, *the conditions*

$$\|x\| = \|y\| = 1 , \qquad \|x - y\| \geqslant \epsilon ,$$

imply

$$\left\| \frac{x+y}{2} \right\| \leqslant 1 - \delta .$$

The number :

$$\delta(\epsilon) = \inf \left\{ 1 - \left\| \frac{x+y}{2} \right\| ; \ \|x\| = \|y\| = 1 , \quad \|x - y\| \geqslant \epsilon \right\}$$

is called the modulus of convexity of E. It is clear on the definition that E is uniformly convex if and only if, for every $\epsilon > 0$, $\delta(\epsilon) > 0$. It is also clear that, if $\epsilon_1 < \epsilon_2$, $\delta(\epsilon_1) < \delta(\epsilon_2)$, and that $\delta(0) = 0$.

The most immediate example of a uniformly convex space is the Hilbert space. Indeed, the Parallelogram Identity

$$\left\| \frac{x+y}{2} \right\|^2 = \frac{1}{2} (\|x\|^2 + \|y\|^2) - \left\| \frac{x-y}{2} \right\|^2$$

shows that if $\|x\| = \|y\| = 1$, $\|x - y\| \geqslant \epsilon$, then :

$$\left\| \frac{x+y}{2} \right\|^2 \leqslant 1 - \frac{\epsilon^2}{4} \text{ , and therefore, } \delta(\epsilon) = 1 - \sqrt{1 - \frac{\epsilon^2}{4}} \text{ .}$$

Uniform convexity implies strict convexity, and, at least formally, is strictly stronger, since it assumes the differences $1 - \left\| \frac{x+y}{2} \right\|$ to be uniformly bounded from below, for all x, y, $\|x\| = \|y\| = 1$, $\|x - y\| \geqslant \epsilon$. In fact, uniform convexity is really stronger, as the following example shows :

Take $(\prod_{n \in \mathbb{N}} E_n)_p$, with $1 < p < +\infty$, where all the E_n's are uniformly convex. Then, by chap. I, § 3, prop. 3, $(\prod_{n \in \mathbb{N}} E_n)_p$ is strictly convex. But, if the E_n's are less and less uniformly convex, that is, if for every $\epsilon > 0$, $\delta_{E_n}(\epsilon) \xrightarrow[n \to +\infty]{} 0$, then $(\prod_{n \in \mathbb{N}} E_n)_p$ will not be uniformly convex. As examples of E_n's , we shall see later that we can take $E_n = \ell_{p_n}$, where $1 < p_n < +\infty$, and p_n tends to 1 or to $+\infty$, when $n \to +\infty$.

The conditions $\|x\| = \|y\| = 1$ may be replaced, in the definition, by $\|x\| \leqslant 1$ and $\|y\| \leqslant 1$. This will allow us to give homogeneous characterizations of uniform convexity :

PROPOSITION 1. - *Let* p , $1 < p < +\infty$. E *is uniformly convex if and only if, for every* $\epsilon > 0$, *there is a number* $\delta_p(\epsilon) > 0$ *such that,*

for all $x,y \in E$, *the conditions*

$$(1) \quad \begin{cases} \|x\| \leqslant 1 , \quad \|y\| \leqslant 1 , \quad \|x - y\| \geqslant \epsilon \\ \text{imply} \\ \left\| \frac{x+y}{2} \right\|^p \leqslant (1 - \delta_p(\epsilon)) \left[\frac{1}{2} (\|x\|^p + \|y\|^p) \right] . \end{cases}$$

Therefore, E *is uniformy convex if and only if, for all* $x,y \in E$,

$$(2) \quad \left\| \frac{x+y}{2} \right\|^p \leqslant \left[1 - \delta_p \left(\frac{\|x-y\|}{\sup(\|x\|,\|y\|)} \right) \right] \frac{1}{2} \left[(\|x\|^p + \|y\|^p) \right] .$$

PROOF. - We shall first prove a lemma of geometric nature.

LEMMA 2. - *Let* x, y *be two distincts points of norm* 1 , *in* E , *uniformly convex. Set* $\epsilon' = \|x - y\|$. *Then, for all* t , $0 \leqslant t \leqslant 1$,

$$\left\| \frac{x + ty}{2} \right\| \leqslant \frac{1 - t}{2} + t(1 - \delta(\epsilon')) = \frac{1 + t}{2} - t\delta(\epsilon') .$$

PROOF OF LEMMA 2. - Let z be the intersection of x,y and the half-line $(0, \frac{x + ty}{2})$, (see picture). Put $z = \lambda \frac{x + ty}{2}$, and $z = \mu x + (1 - \mu) \frac{x + y}{2}$. Then :

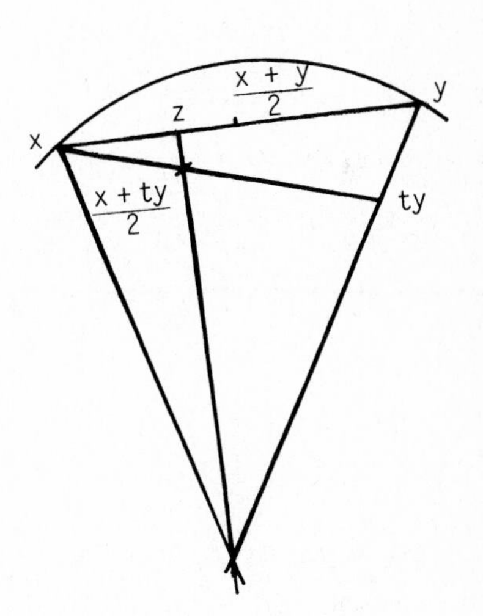

(figure 2)

$$\lambda \frac{x + ty}{2} = \mu x + (1 - \mu) \frac{x + y}{2}$$

$$x\left[\frac{\lambda}{2} - \mu - \frac{1 - \mu}{2}\right] = y\left[\frac{1 - \mu}{2} - \frac{t\lambda}{2}\right]$$

therefore :

$$\begin{cases} \frac{\lambda}{2} - \mu - \frac{1 - \mu}{2} = 0 \\ \frac{1 - \mu}{2} - \frac{t\lambda}{2} = 0 \end{cases}$$

or

$$\begin{cases} \lambda = \mu + 1 \\ 1 - \mu = t\lambda \end{cases} ,$$

and $\lambda = \frac{2}{1 + t}$, $\mu = \frac{1 - t}{1 + t}$.

But, since $z = \mu x + (1 - \mu) \frac{x + y}{2}$,

$$\|z\| \leqslant \frac{1 - t}{1 + t} + (1 - \frac{1 - t}{1 + t})(1 - \delta(\epsilon')) = \frac{1 - t}{1 + t} + \frac{2t}{1 + t}(1 - \delta(\epsilon'))$$

$$= \frac{1 + t - 2t\delta(\epsilon')}{1 + t}$$

that is

$$\|z\| \leqslant 1 - \frac{2t}{1+t}\,\delta(\epsilon') .$$

Consequently :

$$\left\| \frac{x + ty}{2} \right\| = \lambda^{-1}\|z\| = \frac{1+t}{2}\left(1 - \frac{2t}{1+t}\,\delta(\epsilon')\right) = \frac{1+t}{2} - t\,\delta(\epsilon')) ,$$

and the lemma is proved.

Let us come back to the proof of proposition 1. Let p , $1 < p < +\infty$.

Since the conclusion (2) is homogeneous, we may assume $\|x\| = 1$, $\|y\| \leqslant 1$. We set $\widetilde{y} = \frac{y}{\|y\|}$, and $t = \|y\|$. Then $0 \leqslant t \leqslant 1$. We assume $\|x - y\| \geqslant \epsilon$, and set $\epsilon' = \|x - \widetilde{y}\|$. By lemma 2, applied to $x, \widetilde{y}$:

$$\left\| \frac{x + t\widetilde{y}}{2} \right\| \leqslant \frac{1+t}{2} - t\,\delta(\epsilon')$$

and therefore

$$\frac{\left\| \frac{x+y}{2} \right\|^p}{\frac{1}{2}(\|x\|^p + \|y\|^p)} \leqslant \frac{\left(\frac{1+t}{2} - t\,\delta(\epsilon')\right)^p}{\frac{1}{2}(1 + t^p)} \quad ;$$

we call $\varphi(t)$ this last quantity.

Now, two cases can occur :

- If $\epsilon' < \frac{\epsilon}{2}$, then $\|x - \widetilde{y}\| < \frac{\epsilon}{2}$, we have

$$1 - t = \|y - \widetilde{y}\| \geqslant \|y - x\| - \|x - \widetilde{y}\| \geqslant \epsilon - \frac{\epsilon}{2} = \frac{\epsilon}{2}$$

and $t \leqslant 1 - \frac{\epsilon}{2}$. We have :

$$\varphi(t) \leqslant \frac{\left(\frac{1+t}{2}\right)^p}{\frac{1}{2}(1 + t^p)} ,$$

and we use the following lemma :

LEMMA 3. - *The real function*

$$\varphi_1(t) = \frac{(\frac{1+t}{2})^p}{\frac{1}{2}(1+t^p)}$$

is strictly increasing for $0 \leqslant t \leqslant 1$ *and attains its maximum for* $t = 1$.

PROOF. - Just compute the derivative.

Therefore, in this case, we have

$$\varphi(t) \leqslant \frac{(1-\frac{\epsilon}{4})^p}{\frac{1}{2}(1+(1-\frac{\epsilon}{2})^p)},$$

and

$$1-\varphi(t) \geqslant 1 - \frac{(1-\frac{\epsilon}{4})^p}{\frac{1}{2}[1+(1-\frac{\epsilon}{2})^p]}$$

- If $\epsilon' > \frac{\epsilon}{2}$, then $\delta(\epsilon') \geqslant \delta(\frac{\epsilon}{2})$, and

$$\varphi(t) \leqslant \frac{(\frac{1+t}{2} - t\delta(\frac{\epsilon}{2}))^p}{\frac{1}{2}(1+t^p)} = \psi(t).$$

We put δ instead of $\delta(\frac{\epsilon}{2})$.

The maximum of this last quantity is attained for $t^{p-1} = 1-2\delta$, or $t = (1-2\delta)^{1/p-1}$. Its value is :

$$\psi\left[(1-2\delta)^{1/p-1}\right] \leqslant \frac{\left[\frac{1+(1-2\delta)^{1/p-1}}{2} - (1-2\delta)^{1/p-1}\delta\right]^p}{\frac{1}{2}[1+(1-2\delta)^{p/p-1}]}.$$

In all cases, we obtain

$$\delta_p(\epsilon) \geqslant \min\left\{1 - \frac{(1-\frac{\epsilon}{4})^p}{\frac{1}{2}[1+(1-\frac{\epsilon}{2})^p]},\; 1 - \frac{\left[\frac{1+(1-2\delta(\frac{\epsilon}{2}))^{1/p-1}}{2} - [1-2\delta(\frac{\epsilon}{2})]^{1/p-1}\delta(\frac{\epsilon}{2})\right]^p}{\frac{1}{2}[1+(1-2\delta(\frac{\epsilon}{2}))^{p/p-1}]}\right\}$$

and the proposition is proved. This direct computation has a serious advantage : it allows us to compare the behaviour of $\delta(\epsilon)$ and $\delta_p(\epsilon)$ when $\epsilon \to 0$. The first term in the min is then equivalent to $\frac{p(p-1)\epsilon^2}{32}$, and the second is equivalent to $p \cdot \delta(\frac{\epsilon}{2})$. If we admit the fact that, for every Banach space, $\delta(\epsilon) \leqslant C\epsilon^2$, we find that $\delta_p(\epsilon) \geqslant c_p\,\delta(\epsilon)$, for some constant c_p dependent only on p.

PROPOSITION 4. - *Every subspace and every quotient of a uniformly convex space are also uniformly convex.*

PROOF. - The first fact is obvious. For the second, let F be a quotient of a uniformly convex space E. Let $x, y \in F$, with $\|x-y\| \geqslant \epsilon$, $\|x\| \leqslant 1$, $\|y\| \leqslant 1$. Let $\epsilon' > 0$. We may find representants ξ, η of x, y respectively, in E, with $\|\xi\|_E \leqslant 1+\epsilon'$, $\|\eta\|_E \leqslant 1+\epsilon'$. Since $\xi - \eta$ represents $x - y$, we have $\|\xi-\eta\|_E \geqslant \epsilon$. Put $\xi' = \frac{\xi}{1+\epsilon'}$, $\eta' = \frac{\eta}{1+\epsilon'}$.

Then $\left\| \frac{\xi'+\eta'}{2} \right\|_E \leqslant 1 - \delta(\frac{\epsilon}{1+\epsilon'})$, where $\delta(\epsilon)$ is the modulus of convexity of E. Therefore, we obtain :

$$\left\| \frac{\xi+\eta}{2} \right\|_E \leqslant (1+\epsilon')(1-\delta(\frac{\epsilon}{1+\epsilon'}))$$

and

$$\left\| \frac{x+y}{2} \right\| = \inf_{\substack{\xi \text{ rep. } x \\ \eta \text{ rep. } y}} \left\| \frac{\xi+\eta}{2} \right\|_E \leqslant (1+\epsilon')(1-\delta(\frac{\epsilon}{1+\epsilon'})) ,$$

for all $\epsilon' > 0$.

The modulus of convexity $\delta_F(\epsilon)$ is therefore at least

$$\delta_F(\epsilon) \geqslant \lim_{\epsilon' \to 0} \delta(\frac{\epsilon}{1+\epsilon'}) \geqslant \delta(\frac{\epsilon}{2}) .$$

We now give a first application of the notion of uniform convexity :

PROPOSITION 5. - *If* E *is uniformly convex, there is a best approximation projection on every closed convex subset* C *of* E.

(This projection needs not be linear, even if $C = F$ is a closed subspace).

PROOF. - Let C be a closed convex set in E ; we want to show that there is a projection P, from E onto C, such that, for all $x \in E$, $\|x - Px\| = \inf_{z \in C} \|x - z\|$.

We have seen (second part, chap. I) that this property was true in Hilbert spaces.

Let $x \in E$, $x \notin C$. We put, for $n \geqslant 1$,

$$A_n = \{z \in C \ ; \ d(x, z) \leqslant d(x, C) + \frac{1}{n}\}$$

(recall that $d(x, C) = \inf_{z \in C} \|x - z\|$; we put $d = d(x, C)$).

By proposition 1, we can write, if z_1, z_2 are two points of A_n :

$$\left\| x - \frac{z_1 + z_2}{2} \right\|^2 \leqslant$$

$$\leqslant \left[1 - \delta_2\left(\frac{\|z_1 - z_2\|}{\max(\|x - z_1\|, \|x - z_2\|)}\right)\right] \frac{1}{2}(\|x - z_1\|^2 + \|x - z_2\|^2) .$$

But $\frac{z_1 + z_2}{2} \in C$, and therefore $\left\| x - \frac{z_1 + z_2}{2} \right\|^2 \geqslant d^2$. Thus we obtain :

$$d^2 \leqslant \left[1 - \delta_2\left(\frac{\|z_1 - z_2\|}{\max(\|x - z_1\|, \|x - z_2\|)}\right)\right] (d + \frac{1}{n})^2 .$$

When $n \to +\infty$, this implies that $\sup_{z_1, z_2 \in A_n} \|z_1 - z_2\| \xrightarrow[n \to +\infty]{} 0$, that is, the diameter of A_n tends to zero. By completeness, there is a unique point y which belongs to $\bigcap_{n \geqslant 1} A_n$. It is immediately checked that y is the point Px we were looking for.

As we already mentioned, this projection P needs not be linear, even if $C = F$ is a subspace. Indeed, we have built (second part, chapter III) an example of an uncomplemented subspace of ℓ_p $(1 < p < +\infty)$, and we shall see below that ℓ_p $(1 < p < +\infty)$ is uniformly convex.

Uniform convexity is obviously a metric property, and is not preserved if the norm is replaced by an equivalent one. For example, on ℓ_2 , we can consider the norm (if $x = (x_n)_{n \in \mathbb{N}}$) :

$$[\![x]\!] = \max\left(2|x_0| \, , \, (\sum_{n \in \mathbb{N}} |x_i|^2)^{1/2}\right).$$

This norm is equivalent to the usual norm on ℓ_2 , but, if we take :

$$a_1 = (\tfrac{1}{2}, 0, 0, \ldots)$$

$$a_2 = (\tfrac{1}{2}, \tfrac{1}{2}, 0, \ldots)$$

we have $[\![a_1]\!] = [\![a_2]\!] = 1$, $[\![a_1 - a_2]\!] = \frac{1}{2}$, $\left[\!\!\left[\frac{a_1 + a_2}{2}\right]\!\!\right] = 1$. The space fails now to be strictly convex, and, a fortiori, to be uniformly convex.

We are therefore brought to the following question : does uniform convexity imply properties which are invariant under isomorphism ? A first answer, which is very important, is the following :

PROPOSITION 6. - *A uniformly convex Banach space is reflexive.*

PROOF. - We shall give two different proofs of this result.

1°) Let $z \in E^{**}$ with $\|z\| = 1$. Since $\mathscr{B}_E$ is $\sigma(E^{**}, E^*)$ dense in E^{**} , we can find a family $(x_i)_{i \in I}$ of elements of $\mathscr{B}_E$, and a filter $\mathscr{F}$ on I , such that $(x_i)_{i \in I}$ converges to z , for $\sigma(E^{**}, E^*)$. Let $\epsilon > 0$. We can find $f \in E^*$, with $\|f\| = 1$ and $< f,z > \; > 1 - \delta(\epsilon)$. By definition of the convergence of $(x_i)_{i \in I}$ for the filter $\mathscr{F}$, there is a set $X \in \mathscr{F}$ such that for all $i \in X$, $f(x_i) > 1 - \delta(\epsilon)$.

Therefore, if $i,j \in X$, $f(\frac{x_i + x_j}{2}) > 1 - \delta(\epsilon)$, and $\left\| \frac{x_i + x_j}{2} \right\| > 1 - \delta(\epsilon)$.

Since E is uniformly convex, this implies $\|x_i - x_j\| \leqslant \epsilon$. So we have obtained : for every $\epsilon > 0$, there is $X \in \mathscr{F}$ such that, for all $i,j \in \mathscr{F}$, $\|x_i - x_j\| \leqslant \epsilon$. This says that $\mathscr{F}$ is a Cauchy filter on I . Since E is complete, the family $(x_i)_{i\in I}$ converges to a point $x \in E$, and $z = x \in E$. So, E is reflexive.

Let us observe that the notation $\operatorname{Lim\,inf}_{i,j\to+\infty} \|x_i + x_j\|$ has a sense when $(x_i)_{i\in I}$ is a *sequence*, but is meaningless in the present case (one should define and use the product $\mathscr{F}\times\mathscr{F}$).

2°) If E is not reflexive, we can find a norm-one sequence $(x_n)_{n\in\mathbb{N}}$ satisfying condition J, for some $\theta > 0$, that is, for all $k \geqslant 1$:

$$\operatorname{dist}(\operatorname{conv}(x_1,\ldots,x_k),\ \operatorname{conv}(x_{k+1},\ldots)) > \theta .$$

If E is uniformly convex, we obtain, for all $n \geqslant 1$:

$$\left\| \frac{x_1 + x_2}{2} \right\| \leqslant 1 - \delta(\theta),\ldots, \left\| \frac{x_{2n-1} + x_{2n}}{2} \right\| \leqslant 1 - \delta(\theta) .$$

And, since

$$\left\| \frac{x_1 + x_2}{2} - \frac{x_3 + x_4}{2} \right\| \geqslant \theta \quad ,$$

we get :

$$\left\| \frac{x_1 + x_2}{2(1-\delta(\theta))} - \frac{x_3 + x_4}{2(1-\delta(\theta))} \right\| \geqslant \frac{\theta}{1-\delta(\theta)} \geqslant \theta \quad ,$$

which implies

$$\left\| \frac{x_1 + x_2 + x_3 + x_4}{4(1-\delta(\theta))} \right\| \leqslant 1 - \delta(\theta) \ ,$$

or

$$\left\| \frac{x_1 + x_2 + x_3 + x_4}{4} \right\| \leqslant (1 - \delta(\theta))^2 .$$

Using this procedure again, we get

$$\left\| \frac{x_1 + \ldots + x_{2^{n-1}}}{2^{n-1}} \right\| \leqslant (1 - \delta(\theta))^{n-1}$$

and also :

$$\left\| \frac{x_{2^{n-1}+1} + \dots + x_{2^n}}{2^{n-1}} \right\| \leqslant (1 - \delta(\theta))^{n-1}$$

and therefore :

$$\theta \leqslant \left\| \frac{x_1 + \dots + x_{2^{n-1}}}{2^{n-1}} - \frac{x_{2^{n-1}+1} + \dots + x_{2^n}}{2^{n-1}} \right\| \leqslant 2(1 - \delta(\theta))^{n-1} .$$

But $\delta(\theta)$ is strictly positive, and this inequality is impossible for n large enough.

In a uniformly convex space, weak convergence plus convergence of norms imply strong convergence. Indeed, just repeating the arguments of the first proof of the previous proposition, we obtain immediately :

PROPOSITION 7. - *Let* $(x_i)_{i \in I}$ *be a family of elements of* E *, uniformly convex space. Let* $\mathscr{F}$ *be a filter on* I *. We assume that* $(x_i)_{i \in I}$ *converges weakly to an element* $x \in E$ *, and that* $(\|x_i\|)_{i \in I}$ *converges to* $\|x\|$ *. Then* $(x_i)_{i \in I}$ *converges to* x *in norm.*

We shall say that a Banach space is uniformly convexifiable if it is isomorphic to a uniformly convex space, that is, if it can be endowed with an equivalent uniformly convex norm. The previous proposition says that a uniformly convexifiable space must be reflexive. But reflexivity is not sufficient : we shall investigate this question in detail in the fourth part. We shall now give some examples of uniformly convex spaces.

PROPOSITION 8. - *If* $1 < p < +\infty$ *, the spaces* $L_p\,(\Omega, \mathscr{A}, \mu)$ *are uniformly convex* (μ *being any measure on* $(\Omega, \mathscr{A})$).

PROOF. - We shall distinguish two cases : $p < 2$ or $p \geqslant 2$.

1°) $p \geqslant 2$.

We shall use the following lemma, which is a consequence of a Clarkson's Inequality :

LEMMA 9. - *If* $p \geqslant 2$ *, for all* $a, b \in \mathbb{C}$ *, one has :*

$$|a + b|^p + |a - b|^p \leqslant 2^{p-1}(|a|^p + |b|^p) .$$

PROOF. - We can write

$$(|a+b|^p + |a-b|^p)^{1/p} \leqslant (|a+b|^2 + |a-b|^2)^{1/2} = \sqrt{2}(|a|^2 + |b|^2)^{1/2} .$$

But Hölder's Inequality, for $\frac{2}{p} + \frac{p-2}{p} = 1$, gives :

$$|a|^2 + |b|^2 \leqslant (|a|^p + |b|^p)^{2/p} (1+1)^{p-2/p} = (|a|^p + |b|^p)^{2/p} 2^{p-2/p} ,$$

from which the lemma follows.

Let now $f,g \in L_p(\Omega,\mu)$. For every $\omega \in \Omega$, we can write, using the lemma :

$$|f(\omega) + g(\omega)|^p + |f(\omega) - g(\omega)|^p \leqslant 2^{p-1}(|f(\omega)|^p + |g(\omega)|^p)$$

and, integrating each member, we get :

$$\|f + g\|_p^p + \|f - g\|_p^p \leqslant 2^{p-1}(\|f\|_p^p + \|g\|_p^p)$$

Consequently, if $\|f\|_p \leqslant 1$, $\|g\|_p \leqslant 1$, and $\|f - g\|_p \geqslant \epsilon$, we obtain

$$\|f + g\|_p^p \leqslant 2^p - \epsilon^p ,$$

that is

$$\left\| \frac{f+g}{2} \right\|_p^p \leqslant 1 - \left(\frac{\epsilon}{2}\right)^p ,$$

which gives

$$\delta_{L_p}(\epsilon) \geqslant 1 - \left(1 - \left(\frac{\epsilon}{2}\right)^p\right)^{1/p} ,$$

and this last quantity is equivalent to $\frac{1}{p}\left(\frac{\epsilon}{2}\right)^p$ when $\epsilon \to 0$.

It is easy to see that one has in fact $\delta(\epsilon) = 1 - \left[1 - \left(\frac{\epsilon}{2}\right)^p\right]^{1/p}$: if $\Omega = [0,1]$, take

$f(t) = 1$ for all t ,

$g(t) = 1$ if $0 \leqslant t \leqslant 1 - \left(\frac{\epsilon}{2}\right)^p$

and -1 if $1 - (\frac{\epsilon}{2})^p < t \leqslant 1$.

Then $\|f\|_p = \|g\|_p = 1$, $\|f - g\| = \epsilon$, $\left\| \frac{f+g}{2} \right\|_p = (1 - (\frac{\epsilon}{2})^p)^{1/p}$: the estimate given by the previous computation is the best possible.

2°) $1 < p \leqslant 2$.

We first need several lemmas :

LEMMA 10. - *Let* $q \geqslant 2$. *Let* $u_1, \dots, u_n$ *a finite sequence of elements in* L_q . *Then :*

$$\left\| \left(\sum_{j=1}^{n} |u_j|^2 \right)^{1/2} \right\|_q \leqslant \left(\sum_{j=1}^{n} \|u_j\|_q^2 \right)^{1/2} .$$

PROOF. - For any $r \geqslant 1$, we put :

$$M^{(r)} = \sup \left\{ \left\| \left(\sum_j |v_j|^r \right)^{1/r} \right\|_q \; ; \; (v_j) \text{ finite sequence of elements in } L_q \text{, with } \left(\sum \|v_j\|_q^r \right)^{1/r} = 1 \right\} .$$

Putting $v_j = a_j^{1/r} w_j$, with $a_j \geqslant 0$, $\|w_j\|_q = 1$, $\sum_j a_j = 1$, we obtain :

$$M^{(r)} = \sup \left\{ \left\| \left(\sum_j a_j |v_j|^r \right)^{1/r} \right\|_q \; ; \; (v_j) \text{ finite sequence in } L_q \text{, with } \|v_j\| = 1 \text{, } a_j \geqslant 0 \text{, } \sum a_j = 1 \right\} .$$

We shall show that $M^{(r)}$ is increasing.

Let $r \leqslant r'$. Let θ , with $0 < \theta < 1$, such that

$$\frac{1}{r} = \frac{1-\theta}{r'} + \frac{\theta}{1} .$$

Then, by Hölder's Inequality :

$$(\sum_j a_j|v_j|^r)^{1/r} \leqslant \left[(\sum a_j|v_j|^{r'})^{1/r'}\right]^{1-\theta} \left[\sum a_j|v_j|\right]^{\theta} ,$$

and, since (also by Hölder's Inequality) we have

$$\|h_1^{1-\theta} \cdot h_2^{\theta}\|_q \leqslant \|h_1\|_q^{1-\theta} \cdot \|h_2\|_q^{\theta}$$

(for every $h_1,h_2 \in L_q$, every θ , $0 \leqslant \theta \leqslant 1$)

we get :

$$\|(\sum a_j|v_j|^r)^{1/r}\|_q \leqslant \|(\sum a_j|v_j|^{r'})^{1/r'}\|_q^{1-\theta} \cdot \|\sum a_j|v_j|\|_q^{\theta} .$$

We deduce that :

$$M^{(r)} \leqslant (M^{(r')})^{1-\theta} \cdot (M^{(1)})^{\theta} .$$

But $M^{(1)} = \sup\{\|\sum a_j|v_j|\|_q \ ; \ \sum a_j = 1 , \ \|v_j\| = 1 \} = 1$,

and $M^{(r')} \geqslant 1$ (take only v_1 , with $\|v_1\| = 1$, $a_1 = 1$, other a_i's = 0).

So $M^{(r)} \leqslant M^{(r')}$, and our claim is proved.

But $M^{(q)} = 1$, since $\|(\sum_j |v_j|^q)^{1/q}\|_q = (\sum_j \|v_j\|_q^q)^{1/q}$. So, if $q \geqslant 2$, $M^{(2)} \leqslant M^{(q)} = 1$, and

$$\|(\sum_{j=1}^{n} |u_j|^2)^{1/2}\|_q \leqslant (\sum_{j=1}^{n} \|u_j\|_q^2)^{1/2} .$$

This property is known as the "2-convexity" of L_q , for $q \geqslant 2$.

LEMMA 11. - *If* $1 \leqslant p \leqslant 2$, *for every finite sequence* (f_j) *of elements in* L_p , *we have*

$$(\sum \|f_j\|_p^2)^{1/2} \leqslant \|(\sum_j |f_j|^2)^{1/2}\|_p .$$

This property is known as the "2-concavity" of L_p , $p \leqslant 2$.

PROOF. - Let $f_1,\dots,f_n \in L_p$. Since the dual of $(\prod_{i=1}^{n} E_i)_2$, with $E_i = L_p$ for $i = 1,\dots,n$, is isometric to $(\prod_{i=1}^{n} E_i^\star)_2$ (first part, chap. III, exercise 3), we have :

$$\left(\sum_1^n \|f_i\|_p^2\right)^{1/2} = \sup\left\{ \left|\int \sum_{i=1}^n f_i u_i \, d\mu\right| \ ; \ \sum_1^n \|u_i\|_q^2 \leqslant 1 \right\}$$

$$\leqslant \sup\left\{ \left|\int \Sigma f_i u_i \, d\mu\right| \ ; \ \left\|\left(\sum_{i=1}^n |u_i|^2\right)^{1/2}\right\|_q \leqslant 1 \right\} .$$

(by lemma 10)

$$\leqslant \sup\left\{ \int (\Sigma |f_i|^2)^{1/2} (\Sigma |u_i|^2)^{1/2} d\mu \ ; \ \left\|\left(\sum_{i=1}^n |u_i|^2\right)^{1/2}\right\|_q \leqslant 1 \right\}$$

$$\leqslant \left\|\left(\Sigma |f_i|^2\right)^{1/2}\right\|_p ,$$

and the lemma is proved.

In particular, we obtain, for any couple f_1, f_2 of elements in L_p, $p \leqslant 2$:

$$\left(\|f_1\|_p^2 + \|f_2\|_p^2\right)^{1/2} \leqslant \left\|\left(|f_1|^2 + |f_2|^2\right)^{1/2}\right\|_p .$$

LEMMA 12. - *Let* $p,q \in \mathbb{R}$, *with* $1 < p < +\infty$, $q \geqslant 2$. *There exists a constant* C, *depending on* p, q, *such that, for every* $s,t \in \mathbb{R}$:

$$\left(\left|\frac{s-t}{C}\right|^q + \left|\frac{s+t}{2}\right|^q \right)^{1/q} \leqslant \left(\frac{|s|^p + |t|^p}{2} \right)^{1/p} .$$

PROOF. - We may assume $s = 1$, $-1 \leqslant t < 1$. We consider the function $\varphi(t) = \left(\frac{1+|t|^p}{2}\right)^{q/p} - \left(\frac{1+t}{2}\right)^q$. This function is positive on $[-1, 1[$, and satisfies $\varphi''(1) > 0$, $\varphi(1) = \varphi'(1) = 0$. Therefore, $\frac{\varphi(t)}{(1-t)^2}$, and a fortiori $\frac{\varphi(t)}{(1-t)^q}$, are bounded from below on $[-1,+1[$. The lemma follows.

We can now prove the uniform convexity of L_p, $1 < p \leqslant 2$.

Let $f,g \in L_p(\Omega, \mu)$, with $\|f\|_p = \|g\|_p = 1$, $\|f - g\|_p \geqslant \epsilon$. Using lemma 12, with $q = 2$, we can write, for every $\omega \in \Omega$:

$$\left(\left| \frac{f(\omega) - g(\omega)}{C} \right|^2 + \left| \frac{f(\omega) + g(\omega)}{2} \right|^2 \right)^{1/2} \leq \left(\frac{|f(\omega)|^p + |g(\omega)|^p}{2} \right)^{1/p} .$$

Therefore :

$$\left\| \left(\left| \frac{f-g}{C} \right|^2 + \left| \frac{f+g}{2} \right|^2 \right)^{1/2} \right\|_p \leq \left(\frac{\|f\|_p^p + \|g\|_p^p}{2} \right)^{1/2} = 1 .$$

But we have seen that

$$\left\| \left(\left| \frac{f-g}{C} \right|^2 + \left| \frac{f+g}{2} \right|^2 \right)^{1/2} \right\|_p \geq \left(\left\| \frac{f-g}{C} \right\|_p^2 + \left\| \frac{f+g}{2} \right\|_p^2 \right)^{1/2} .$$

So we get

$$\left\| \frac{f-g}{C} \right\|_p^2 + \left\| \frac{f+g}{2} \right\|_p^2 \leq 1 ,$$

and

$$\left\| \frac{f+g}{2} \right\|_p \leq 1 - \frac{\epsilon^2}{C} ,$$

which means that $L_p(\Omega, \mu)$ is uniformly convex, with $\delta_{L_p}(\epsilon) \geq \frac{\epsilon^2}{C}$.

The constant C has not been computed (it could be, from lemma 12), but the exponent 2, for ϵ^2 , is certainly the best possible : we have seen (second part, chapter VI) that $L_p[0,1]$ contained subspaces isomorphic to ℓ_2 , and we know that $\delta_{\ell_2}(\epsilon) = 1 - \sqrt{\frac{1-\epsilon^2}{4}}$.

Let us now come back to our example of $(\prod_{n \in \mathbb{N}} E_n)_p$, $1 < p \leq +\infty$. One shows easily (see exercise 4) that if the E_n's are uniformly convex, and if, for every $\epsilon > 0$, $\inf_{n \in \mathbb{N}} \delta_{E_n}(\epsilon) > 0$ (that is to say : the moduli of E_n's are bounded from below), then $(\prod_{n \in \mathbb{N}} E_n)_p$ is also uniformly convex. Conversely, if $(\prod_{n \in \mathbb{N}} E_n)_p = E$ is uniformly convex, then each E_n must be, and must verify :

$$\delta_{E_n}(\epsilon) \geq \delta_E(\epsilon) \qquad \text{for every} \quad \epsilon > 0 ,$$

since each E_n is isometric to a subspace of $(\prod_{n \in \mathbb{N}} E_n)_p$.

So if we take $E_n = \ell_{p_n}$ $1 < p_n < +\infty$, $p_n \xrightarrow[n\to+\infty]{} +\infty$ or $p_n \xrightarrow[n\to+\infty]{} 1$, then $(\prod_{n\in\mathbb{N}} E_n)_{n\in\mathbb{N}}$ is not uniformly convex. But this space is strictly convex, and is reflexive. (See first part, chap. III, exercise 3).

We shall now study the dual property of uniform convexity :

§ 2. UNIFORM SMOOTHNESS AND UNIFORM FRECHET DIFFERENTIABILITY.

Let E be a Banach space. We set, for $\tau > 0$:

$$\rho_E(\tau) = \sup\{\tfrac{1}{2}(\|x + y\| + \|x - y\|) - 1 \ ; \quad \|x\| = 1 \ , \quad \|y\| \leq \tau\}$$

The function $\rho_E(\tau)$ will be called *modulus of smoothness* of E ; we shall say that E is *uniformly smooth* if $\rho_E(\tau)/_\tau \xrightarrow[\tau\to 0]{} 0$.

The meaning of the modulus $\rho_E(\tau)$ and of the condition $\rho_E(\tau)/_\tau \xrightarrow[\tau\to 0]{} 0$ is not as clear as the meaning of uniform convexity.

In order to get acquainted with it, we shall first prove that a uniformly smooth space is smooth, which is a good thing for the terminology.

Suppose that E is not smooth, that is, for some $x_0 \in E$ with $\|x_0\| = 1$, there are two linear functionals f, g , $f \neq g$, $\|f\| = \|g\| = 1$, with $f(x_0) = g(x_0) = 1$. Take $y_0 \in E$, with $\|y_0\| = 1$, $y_0 \in \operatorname{Ker} \frac{f+g}{2}$, $y_0 \notin \ker f$. Then, for every $\tau > 0$:

$$\tfrac{1}{2}(\|x_0 + \tau y_0\| + \|x_0 - \tau y_0\|) - 1 \geq \tfrac{1}{2}\,[\,(f(x_0) + \tau f(y_0)) + (g(x_0) - \tau g(y_0))\,] - 1$$

$$\geq \tfrac{\tau}{2}(f(y_0) - g(y_0))$$

$$\geq \tau f(y_0) \ , \quad \text{since} \quad g(y_0) = -f(y_0) \ .$$

This implies that

$\rho_E(\tau)/_\tau \geq f(y_0)$, and $\rho_E(\tau)/_\tau$ does not tend to zero when $\tau \to 0$,

and E is not uniformly smooth.

If E is uniformly smooth, we can write for all $t \in \mathbb{R}$, all $x,y \in E$ with $\|x\| = 1$, $\|y\| \leqslant 1$:

$$(1) \qquad \frac{1}{2}(\|x + ty\| + \|x - ty\|) - 1 \leqslant \rho(|t|) , \qquad \text{with } \frac{\rho(|t|)}{|t|} \xrightarrow[t \to 0]{} 0 .$$

If we put $\rho(\tau) = \tau\epsilon(\tau)$, we obtain :

$$(2) \qquad \|x + ty\| + \|x - ty\| \leqslant 2 + 2|t|\epsilon(|t|) \qquad \epsilon(|t|) \xrightarrow[t \to 0]{} 0$$

Like smoothness, uniform smoothness can be characterized by a differentiability property of the norm. Let us give two definitions :

We say that the norm of E is *Fréchet-differentiable* if it is Gâteaux-differentiable, and if, for every $x \neq 0$, with $\|x\| \leqslant 1$,

$$\lim_{t \to 0} \sup_{\substack{h \in E \\ \|h\| \leqslant 1}} \left| \frac{\|x + th\| - \|x\|}{t} - G_x(h) \right| = 0$$

In other words, there exists a function $\epsilon_x(\tau)$, $\epsilon_x(\tau) \xrightarrow[\tau \to 0]{} 0$ such that, for all $h \in E$, with $\|h\| \leqslant 1$,

$$|\|x + th\| - \|x\| - t\, G_x(h)| \leqslant |t|\, \epsilon_x(|t|) .$$

The difference with the Gâteaux-differentiability is that the convergence of $\frac{\|x + th\| - \|x\|}{t}$ to $G_x(h)$ is uniform in h, $\|h\| \leqslant 1$, that is, uniform in all directions. But it depends on the point x.

We say that the norm is *uniformly Fréchet-differentiable* if there is a function $\epsilon(\tau)$, $\epsilon(\tau) \xrightarrow[\tau \to 0]{} 0$, such that, for all $x \in E$, $x \neq 0$, $\|x\| \leqslant 1$, for all $h \in E$, $\|h\| \leqslant 1$,

$$\left| \frac{\|x + th\| - \|x\|}{t} - G_x(h) \right| \leqslant \epsilon(|t|) .$$

In other words, the convergence of $\frac{\|x + th\| - \|x\|}{t}$ to $G_x(h)$ is now uniform in x, $\|x\| \leqslant 1$ and h, $\|h\| \leqslant 1$. This can also be written :

$$|\,\|x + th\| - \|x\| - t\,G_x(h)| \leqslant |t|\epsilon(|t|)\ , \qquad \epsilon(|t|) \xrightarrow[t \to 0]{} 0\ .$$

PROPOSITION 1. - E *is uniformly smooth if and only if its norm is uniformly Fréchet-differentiable.*

PROOF.

a) Let us assume the norm to be uniformly Fréchet-differentiable. We have, if $\|x\| = 1$, $\|h\| \leqslant 1$:

$$|\,\|x + th\| - \|x\| - t\,G_x(h)| \leqslant |t|\epsilon(|t|)$$

$$|\,\|x - th\| - \|x\| + t\,G_x(h)| \leqslant |t|\epsilon(|t|)$$

and consequently

$$0 \leqslant \|x + th\| + \|x - th\| - 2\|x\| \leqslant 2|t|\epsilon(|t|)\ ,$$

from which follows :

$$\sup\{\tfrac{1}{2}(\|x + th\| + \|x - th\|) - 1\ ;\quad \|x\| = 1\ ,\ \|y\| = \|th\| \leqslant \tau\} \leqslant \tau\,\epsilon(\tau)\ ,$$

and $\rho_E(\tau)/_\tau \leqslant \epsilon(\tau) \xrightarrow[\tau \to 0]{} 0$, and E is uniformly smooth.

b) Conversely, let us assume E to be uniformly smooth. First, since uniform smoothness implies smoothness, E has a Gâteaux-differentiable norm, and G_x exists. Moreover we know (from the convexity of $t \to \|x + th\|$) that, for every $t > 0$:

$$\frac{\|x - th\| - \|x\|}{-t} \leqslant G_x(h) \leqslant \frac{\|x + th\| - \|x\|}{t}\ .$$

We have, since E is uniformly smooth :

$$|\,\|x + th\| - \|x\| + \|x - th\| - \|x\|\,| \leqslant |t|\epsilon(|t|)\ ,$$

which implies, for $t \geqslant 0$:

$$\frac{\|x + th\| - \|x\|}{-t} + \frac{\|x - th\| - \|x\|}{-t} \geqslant - \epsilon(t)$$

that is

$$G_x(h) \geqslant \frac{\|x + th\| - \|x\|}{t} - \epsilon(t) ,$$

and

$$\left| G_x(h) - \frac{\|x + th\| - \|x\|}{t} \right| < \epsilon(t) .$$

The same way, we obtain, also for $t > 0$:

$$\left| G_x(h) - \frac{\|x - th\| - \|x\|}{-t} \right| < \epsilon(t)$$

and finally, for all $t \in \mathbb{R}$:

$$\left| G_x(h) - \frac{\|x + th\| - \|x\|}{t} \right| < \epsilon(|t|) ,$$

which means that the norm is uniformly Fréchet-differentiable.

Uniform smoothness is the dual notion of uniform convexity :

PROPOSITION 2. - E *is uniformly convex if and only if* $E^\star$ *is uniformly smooth ;* E *is uniformly smooth if and only if* $E^\star$ *is uniformly convex.*

PROOF. - All we have to do is to establish the following two implications :

a) E uniformly convex implies $E^\star$ uniformly smooth ,

b) E uniformly smooth implies $E^\star$ uniformly convex.

Indeed, b) implies that E is reflexive, and one gets the other implications by applying these two to $E^\star$, $E^{\star\star} = E$ instead of E , $E^\star$.

The implication a) will follow from the following lemma (*Lindenstrauss' duality formula*) :

LEMMA 3. - *For every Banach space* E , *every* $\tau > 0$,

$$\rho_{E^\star}(\tau) = \sup_{0 \leqslant \epsilon \leqslant 2} \left(\frac{\tau\epsilon}{2} - \delta_E(\epsilon) \right) .$$

PROOF OF LEMMA 3. - Let us show first, that for every $\epsilon > 0$ and every $\tau > 0$,

$$(1) \quad \delta_E(\epsilon) + \rho_{E^\star}(\tau) \geqslant \frac{\tau\epsilon}{2} .$$

Let $x,y \in E$, with $\|x\| = \|y\| = 1$, $\|x - y\| \geqslant \epsilon$, and let $f,g \in E^\star$, with $\|f\| = \|g\| = 1$, $f(x + y) = \|x + y\|$, $g(x - y) = \|x - y\|$. Then :

$$\begin{aligned} 2\,\rho_{E^\star}(\tau) &\geqslant \|f + \tau g\| + \|f - \tau g\| - 2 \\ &\geqslant (f + \tau g)(x) + (f - \tau g)(y) - 2 \\ &\geqslant f(x + y) + \tau g(x - y) - 2 \\ &\geqslant \|x + y\| + \tau\epsilon - 2 , \end{aligned}$$

that is

$2 - \|x + y\| \geqslant \tau\epsilon - 2\,\rho_{E^\star}(\tau)$, from which (1) follows.

But (1) implies that

$$\rho_{E^\star}(\tau) \geqslant \sup_{0 \leqslant \epsilon \leqslant 2} \left(\frac{\tau\epsilon}{2} - \delta_E(\epsilon) \right) .$$

Now, let $f,g \in E^\star$ with $\|f\| = 1$, $\|g\| = \tau$. Let $\eta > 0$. We can find $x,y \in E$ with

$$\|x\| = \|y\| = 1 , \qquad (f + g)(x) \geqslant \|f + g\| - \eta$$

$$(f - g)(y) \geqslant \|f - g\| - \eta$$

Therefore, we have :

$$\|f + g\| + \|f - g\| \leqslant f(x) + g(x) + f(y) - g(y) + 2\eta$$

$$\leqslant f(x + y) + g(x - y) + 2\eta$$

$$\leqslant \|x + y\| + \tau\|x - y\| + 2\eta .$$

Set $\epsilon = x - y$. Then $\left\| \frac{x + y}{2} \right\| \leqslant 1 - \delta_E(\epsilon)$, and

$$\|f + g\| + \|f - g\| \leqslant 2 - 2\,\delta_E(\epsilon) + \tau\epsilon + 2\eta$$

$$\leqslant 2 + 2\left(\frac{\tau\epsilon}{2} - \delta_E(\epsilon)\right) + 2\eta .$$

Consequently, we obtain :

$$\|f + g\| + \|f - g\| \leqslant 2 + 2 \sup_{0 \leqslant \epsilon \leqslant 2} \left(\frac{\tau\epsilon}{2} - \delta_E(\epsilon)\right) + 2\eta ,$$

from which follows

$$\rho_{E^\star}(\tau) \leqslant \sup_{0 \leqslant \epsilon \leqslant 2} \left(\frac{\tau\epsilon}{2} - \delta_E(\epsilon)\right) ,$$

and the formula is established. Let us observe that it does not require any hypothesis on E or on $E^\star$.

We shall now show that $E^\star$ is uniformly smooth when E is uniformly convex. For this, we must show that if $\delta_E(\epsilon) > 0$ for every $\epsilon > 0$, then $\rho_{E^\star}(\tau)/\tau \xrightarrow[\tau \to 0]{} 0$.

By the lemma, we have

$$\frac{\rho(\tau)}{\tau} = \sup_{0 \leqslant \epsilon \leqslant 2} \left(\frac{\epsilon}{2} - \frac{\delta(\epsilon)}{\tau} \right) .$$

The second member is a decreasing function of $\tau > 0$

Let us assume that $\lim_{\tau \to 0} \frac{\rho(\tau)}{\tau} = a > 0$. For a sequence of $\tau \to 0$, let ϵ_τ be such that :

$$\frac{\epsilon_\tau}{2} - \frac{\delta(\epsilon_\tau)}{\tau} > \frac{a}{2} .$$

Then

$$\frac{\epsilon_\tau}{2} - \frac{a}{2} > \frac{\delta(\epsilon_\tau)}{\tau} > 0$$

and therefore $\epsilon_\tau > a$, and the sequence ϵ_τ cannot tend to zero. But $\frac{\tau\epsilon_\tau}{2} - \frac{a\tau}{2} > \delta(\epsilon_\tau)$, which implies $\delta(\epsilon_\tau) \xrightarrow[\tau \to 0]{} 0$, and E is not uniformly convex.

Let us observe that the converse implication also follows from the lemma, on an obvious way : if there is an $\epsilon_0 > 0$ such that $\delta(\epsilon_0) = 0$, then $\rho(\tau)/_\tau \geqslant \frac{\epsilon_0}{2}$ for all τ , and $E^\star$ is not uniformly smooth.

For the second part of the proposition, one establishes the other *duality formula* :

$$\rho_E(\tau) = \sup_{0 \leqslant \epsilon \leqslant 2} \left(\frac{\tau\epsilon}{2} - \delta_{E^\star}(\epsilon) \right) ,$$

which is left to the reader.

PROPOSITION 4. - *If* E *is uniformly smooth, the application* $x \to G_x$ *is continuous, from* $E - \{0\}$ *endowed with its norm, into* $E^\star$ *endowed with its norm.*

PROOF. - Let $(x_n)_{n \in \mathbb{N}}$ be a sequence of points in E , converging in norm to a point $x \in E$. We want to show that $G_{x_n} \xrightarrow[n \to +\infty]{} G_x$ in $E^\star$, for the norm. But we know already (chap. I, § 2, prop. 1) that $G_{x_n} \xrightarrow[n \to +\infty]{} G_x$ for $\sigma(E^\star, E)$. We know also (previous proposition) that $E^\star$ is uniformly convex, and we know finally that $\|G_{x_n}\| = \|G_x\| = 1$, for all $n \in \mathbb{N}$. The proposition follows now from § 1, prop. 7.

One may wonder when the mapping $x \to G_x$ is continuous from $E - \{0\}$ endowed with $\sigma(E, E^\star)$ into $E^\star$, $\sigma(E^\star, E)$. Of course, E has to be smooth, in order that this mapping can be defined. It is continuous, with these topologies, if E is a Hilbert space, or, if $E = \ell_p$, $1 < p < +\infty$, since, in this last case, the norming functional f_x is computed from x coordinate-wise. But this is not the case in $L_p([0,1], dt)$.

EXERCISES ON CHAPTER II.

EXERCISE 1. - Let E be a Banach space, $\|\cdot\|$ its norm. We assume that there is a norm $[\![\cdot]\!]$, defined on E, satisfying $[\![\cdot]\!] \leqslant C \|\cdot\|$, and uniformly convex. Let p with $1 < p < +\infty$. We set

$$\|\cdot\|_1 = (\|\cdot\|^p + [\![\cdot]\!]^p)^{1/p} .$$

Show that $\|\cdot\|_1$ is a uniformly convex norm, which is equivalent to the original norm of E.

EXERCISE 2. - Show that on the unit sphere of a uniformly convex space, the topologies of the norm and $\sigma(E, E^\star)$ coïncide.

EXERCISE 3. - Let E be a Banach space, and $x_1,\dots,x_n$ a finite number of points in E. We define, for all $z \in E$:

$$\varphi(z) = \frac{1}{n} \sum_1^n \|z - x_i\|^2 .$$

a) Show that φ is a convex function and that, if E is reflexive, φ attaints its minimum.

b) Show that if E is reflexive and strictly convex, there is a unique point, called σ, at which φ attains its minimum.

c) Show that if E is uniformly convex, φ increases quickly in a neighbourhood of σ, that is, for all $z \in E$,

$$\varphi(z) \geqslant \varphi(\sigma) + \|z - \sigma\|^2 \delta_E(\|z - \sigma\|) .$$

EXERCISE 4. - Let $(E_n)_{n\in\mathbb{N}}$ be a sequence of uniformly convex spaces with, for every $\epsilon > 0$, $\inf_n \delta_{E_n}(\epsilon) = \delta(\epsilon) > 0$.

Show that, for every p , $1 < p < +\infty$, $(\prod_{n\in\mathbb{N}} E_n)_p$ is uniformly convex. [Take $x = (x_n)_{n\in\mathbb{N}}$, $y = (y_n)_{n\in\mathbb{N}}$ with

$$\sum_n \|x_n\|_{E_n}^p = 1 \ , \quad \sum_n \|y_n\|_{E_n}^p = 1 \ , \quad \sum_n \|x_n - y_n\|_{E_n}^p = \epsilon^p \ .$$

Put

$$M = \left\{ n \in \mathbb{N} \ , \quad \|x_n - y_n\|_{E_n}^p \geq \frac{\epsilon^p}{4} (\|x_n\|_{E_n}^p + \|y_n\|_{E_n}^p) \right\}.$$

Show that

$$\sum_{n\in M} (\|x_n\|_{E_n}^p + \|y_n\|_{E_n}^p) \geq \frac{\epsilon^p}{2^{p+1}} \ ,$$ and use prop. 1, § 1].

EXERCISE 5. - Let E be a Banach space. For every C , bounded closed convex subset of E , we call $D(C)$ the diameter of C , that is :

$$D(C) = \sup \{\|x - y\| \ ; \ x,y \in C\} \ ,$$

and for each $x \in C$, we put

$$r_x(C) = \sup \{\|x - y\| \ ; \ y \in C\} \ .$$

We say that E has a normal structure if, for every C , containing more than one point, there is a $x \in C$, with $r_x(C) < D(C)$.

Show that if E is uniformly convex, it has a normal structure.

EXERCISE 6. - Let E be a reflexive Banach space, and C be a bounded closed convex subset of E . Let T be a (non-linear) contraction, from C into C , that is an application satisfying

$$\|Tx - Ty\| \leq \|x - y\| \qquad \text{for all } x,y \in C$$ (see exercise 7, p. 77).

1°) Using Zorn's Axiom, show that there is $C_1 \subset C$ with $TC_1 \subset C_1$, which is minimal, that is, if $C_2 \subset C_1$ satisfies $TC_2 \subset C_2$, then $C_2 = C_1$.

2°) Show that $\overline{\text{conv}}\, TC_1 = C_1$.

3°) Let $\varphi : C_1 \to \mathbb{R}_+$ a convex, lower semi-continuous function, such that $\varphi(Tx) \leqslant \varphi(x)$ for all $x \in C_1$. Show that φ is constant.

(Show that φ reaches its minimum, and consider the set of points at which this minimum is reached).

4°) Show that $r_x(C_1)$, defined in the previous exercise, satisfies

$$r_x(C_1) = D(C_1) \qquad \text{for all } x \in C_1 .$$

5°) Show that if E has normal structure (see previous exercise), C_1 is reduced to a single point $\{a\}$, and that $Ta = a$.

6°) Deduce from the previous exercise that, in every uniformly convex space, any contraction T defined on a closed convex bounded set C has a fixed point.

EXERCISE 7. - On ℓ_2 , consider the norm (for $\lambda \geqslant 0$) :

$$\|x\|_{(\lambda)} = \max(\|x\|_2 , \lambda\|x\|_\infty) , \qquad \text{for } x = (x_n)_{n\in\mathbb{N}} \in \ell_2 .$$

1) Show that $\|\cdot\|_{(\lambda)}$ is equivalent to the ℓ_2-norm.

2) Show that ℓ_2 , equipped with $\|\cdot\|_{(\lambda)}$, has a normal structure if $\lambda < \sqrt{2}$.

3) Take $\lambda = \sqrt{2}$. Considering

$$C = \{x = (x_n)_{n\in\mathbb{N}} \in \ell^2 , \quad x_n \geqslant 0 \quad \forall n , \quad \Sigma\, x_n^2 \leqslant 1\} ,$$

show that ℓ_2 equipped with $\|\cdot\|_{(\lambda)}$ does not have a normal structure.

REFERENCES ON CHAPTER II.

The homogeneous characterization of uniform convexity (statement of prop. 1, § 1) is given in KÖTHE [31] but his proof does not allow to compare δ and δ_p . Our proof establishes that they are of the same order ; as far as we know, it is new. The computation showing the uniform convexity of L_p-spaces is taken from G. KÖTHE [31] for $p \geq 2$, and from LINDENSTRAUSS-TZAFRIRI [34] for $1 < p \leq 2$. The first proof of the fact that every uniformly convex space is reflexive, and the Lindenstrauss' duality formula, can be found in LINDENSTRAUSS-TZAFRIRI [34].

Exercise 3 is inspired by an idea introduced by B. MAUREY and the author [9].

PART 4

THE GEOMETRY OF SUPER-REFLEXIVE BANACH SPACES

The general idea in this fourth and last part is to investigate the following question : under what conditions does a Banach space admit an equivalent uniformly convex norm ? Such spaces are called uniformly convexifiable. We give several necessary and sufficient conditions, of geometric nature, for a Banach space to be uniformly convexifiable. Finally, we make a systematic investigation of this class of Banach spaces.

CHAPTER I

FINITE REPRESENTABILITY AND SUPER-PROPERTIES OF BANACH SPACES

§ 1. FINITE-REPRESENTABILITY AND ULTRAPOWERS OF BANACH SPACES.

Let E and F be two Banach spaces. We shall say that F is *finitely representable* in E (in short : F f.r. E) if :

For every $\epsilon > 0$, every finite-dimensional subspace F° of F , there exists a finite-dimensional subspace E° of E , with $\dim F^\circ = \dim E^\circ$, and an isomorphism T , from F° onto E° , with

$$(1) \qquad \|T\| \cdot \|T^{-1}\| \leq 1 + \epsilon \ .$$

This definition means that the finite-dimensional subspaces of F can be found in E , with an approximation as good as one wants. But of course, E may be much larger.

Condition (1) is given under this form, because it allows the replacement of T by λT , $\lambda \in \mathbb{K}$, without being modified. But if we find, for all $\eta > 0$, a T with :

$$(2) \qquad (1 - \eta)\|x\| \leq \|Tx\| \leq (1 + \eta)\|x\| \qquad , \text{ for all } x \in F^\circ ,$$

then $\|T\| \cdot \|T^{-1}\| \leq \frac{1 + \eta}{1 - \eta}$, and F will be finitely representable in E .

Of course, E is f.r. in itself, and every subspace of E is also f.r. in E .

One checks easily that the relation "f.r." is transitive : if G f.r. F, F f.r. E , then G f.r. E .

To illustrate our definition, we shall now treat an example.

EXAMPLE. - Let $E = (\prod_{n\geq 1} \ell_1^{(n)})_2$ (see second part, chap. III, for the definition of this space). Then ℓ_1 is finitely representable in E.

PROOF. - Let $\epsilon > 0$, and $F°$ be a finite-dimensional subspace of ℓ_1. Let $f_1,\dots,f_m$ $(m = \dim F°)$ be an algebraic basis of $F°$, with $\|f_i\|_1 = 1$, $i = 1,\dots,m$. Let $\delta > 0$ be the constant such that

$$\delta \sum_1^m |a_i| \leq \Big\| \sum_1^m a_i f_i \Big\|_1 ,$$

for every sequence of scalars $a_1,\dots,a_m$.

For $i = 1,\dots,m$, choose $f_i' \in \ell_1$, with only finitely many non-zero coefficients, such that $\|f_i - f_i'\|_1 < \alpha$, where α satisfies

$$\frac{1 + \frac{\alpha}{\delta}}{1 - \frac{\alpha}{\delta}} < 1 + \epsilon .$$

Then we obtain :

$$\Big\| \sum_1^m a_i (f_i - f_i') \Big\|_1 \leq \Big(\sum_1^m |a_i| \Big)\alpha$$

and thus

$$\Big\| \sum_1^m a_i f_i' \Big\|_1 \leq \Big\| \sum_1^m a_i f_i \Big\|_1 + \alpha \sum_1^m |a_i| \leq (1 + \tfrac{\alpha}{\delta}) \Big\| \sum_1^m a_i f_i \Big\|_1$$

and also

$$\Big\| \sum_1^m a_i f_i' \Big\|_1 \geq (1 - \tfrac{\alpha}{\delta}) \Big\| \sum_1^m a_i f_i \Big\|_1$$

that is

$$(1 - \tfrac{\alpha}{\delta}) \Big\| \sum_1^m a_i f_i \Big\|_1 \leq \Big\| \sum_1^m a_i f_i' \Big\|_1 \leq \Big\| \sum_1^m a_i f_i \Big\|_1 (1 + \tfrac{\alpha}{\delta})$$

This means that the operator U, from $F°$ onto $\operatorname{span}\{f_1',\dots,f_m'\}$, defined by $Uf_1 = f_1',\dots,Uf_m = f_m'$, satisfies $\|U\| \leq 1 + \frac{\alpha}{\delta}$, $\|U^{-1}\| \leq (1 - \frac{\alpha}{\delta})^{-1}$.

Let n_k be the index of the last non-zero term in f'_k, $k = 1,\dots,m$, and $n = \max_{1 \leq k \leq m} n_k$. Then $\mathrm{span}\{f'_1,\dots,f'_m\}$ is (isometrically) a subspace of $\ell_1^{(n)}$, which, in its turn, is a subset of E. Therefore U, composed with these embeddings, gives the desired isomorphism.

The fact that we used the ℓ_2-norm to define E is of no importance, and any norm on the product $\prod_{n \geq 1} \ell_1^{(n)}$ would lead to the same result. But, with the ℓ_2-norm, or, more generally, with an ℓ_p-norm, $(1 < p < +\infty)$, E is reflexive (first part, chap. III, exercise 3). We obtain, therefore, an example of a non reflexive space (here, ℓ_1), which is finitely representable in a reflexive one. Reflexivity is not preserved under finite-representability.

For uniform convexity, the situation is different :

PROPOSITION 1. - *If* E *is uniformly convex and* F *finitely representable in* E, *then* F *is uniformly convex.*

PROOF. - Let $x,y \in F$, with $\|x\| = \|y\| = 1$, $\|x - y\| \geq \epsilon$. For each $\eta > 0$, let $E°$ be a two-dimensional subspace of E and T be an isomorphism from $\mathrm{span}\{x,y\}$ onto $E°$, with $\|T^{-1}\| \leq 1$, $\|T\| \leq 1 + \eta$. Put

$$x' = Tx , \quad y' = Ty .$$

Then :

$$\|x'\| \leq 1 + \eta , \quad \|y'\| \leq 1 + \eta$$

$$\|x' - y'\| = \|T(x - y)\| \geq \|x - y\| \geq \epsilon .$$

Since E is uniformly convex, we obtain :

$$\left\| \frac{1}{2}\left(\frac{x'}{1+\eta} + \frac{y'}{1+\eta}\right) \right\| \leq 1 - \delta_E\left(\frac{\epsilon}{1+\eta}\right)$$

or

$$\left\| \frac{x' + y'}{2} \right\| \leq (1 + \eta)\left(1 - \delta_E\left(\frac{\epsilon}{1+\eta}\right)\right) .$$

Finally, for every $\eta > 0$,

$$\left\| \frac{x + y}{2} \right\| \leqslant \left\| \frac{x' + y'}{2} \right\| \leqslant (1 + \eta)(1 - \delta_E(\frac{\epsilon}{1 + \eta})) ,$$

and this proves that F is uniformly convex, and that

$$\delta_F(\epsilon) \geqslant \lim_{\eta \to 0} \delta_E(\frac{\epsilon}{1 + \eta}) \geqslant \delta_E(\frac{\epsilon}{2}) .$$

(The previous limit is not necessarily equal to $\delta_E(\epsilon)$, since δ_E needs not be continuous. But it is certainly greater than $\delta_E(\frac{\epsilon}{1 + \alpha})$, for every $\alpha > 0$). Our proposition is proved.

We shall now give a generic way of producing Banach spaces which are finitely representable in a given Banach space E . The key for this construction is the notion of Ultrapower of a Banach space.

Let $\mathcal{U}$ be a non-trivial ultrafilter on $\mathbb{N}$ (recall that an ultra-filter is said to be *trivial* if it consists in all subsets of $\mathbb{N}$ which contain a given integer k_0).

We consider the product $E^{\mathbb{N}}$ and, in it, the subspace $\mathcal{F}$ of bounded sequences :

$$\mathcal{F} = \{\bar{x} \in E^{\mathbb{N}} , \bar{x} = (x_n)_{n \in \mathbb{N}} ; \sup_{n \in \mathbb{N}} \|x_n\| < + \infty \} .$$

On this subspace, the application

$$\bar{x} = (x_n)_{n \in \mathbb{N}} \longrightarrow \operatorname{Lim}_{\mathcal{U}} \|x_n\|$$

defines a semi-norm.

(If the ultrafilter was trivial, it would just be $(x_n)_{n \in \mathbb{N}} \to \|x_{k_0}\|$).

The kernel of this semi-norm is the subspace $\mathcal{N}$:

$$\mathcal{N} = \{\bar{x} \in E^{\mathbb{N}} , \bar{x} = (x_n)_{n \in \mathbb{N}} ; \operatorname{Lim}_{\mathcal{U}} \|x_n\| = 0\} .$$

We consider the quotient $\mathcal{F}/\mathcal{N}$, which we call $E^{\mathbb{N}}/\mathcal{U}$ or, more simply, $\tilde{E}$.

On this quotient, the application

$$\tilde{x} \to \operatorname{Lim}_{\mathcal{U}} \|x_n\| ,$$

if $\tilde{x}$ is the class of $(x_n)_{n \in \mathbb{N}}$, is a norm.

We recall that (by definition of a quotient), if $(x_n)_{n \in \mathbb{N}}$, $(x'_n)_{n \in \mathbb{N}}$ are two representants of a class $\tilde{x} \in \tilde{E}$, then $\operatorname{Lim}_{\mathcal{U}} \|x_n - x'_n\| = 0$, and $\operatorname{Lim}_{\mathcal{U}} \|x_n\| = \operatorname{Lim}_{\mathcal{U}} \|x'_n\|$ (the first fact implies the second !).

If E is finite dimensional, the closed balls are compact, and for every bounded sequence in E, $(x_n)_{n \in \mathbb{N}}$, the limit $\operatorname{Lim}_{\mathcal{U}} x_n$ exists in E, for the norm. If $x = \operatorname{Lim}_{\mathcal{U}} x_n$, then $\|x\| = \operatorname{Lim}_{\mathcal{U}} \|x_n\|$, and $\tilde{E}$ is isometric to E : one gets nothing more by this process when E is finite dimensional.

PROPOSITION 2. - $\tilde{E}$ *is a Banach space.*

PROOF. - Let $(f^{(n)})_{n \in \mathbb{N}}$ be a Cauchy sequence in $\tilde{E}$. In order to show that it converges, it suffices to show that it has a converging subsequence, Therefore, we may assume $\left\| f^{(n)} - f^{(n-1)} \right\|_{\tilde{E}} < \frac{1}{2^n}$, for all $n \geqslant 1$. We put :

$$u^{(1)} = f^{(1)} , \quad u^{(2)} = f^{(2)} - f^{(1)}, \ldots, u^{(n)} = f^{(n)} - f^{(n-1)}, \ldots,$$

then $f^{(n)} = \sum_{j=1}^{n} u^{(j)}$, $\quad \|u^{(n)}\|_{\tilde{E}} < \frac{1}{2^n}$ for all $n \geqslant 1$, and we want to show that the series $\sum_{1}^{\infty} u^{(n)}$ is convergent in $\tilde{E}$.

For every $n \geqslant 1$, let us choose a representant $(u_i^{(n)})_{i \in \mathbb{N}}$ of $u^{(n)}$ (that is, an element in the class $u^{(n)}$). This can be done with $\left\| u_i^{(n)} \right\| < \frac{1}{2^n}$ for all $i \in \mathbb{N}$. Indeed, if $(v_i^{(n)})_{i \in \mathbb{N}}$ is any representant of $u^{(n)}$, we have $\operatorname{Lim}_{\mathcal{U}} \left\| v_i^{(n)} \right\| < \frac{1}{2^n}$, and we choose $u_i^{(n)} = v_i^{(n)}$ if $\left\| v_i^{(n)} \right\| < \frac{1}{2^n}$, and $u_i^{(n)} = 0$ otherwise.

The series $\sum_{j=1}^{\infty} u_i^{(j)}$ converges normally in E ; we call $u_i = \sum_{j=1}^{\infty} u_i^{(j)}$ its sum, for all $i \in \mathbb{N}$. Then $u_i \in E$. Let u be the class, in $\widetilde{E}$, of $(u_i)_{i \in \mathbb{N}}$. We shall see that the series $\sum_{n=1}^{\infty} u^{(n)}$ converges to u .

Indeed, we have, for all $i \in \mathbb{N}$, all $n \geqslant 1$:

$$\left\| \sum_{j=1}^{n} u_i^{(j)} - u_i \right\| = \left\| \sum_{j \geqslant n+1} u_i^{(j)} \right\| \leqslant \sum_{j \geqslant n+1} \left\| u_i^{(j)} \right\| \leqslant \frac{1}{2^n} .$$

But $\left(\sum_{j=1}^{n} u_i^{(j)} \right)_{i \in \mathbb{N}}$ is a representant of $\sum_{j=1}^{n} u^{(j)}$. Therefore we obtain :

$$\left\| \sum_{j=1}^{n} u^{(j)} - u \right\|_{\widetilde{E}} \leqslant \frac{1}{2^n} ,$$

and the conclusion follows.

PROPOSITION 3. - $\widetilde{E}$ *contains a subspace isometric to* E .

PROOF. - To every $x \in E$, we can associate the constant sequence $x_n = x$, for all $n \in \mathbb{N}$, and $\dot{x}$ the class of $(x_n)_{n \in \mathbb{N}}$ in $\widetilde{E}$. Then the application $x \to \dot{x}$ is an isometry, since

$$\|\dot{x}\|_{\widetilde{E}} = \lim_{\mathcal{U}} \|x_n\| = \|x\| .$$

PROPOSITION 4. - $\widetilde{E}$ *is finitely representable in* E .

PROOF. - Let $\widetilde{E}°$ be a finite-dimensional subspace of $\widetilde{E}$, ant let $\epsilon > 0$. In $\widetilde{E}°$, we choose an algebraic basis $(z^{(1)},\ldots,z^{(m)})$ $(m = \dim \widetilde{E}°)$, with $\|z^{(1)}\|_{\widetilde{E}} = \ldots = \|z^{(m)}\|_{\widetilde{E}} = 1$.

Since $\widetilde{E}°$ is finite-dimensional, there is a $\delta > 0$ such that for every sequence $a_1,\ldots,a_m$ of scalars :

$$(1) \qquad \delta \sum_1^m |a_j| \leqslant \left\| \sum_1^m a_j \, z^{(j)} \right\|_{\widetilde{E}} \leqslant \sum_1^m |a_j| .$$

The unit sphere of $\ell_1^{(m)}$ is compact, therefore, for every $\eta > 0$, we can find a η-net in this unit sphere, that is, a finite number L of finite sequences $(a_1^{(\ell)},\ldots,a_m^{(\ell)})_{\ell=1\ldots L}$, such that $\sum_{j=1}^m |a_j^{(\ell)}| = 1$ for every $\ell = 1,\ldots,L$, and every sequence $a_1,\ldots,a_m$ with $\sum_1^m |a_j| = 1$ is at distance at most η of one of the $(a_j^{(\ell)})_{1 \leqslant j \leqslant m}$, $\ell = 1,\ldots,L$. In the sequel, we choose η such that $\frac{\delta + 3\eta(1+\eta)}{\delta - 3\eta(1+\eta)} < 1 + \epsilon$. We also assume that the canonical basis of $\ell_1^{(m)}$ is included in the η-net.

For each $j = 1,\ldots,m$, let $(z_i^{(j)})_{i \in \mathbb{N}}$ be a representant of $z^{(j)}$. For each $\ell = 1,\ldots,L$, we have by definition of $\widetilde{E}$:

$$\operatorname*{Lim}_{\mathscr{U}} \left\| \sum_{j=1}^m a_j^{(\ell)} z_i^{(j)} \right\|_E = \left\| \sum_{j=1}^m a_j^{(\ell)} z^{(j)} \right\|_{\widetilde{E}} .$$

Therefore, we can find an index i_o for which, for every $\ell = 1,\ldots,L$:

$$\left| \left\| \sum_{j=1}^m a_j^{(\ell)} z_{i_o}^{(j)} \right\|_E - \left\| \sum_{j=1}^m a_j^{(\ell)} z^{(j)} \right\|_{\widetilde{E}} \right| < \eta$$

and, in particular :

$$\| z_{i_o}^{(j)} \| \leqslant 1 + \eta , \qquad j = 1,\ldots,m .$$

If now $(a_j)_{j=1,\ldots,m}$ is any sequence with $\sum_{j=1}^m |a_j| = 1$, since for some ℓ $(1 \leqslant \ell \leqslant L)$, we have $\sum_{j=1}^m |a_j - a_j^{(\ell)}| < \eta$, we obtain :

$$\left| \left\| \sum_{j=1}^m a_j z_{i_o}^{(j)} \right\|_E - \left\| \sum_{j=1}^m a_j z^{(j)} \right\|_{\widetilde{E}} \right| < \eta + \sum_{j=1}^m |a_j - a_j^{(\ell)}| \, \| z_{i_o}^{(j)} \|_E$$

$$+ \sum_{j=1}^m |a_j - a_j^{(\ell)}| \, \| z^{(j)} \|_{\widetilde{E}}$$

$$< \eta + \eta(1+\eta) + \eta < 3\eta(1+\eta) .$$

Since $\sum_{j=1}^{m} |a_j| = 1$, this can also be written, using (1) :

$$\left| \left\| \sum_{j=1}^{m} a_j\, z_{i_o}^{(j)} \right\|_E - \left\| \sum_{j=1}^{m} a_j\, z^{(j)} \right\|_{\widetilde{E}} \right| < \frac{3\eta(1+\eta)}{\delta} \left\| \sum_{j=1}^{m} a_j\, z^{(j)} \right\|_{\widetilde{E}} ,$$

that is

$$\left(1 - \frac{3\eta(1+\eta)}{\delta}\right) \left\| \sum_{j=1}^{m} a_j\, z^{(j)} \right\|_{\widetilde{E}} \leqslant \left\| \sum_{j=1}^{m} a_j\, z_{i_o}^{(j)} \right\|_E$$

$$\leqslant \left(1 + \frac{3\eta(1+\eta)}{\delta}\right) \left\| \sum_{j=1}^{m} a_j\, z^{(j)} \right\|_{\widetilde{E}} .$$

These inequalities imply that the operator T , from $\widetilde{E}_o$ onto $\mathrm{span}\{z_{i_o}^{(1)},\ldots,z_{i_o}^{(m)}\}$, defined by $Tz^{(1)} = z_{i_o}^{(1)},\ldots,Tz^{(m)} = z_{i_o}^{(m)}$ is an isomorphism, with

$$\|T\| \leqslant 1 + \frac{3\eta(1+\eta)}{\delta} , \qquad \|T^{-1}\| \leqslant \frac{1}{1 - \frac{3\eta(1+\eta)}{\delta}} ,$$

and the conclusion follows.

So we obtain a very important class of Banach spaces which are finitely representable in E . This is a class, since we can consider any ultrafilter on $\mathbb{N}$ (though our notation $\widetilde{E}$ does not make this evident). We shall need for the sequel only this type of ultrapowers, but one may define as well more general ultrapowers $E^I/_{\mathcal{U}}$, where $\mathcal{U}$ is an ultrafilter on the set I . With this more general definition, it can be proved that any space F which is finitely representable in E is a subspace of an ultrapower of E .

PROPOSITION 5. - *If* E *is infinite-dimensional, the ultrapower* $E^{\mathbb{N}}/_{\mathcal{U}}$ *is not separable.*

PROOF. - Let $(x_n)_{n\in\mathbb{N}}$ be a sequence in E with $\|x_n\| = 1$, $\|x_n - x_m\| = 1$ if $n \neq m$, for all $n,m \in \mathbb{N}$ (first part, Chap. II, introduction, prop. 1).

We consider an uncountable family $\mathcal{F}_0$ of subsequences of $(x_n)_{n\in\mathbb{N}}$ such that two distinct elements in $\mathcal{F}_0$ have only finitely many terms in common. So, if y_1 , y_2 are two elements in $\mathcal{F}_0$, $y_1 = (x_{n_k})_{k\in\mathbb{N}}$, $y_2 = (x_{n'_k})_{k\in\mathbb{N}}$ then $\|y_1 - y_2\|_{\widetilde{E}} = \lim_{\mathcal{U}} \|x_{n_k} - x_{n'_k}\| = 1$. Since $\mathcal{F}_0$ is uncountable, this proves that $\widetilde{E}$ is not separable.

§ 2. SUPER-PROPERTIES OF BANACH SPACES ; SUPER-REFLEXIVE SPACES.

If $(\mathcal{P})$ is a property defined for Banach spaces, we say that a Banach space E has the property "*Super* $(\mathcal{P})$ " if every Banach space finitely representable in E has the property $(\mathcal{P})$.

Since E itself is finitely representable in E , this implies that E itself has the property $(\mathcal{P})$. According to what we have seen in the previous paragraph, E has super $(\mathcal{P})$ if and only if all its ultrapowers $E^I/_{\mathcal{U}}$ have the property $(\mathcal{P})$.

According to this definition, a Banach space E will be called *super-reflexive* if every Banach space F finitely representable in E is reflexive. Thus, a super-reflexive space is reflexive, but a reflexive space may not be super-reflexive, as the example of $(\prod_{n\geqslant 1} \ell_1^{(n)})_2$ shows.

Another example of super-property is the fact that ℓ_1 is not finitely representable in E . This is a super-property, because, as we saw, the notion of finite-representability is transitive. This super-property is called B-*convexity* (this could be defined with any space as well as ℓ_1 , but is specially interesting with ℓ_1). Obviously, super-reflexivity implies B-convexity, because ℓ_1 is not reflexive.

Also, let us observe the following thing : if $(\mathcal{P}_1)$ and $(\mathcal{P}_2)$ are two properties such that

$$\text{super}(\mathcal{P}_1) \Rightarrow (\mathcal{P}_2) \Rightarrow (\mathcal{P}_1) ,$$

then super$(\mathcal{P}_1)$ is equivalent to super$(\mathcal{P}_2)$. Indeed, if $(\mathcal{P}_2)$ implies $(\mathcal{P}_1)$, then super$(\mathcal{P}_2)$ implies super$(\mathcal{P}_1)$, and if super$(\mathcal{P}_1)$ implies $(\mathcal{P}_2)$, then super(super$(\mathcal{P}_1)$) = super$(\mathcal{P}_1)$ implies super $(\mathcal{P}_2)$.

As an application of these definitions, let us mention that proposition 1 of the previous paragraph means that super-uniform convexity is just uniform convexity. Since we know that a uniformly convex space is reflexive, we obtain that a uniformly convex space is super-reflexive.

§ 3. THE TREE PROPERTIES.

We shall now define a geometric tool which will be very useful for the study of super-reflexive spaces. This tool is the Tree Property, which was introduced by R.C. James.

1) *The Finite Tree Property.*

Let ϵ , with $0 < \epsilon \leqslant 2$. We say that two points $x_1, x_2 \in E$ form a $(1,\epsilon)$-branch if $\|x_1 - x_2\| \geqslant \epsilon$.

Assume that a $(n - 1, \epsilon)$ - branch $(n \geqslant 2)$ has been defined. We shall say that the 2^n points $x_1, \ldots, x_{2^n}$ form a (n,ϵ)-branch if :

a) $\|x_{2i-1} - x_{2i}\| \geqslant \epsilon$ for $i = 1,\ldots,2^{n-1}$,

b) The middlepoints $\dfrac{x_{2i-1} + x_{2i}}{2}$, $i = 1,\ldots,2^{n-1}$, from a $(n - 1, \epsilon)$-branch.

Geometrically, for $n = 3$, a $(3 , \epsilon)$-branch is of this type :

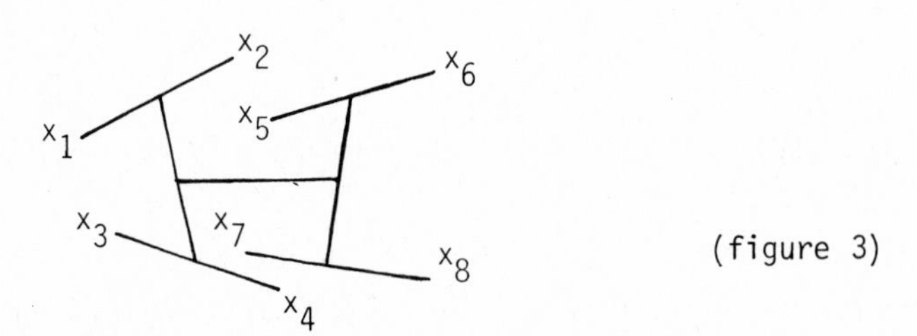

(figure 3)

where every line which appears on the picture represents a length, in norm, at least ϵ .

A different indexation of the points will help understanding the geometric structure.

We start with a point $x_0 \in E$. A $(1 , \epsilon)$-branch is made of two points x_{+1} , x_{-1} , such that $\|x_{+1} - x_{-1}\| \geqslant \epsilon$, and $\dfrac{x_{+1} + x_{-1}}{2} = x_0$.

To obtain a $(2, \epsilon)$-branch, we subdivide x_{+1} into two points $x_{+1,+1}$, $x_{+1,-1}$, with $x_{+1} = \frac{x_{+1,+1} + x_{+1,-1}}{2}$ and $\|x_{+1,+1} - x_{+1,-1}\| \geqslant \epsilon$, and we subdivide x_{-1} into $x_{-1,+1}$, $x_{-1,-1}$, with similar properties. Thus, if $(x_{\epsilon_1,\dots,\epsilon_{n-1}})_{\epsilon_i = \pm 1}$ are the 2^{n-1} points of a $(n-1, \epsilon)$-branch $(\epsilon_1,\dots,\epsilon_{n-1} = \pm 1)$, we obtain a (n, ϵ)-branch by dividing each $x_{\epsilon_1,\dots,\epsilon_{n-1}}$ into $\frac{x_{\epsilon_1,\dots,\epsilon_{n-1},+1} + x_{\epsilon_1,\dots,\epsilon_{n-1},-1}}{2}$, with $\left\|x_{\epsilon_1,\dots,\epsilon_{n-1},+1} - x_{\epsilon_1,\dots,\epsilon_{n-1},-1}\right\| \geqslant \epsilon$.

We now say that E has the *Finite Tree Property* if, for some ϵ, $0 < \epsilon \leqslant 2$, one can find, for every $n \geqslant 1$, a (n, ϵ)-branch contained in the unit ball of E.

EXAMPLE. - The space $\ell_1^{(2^n)}$ contains, in its unit ball, a $(n, 2)$-branch, given by the points of the canonical basis : $e_1,\dots,e_{2^n}$. One checks immediately that $\|e_1 - e_2\|_1 = \|e_3 - e_4\|_1 = \dots = \|e_{2^n-1} - e_{2^n}\|_1 = 2$,

$$\left\|\frac{e_1 + e_2}{2} - \frac{e_3 + e_4}{2}\right\|_1 = \dots = \left\|\frac{e_{2^n-3} + e_{2^n-2}}{2} - \frac{e_{2^n-1} + e_{2^n}}{2}\right\|_1 = 2,$$

and so on.

Therefore, the space $(\prod_{n \geqslant 1} \ell_1^{(n)})_2$ has the finite tree property : for every $n \geqslant 1$, there is a $(n, 2)$-branch in the unit ball of $\ell_1^{(2^n)}$, and therefore in the unit ball of the whole space. But one should observe that the $(n+1, 2)$-branch is not obtained by a division of the $(n, 2)$-branch which is in $\ell_1^{(2^n)}$: one has to go somewhere else in the space (namely in $\ell_1^{(2^{n+1})}$) to find it. The Finite Tree Property does not mean that one starts at some point x_0, and keeps on dividing : it may be necessary to start again somewhere else. On the opposite, we have :

2) A Banach space E is said to have the *Infinite Tree Property* if, for some ϵ, $0 < \epsilon \leqslant 2$, one can find, in the unit ball of E, a family

$(x_{\epsilon_1,\dots,\epsilon_n})_{\substack{n \geq 1 \\ \epsilon_1,\dots,\epsilon_n = \pm 1}}$ such that :

- for all $n \geq 1$, the 2^n points $(x_{\epsilon_1,\dots,\epsilon_n})_{\epsilon_1,\dots,\epsilon_n = \pm 1}$ form a (n, ϵ)-branch ;

- for all $n \geq 1$, $x_{\epsilon_1,\dots,\epsilon_n} = \frac{1}{2}(x_{\epsilon_1,\dots,\epsilon_n,+1} + x_{\epsilon_1,\dots,\epsilon_n,-1})$.

This means that the whole infinite construction is given at once.

EXAMPLE. - The space c_o has the Infinite Tree Property, with $\epsilon = 2$: we take $x_o = 0$, $x_{+1} = e_1$, $x_{-1} = -e_1$, $x_{+1,+1} = e_1 + e_2$, $x_{+1,-1} = e_1 - e_2$, $x_{-1,+1} = -e_1 + e_2$, $x_{-1,-1} = -e_1 - e_2$, and more generally, $x_{\epsilon_1,\dots,\epsilon_n} = \epsilon_1 e_1 + \epsilon_2 e_2 + \dots + \epsilon_n e_n$, where $(e_i)_{i \geq 1}$ is the canonical basis.

A first connexion between the Finite and the Infinite Tree Properties is given by :

PROPOSITION 1. - *If* E *has the Finite Tree Property, then* $\widetilde{E}$ *has the Infinite Tree Property.*

PROOF. - If E has the Finite Tree Property, one can, for some $\epsilon > 0$, find in the unit ball of E, for every $n \geq 1$, points $(x^{(n)}_{\epsilon_1,\dots,\epsilon_n})_{\epsilon_1,\dots,\epsilon_n = \pm 1}$ which form a (n, ϵ)-branch. In $\widetilde{E}$, we shall build a family $(z_{\epsilon_1,\dots,\epsilon_n})_{\substack{n \geq 1 \\ \epsilon_1,\dots,\epsilon_n = \pm 1}}$, forming an infinite tree (for the same ϵ).

These points will be given by their representants $(z^{(i)}_{\epsilon_1,\dots,\epsilon_n})_{i \in \mathbb{N}}$ (now, the index "i" is up ; this is a modification of the notations of § 1). Each $z^{(i)}_{\epsilon_1,\dots,\epsilon_n}$ is defined by :

$$z^{(i)}_{\epsilon_1,\dots,\epsilon_n} = 0 \quad \text{if} \quad i < n .$$

$$z^{(i)}_{\epsilon_1,\dots,\epsilon_n} = \frac{1}{2^{i-n}} \sum_{\substack{\epsilon_{n+1}=\pm 1 \\ \vdots \\ \epsilon_i = \pm 1}} x^{(i)}_{\epsilon_1,\dots,\epsilon_n,\epsilon_{n+1},\dots,\epsilon_i} , \quad \text{if} \quad i \geqslant n ;$$

for each $\epsilon_1,\dots,\epsilon_n = \pm 1$.

To make things simpler, we give explicitly the beginning of this construction :

$$z^{(1)}_{+1} = x^{(1)}_{+1} ,$$

$$z^{(2)}_{+1} = \frac{1}{2} (x^{(2)}_{+1,+1} + x^{(2)}_{+1,-1}) ,$$

$$z^{(3)}_{+1} = \frac{1}{4} (x^{(3)}_{+1,+1,+1} + x^{(3)}_{+1,+1,-1} + x^{(3)}_{+1,-1,+1} + x^{(3)}_{+1,-1,-1,}) ,$$

z_{-1} is defined the same, and for $z_{+1,+1}$, one has :

$$z^{(1)}_{+1,+1} = 0 , \quad z^{(2)}_{+1,+1} = x^{(2)}_{+1,+1} , \quad z^{(3)}_{+1,+1} = \frac{1}{2} (x^{(3)}_{+1,+1,+1} + x^{(3)}_{+1,+1,-1}) ,$$

and so on.

We shall now prove that the family $(z_{\epsilon_1,\dots,\epsilon_n})_{\substack{n \geqslant 1 \\ \epsilon_1,\dots,\epsilon_n = \pm 1}}$ forms an infinite tree, in the unit ball of $\widetilde{E}$. This last fact is obvious : since $\left\|x^{(n)}_{\epsilon_1,\dots,\epsilon_n}\right\|_E \leqslant 1$, for every $n \geqslant 1$, every $\epsilon_1,\dots,\epsilon_n = \pm 1$, we have $\left\|z^{(i)}_{\epsilon_1,\dots,\epsilon_n}\right\|_{\widetilde{E}} \leqslant 1$ for every $i \geqslant 1$, every $n \geqslant 1$, every $\epsilon_1,\dots,\epsilon_n = \pm 1$.

If $i \geqslant n$, we have :

$$\left\| \frac{1}{2^{i-n}} \sum_{\substack{\epsilon_{n+1}=\pm 1 \\ \vdots \\ \epsilon_i = \pm 1}} x^{(i)}_{\epsilon_1,\ldots,\epsilon_{n-1}, +1, \epsilon_{n+1},\ldots,\epsilon_i} - \frac{1}{2^{i-n}} \sum_{\substack{\epsilon_{n+1}=\pm 1 \\ \vdots \\ \epsilon_i = \pm 1}} x^{(i)}_{\epsilon_1,\ldots,\epsilon_{n-1}, -1, \epsilon_{n+1},\ldots,\epsilon_i} \right\| \geqslant \epsilon$$

since this condition appears in the definition of a (i , ϵ)-branch . It implies that, for all $n \geqslant 1$, all $\epsilon_1,\ldots,\epsilon_{n-1} = \pm 1$

$$\left\| z^{(i)}_{\epsilon_1,\ldots,\epsilon_{n-1}, +1} - z^{(i)}_{\epsilon_1,\ldots,\epsilon_{n-1}, -1} \right\| \geqslant \epsilon \qquad \text{for all } i \geqslant n ,$$

and

$$\left\| z_{\epsilon_1,\ldots,\epsilon_{n-1}, -1} - z_{\epsilon_1,\ldots,\epsilon_{n-1}, +1} \right\|_{\widetilde{E}} \geqslant \epsilon \quad .$$

Also, when $i \geqslant n+1$, we have :

$$\frac{1}{2}\left[\frac{1}{2^{i-(n+1)}} \sum_{\substack{\epsilon_{n+2}=\pm 1 \\ \vdots \\ \epsilon_i = \pm 1}} x^{(i)}_{\epsilon_1,\ldots,\epsilon_n, +1, \epsilon_{n+2},\ldots,\epsilon_i} + \frac{1}{2^{i-(n+1)}} \sum_{\substack{\epsilon_{n+2}=\pm 1 \\ \vdots \\ \epsilon_i = \pm 1}} x^{(i)}_{\epsilon_1,\ldots,\epsilon_n, -1, \epsilon_{n+2},\ldots,\epsilon_i} \right]$$

$$= \frac{1}{2^{i-n}} \sum_{\substack{\epsilon_{n+1}=\pm 1 \\ \vdots \\ \epsilon_i = \pm 1}} x^{(i)}_{\epsilon_1,\ldots,\epsilon_n, \epsilon_{n+1},\ldots,\epsilon_i}$$

and therefore

$$\frac{1}{2}\left(z^{(i)}_{\epsilon_1,\ldots,\epsilon_n,+1} + z^{(i)}_{\epsilon_1,\ldots,\epsilon_n,-1}\right) = z^{(i)}_{\epsilon_1,\ldots,\epsilon_n} ,$$

which implies

$$\frac{1}{2}\left(z_{\epsilon_1,\ldots,\epsilon_n,+1} + z_{\epsilon_1,\ldots,\epsilon_n,-1}\right) = z_{\epsilon_1,\ldots,\epsilon_n} ,$$ in $\widetilde{E}$, and our

proposition is proved.

We can now relate super-reflexivity and the Finite Tree Property.

THEOREM 2. (R.C. James). - E *is super-reflexive if and only if it does not have the Finite Tree Property.*

PROOF.

1°) We assume first that E is not super-reflexive. Then, there is a Banach space F , f.r. in E , which is not reflexive. By first part, chap. III, § 1, theorem 6, we can find for some θ , $0 < \theta < 1$, a sequence $(x_k)_{k\in\mathbb{N}}$ of points of norm 1 , in F satisfying condition (J), that is :

$$(1) \quad \mathrm{dist}_F(\mathrm{conv}(x_1,\ldots,x_k), \mathrm{conv}(x_{k+1},\ldots)) > \theta .$$

For every $n \geqslant 1$, let F_n be the subspace of F generated by $x_1,\ldots,x_{2^n}$. Since F is finitely representable in E , we can find a subspace E_n of E $(\dim E_n = \dim F_n)$, and an isomorphism T , from F_n onto E_n , with $\|T\| \leqslant 1$, $\|T^{-1}\| \leqslant 2$. Let $y_1,\ldots,y_{2^n}$ the images of $x_1,\ldots,x_{2^n}$ by T : we have $\|y_i\| \leqslant 1$, $i = 1,\ldots, 2^n$, and, for every $k \leqslant 2^n$,

$$(2) \quad \mathrm{dist}(\mathrm{conv}(y_1,\ldots,y_k), \mathrm{conv}(y_{k+1},\ldots,y_{2^n})) > \frac{\theta}{2} .$$

This last condition implies that the points $(y_1,\ldots,y_{2^n})$ form a $(n, \frac{\theta}{2})$-branch in the unit ball of E : indeed, conditions of the form

$$\left\| \frac{x_1 + \ldots + x_{2^k}}{2^k} - \frac{x_{2^k+1} + \ldots + x_{2^{k+1}}}{2^k} \right\| > \frac{\theta}{2}$$ appear as a consequence of (2).

Instead of $\frac{\theta}{2}$, we might have chosen $\theta(1-\eta)$, for an arbitrary $\eta > 0$. Since θ itself is arbitrary in $]0,1[$. (See th. 6, chap. III, § 1, first part), we obtain :

COROLLARY 3. - *If* E *is not super-reflexive, for every* ϵ, $0 < \epsilon < 1$, *every* $n \geq 1$, *there is a* (n, ϵ)*-branch in the unit ball of* E. *In other terms,* E *has the Finite Tree Property for every* ϵ, $0 < \epsilon < 1$.

2°) Now, we assume that E has the Finite Tree Property. Then any ultrapower $\tilde{E}$ is finitely representable in E (§ 1, prop. 4) and has the Infinite Tree Property (prop. 1). Therefore, to conclude the proof of our theorem, all we have to show is the following

PROPOSITION 4.- *If a Banach space has the Infinite Tree Property, it cannot be reflexive.*

The proof uses a lemma, due to ASPLUND and NAMIOKA [4].

LEMMA 5. - *Let* K *be a convex, separable, weakly compact subset in a Banach space* F. *For every* $\epsilon > 0$, *one can find a closed convex subset* C *of* K, *not equal to* K, *such that the diameter of* $K \setminus C$ *is smaller than* ϵ.

($K \setminus C$ is the complement of C in K ; its diameter is $\sup_{x,y \in K \setminus C} \|x - y\|$).

PROOF OF THE LEMMA. - We call $\mathscr{E}(K)$ the set of extreme points of K (See second part, chap. V), and D the closure of $\mathscr{E}(K)$ in $\sigma(F, F^\star)$. D is a compact set for this topology, therefore a Baire space (first part, chapter I, prop. 1).

Let $\epsilon > 0$. Since K is separable, it may be covered by a countable family of balls B_n, of radius $\frac{\epsilon}{3}$. These balls are also $\sigma(F, F^\star)$-closed, they cover D , and so, for some $n_o \geqslant 1$, the intersection of B_{n_o} with D has non-empty interior (in D , for the weak topology) : there is a set 0 , open for $\sigma(F , F^\star)$, such that $0 \cap D$ is contained in $B_{n_o} \cap D$. Put $U = D \setminus 0$: this is a weakly closed set. Its closed convex hull K_1 is not equal to D , because otherwise all extreme points of K would be in U (Second part, chap. V, § 1, prop. 4), but this is not possible : 0 is open, meets $\overline{D}$, and therefore meets D . Let x_o be an extreme point of K which is not in K_1 .

Let us observe, at this point of the proof, that the set K_1 cannot be taken as the desired C , because nothing says that the diameter of $K \setminus K_1$ is small. This is false in general, as the following picture shows :

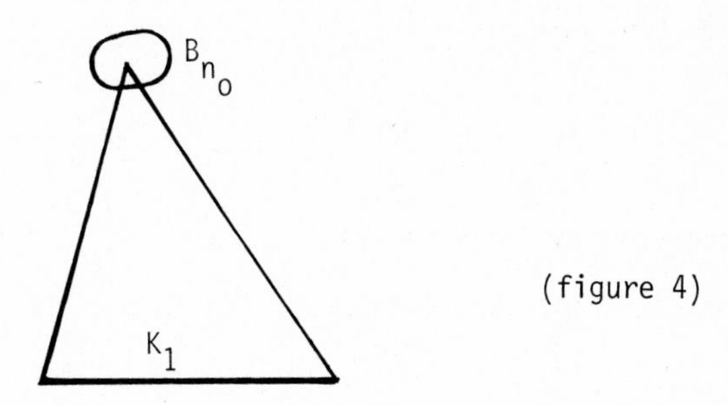

(figure 4)

We put $K_2 = \overline{\text{conv}}(0 \cap D)$. Then $K = \text{conv}(K_1 \cup K_2)$, because every extreme point of K is either in K_1 or in K_2 (recall that $\text{conv}(K_1 \cup K_2)$ is closed, since K_1 and K_2 are compact : it is the image of $K_1 \times K_2 \times [0,1]$ in the application $x \in K_1$, $y \in K_2$, $\lambda \in [0,1] \to \lambda x + (1 - \lambda)y$). Consequently, every point $k \in K$ has a decomposition

$$(1) \quad k = \lambda k_1 + (1 - \lambda)k_2 , \quad 0 \leqslant \lambda \leqslant 1 , \quad k_1 \in K_1 , \quad k_2 \in K_2 .$$

We call A_r the set of points of K which admit a decomposition of the form (1) for some $\lambda \geqslant r$ $(0 \leqslant r \leqslant 1)$.

We shall see that, if r is small enough, A_r is the set C we are looking for.

a) A_r is convex : if $k,k' \in A_r$, then :

$k = \lambda k_1 + (1-\lambda)k_2$, $k' = \lambda' k_1' + (1-\lambda')k_2'$, with $\lambda, \lambda' \in r$, and

$$\frac{k+k'}{2} = \frac{\lambda+\lambda'}{2} \frac{\lambda k_1 + \lambda' k_1'}{\lambda+\lambda'} + \left(1 - \frac{\lambda+\lambda'}{2}\right) \frac{(1-\lambda)k_2 + (1-\lambda')k_2'}{1-\lambda+1-\lambda'}$$

with

$$\frac{\lambda+\lambda'}{2} \geqslant r, \quad \frac{\lambda k_1 + \lambda' k_1'}{\lambda+\lambda'} \in K_1, \quad \frac{(1-\lambda)k_2 + (1-\lambda')k_2'}{1-\lambda+1-\lambda'} \in K_2$$

b) A_r is weakly closed (or strongly : this is the same, since A_r is convex).

Let $k^{(n)}$ be a sequence of points in A_r converging to an element $k \in K$, for the norm. Let

$$k^{(n)} = \lambda_n k_1^{(n)} + (1-\lambda_n)k_2^{(n)}$$

be the corresponding decompositions. Since K_1 and K_2 are weakly compact, by first part, chap. III, § 3, prop. 1 , there is a sequence $(n_j)_{j \in \mathbb{N}}$ for which $k_1^{(n_j)} \xrightarrow[j \to +\infty]{} k_1$, $k_2^{(n_j)} \xrightarrow[j \to +\infty]{} k_2$, $\lambda_{n_j} \to \lambda$, and $k = \lambda k_1 + (1-\lambda)k_2$, with $\lambda \geqslant r$ (in this proof, one could as well use filters, and avoid any appeal to the mentioned proposition).

c) A_r is not equal to K :

Indeed, the point x_0 cannot be in A_r : its only decomposition is $x_0 = 0 \cdot k_1 + x_0$.

d) If $M = \sup\{\|x\|,\ x \in K\}$, then the diameter of $K \setminus A_r$ is smaller than ϵ, if $r \leqslant \frac{\epsilon}{9M}$.

For this, we observe first that, since K_2 is the closed convex hull of $0 \cap D$, K_2 is contained in B_{n_0}. So the diameter of K_2 is at most $\frac{2\epsilon}{3}$.

Let now k, k' be two points in $K \setminus A_r$. They have decompositions :

$$k = \lambda k_1 + (1-\lambda)k_2 \ , \quad k' = \lambda' k_1' + (1-\lambda')k_2' \ , \quad \text{with } \lambda < r \ , \lambda' < r \ ,$$

and therefore :

$$\|k - k'\| \leqslant \lambda \|k_1\| + \lambda' \|k_1'\| + (1-\lambda)\|k_2 - k_2'\| + |\lambda - \lambda'| \, \|k_2'\|$$

$$\leqslant 3rM + \frac{2\epsilon}{3} \leqslant \epsilon \ ,$$

and the lemma is proved.

PROOF OF PROPOSITION 4. - Assume that F has the Infinite Tree Property (for some $\epsilon_0 > 0$), and that F is reflexive. Let K be the closed convex hull of the points of the tree : K satisfies the hypotheses of the lemma (K is separable because the family $(z_{\epsilon_1,\ldots,\epsilon_n})_{\substack{n \geqslant 1 \\ \epsilon_1,\ldots,\epsilon_n = \pm 1}}$ of the points of the tree is countable, and therefore the convex linear combinations with rational coefficients, or with rational real and imaginary parts of the coefficients, form a countable dense set). Let $\epsilon = \frac{\epsilon_0}{3}$, and C be the convex given by the lemma. The set $K \setminus C$ must contain some point of the tree, say $z_{\epsilon_1,\ldots,\epsilon_n}$, otherwise they would all be in C , and then $K = C$. But

$$z_{\epsilon_1,\ldots,\epsilon_n} = \frac{z_{\epsilon_1,\ldots,\epsilon_n,+1} + z_{\epsilon_1,\ldots,\epsilon_n,-1}}{2} \ ,$$

with

$$\begin{cases} \left\| z_{\epsilon_1,\ldots,\epsilon_n} - z_{\epsilon_1,\ldots,\epsilon_n,+1} \right\| \geqslant \frac{\epsilon_0}{2} \ , \\ \left\| z_{\epsilon_1,\ldots,\epsilon_n} - z_{\epsilon_1,\ldots,\epsilon_n,-1} \right\| \geqslant \frac{\epsilon_0}{2} \ . \end{cases}$$

Since the diameter of $K \setminus C$ is smaller than ϵ , neither $z_{\epsilon_1,\ldots,\epsilon_n,+1}$ nor $z_{\epsilon_1,\ldots,\epsilon_n,-1}$ can be in $K \setminus C$: they must be both in C . But then, their middle, $z_{\epsilon_1,\ldots,\epsilon_n}$, must also be in C : this contradiction ends the proof of proposition 4, and of theorem 2.

COROLLARY 6. - *Super-reflexivity is invariant under isomorphisms.*

This means that if a space is isomorphic to a super-reflexive space, it is itself super-reflexive. This is not obvious on the definition because if E and E_1 are isomorphic, if F is finitely representable in E, F needs not be finitely representable in E_1. But the Finite Tree Property is obviously invariant under isomorphisms, and the corollary follows.

From first part, chap. III, § 1, theorem 6, we deduce, just applying the definition of finite representability, the following characterizations of non-super-reflexivity :

PROPOSITION 7. - *For a Banach space* E, *the following properties are equivalent :*

1) E *is not super-reflexive.*

2) *For every* $n \geq 1$, *every* θ, $0 < \theta < 1$ *there is a sequence* $x_1^{(n)},\dots,x_n^{(n)}$ *in the unit ball of* E, *such that, for every* k, $1 \leq k \leq n$,

$$\operatorname{dist}(\operatorname{conv}(x_1^{(n)},\dots,x_k^{(n)}), \operatorname{conv}(x_{k+1}^{(n)},\dots,x_n^{(n)})) > \theta .$$

3) *For every* $n \geq 1$, *every* θ, $0 < \theta < 1$, *there are sequences* $x_1^{(n)},\dots,x_n^{(n)}$, *in the unit ball of* E, *and* $f_1^{(n)},\dots,f_n^{(n)}$ *in the unit ball of* $E^\star$ *with*

$$\begin{aligned} f_n^{(n)}(x_k^{(n)}) &= \theta \qquad \text{if } m \leq k \leq n \\ &= 0 \qquad \text{if } k < m \leq n . \end{aligned}$$

PROOF.

3) $\Rightarrow$ 2) as in the proof of theorem 6, first part, chap. III, § 1.

2) $\Rightarrow$ 1) was already observed in the proof of theorem 2 above.

Let us show that 1) $\Rightarrow$ 3). If E is not super-reflexive, there is F, finitely representable in E, which is not reflexive. Then, for every θ, $0 < \theta < 1$, there are sequences $(x_n)_{n\in\mathbb{N}}$ in $\mathscr{B}_F$, $(f_n)_{n\in\mathbb{N}}$ in $\mathscr{B}_{F^\star}$, with

$$f_m(x_k) = \theta \quad \text{if } m \leqslant k$$
$$= 0 \quad \text{if } m > k .$$

Let $\epsilon > 0$, and let $F_n = \operatorname{span}\{x_1,\ldots,x_n\}$. There is a subspace E_n of E , and an isomorphism T from F_n onto E_n , with

$$\|T\| \leqslant 1 + \epsilon , \qquad \|T^{-1}\| = 1 .$$

Then we put $x_k^{(n)} = Tx_k$, $f_k^{(n)} = f_k \circ T^{-1}$. We have

$$f_m^{(n)}(x_k^{(n)}) = \theta \quad \text{if } m \leqslant k$$
$$= 0 \quad \text{if } m > k$$

and

$$\|f_m^{(n)}\|_{E^\star} \leqslant \|T^{-1}\| = 1$$

$$\|x_k^{(n)}\|_E \leqslant \|T\| \leqslant 1 + \epsilon$$

and therefore we obtain the proposition if we start with $\theta(1 + \epsilon)$ instead of θ and take $\dfrac{x_k^{(n)}}{1 + \epsilon}$ instead of $x_k^{(n)}$.

REMARK. - As we already observed, we can replace the words "for every θ , $0 < \theta < 1$ " by "for some θ , $0 < \theta < 1$ " , in 2) and 3). Also, obviously, we can put $f_m^{(n)}(x_k) \geqslant \theta$ if $m \leqslant k \leqslant n$, instead of $f_m^{(n)}(x_k) = \theta$, in 3).

COROLLARY 8. - $E^\star$ *is super-reflexive if and only if* E *is super-reflexive.*

PROOF.

1) Assume first that E is not super-reflexive. Then, by 3) of the previous proposition, for every $n \geqslant 1$, every θ , $0 < \theta < 1$, there are $x_1^{(n)},\ldots,x_n^{(n)}$ in $\mathscr{B}_E$, $f_1^{(n)},\ldots,f_n^{(n)}$ in $\mathscr{B}_{E^\star}$, with

$$f_m^{(n)}(x_k^{(n)}) = \theta \quad \text{if } m \leqslant k \leqslant n$$
$$= 0 \quad \text{if } k < m \leqslant n$$

The points $(x_i^{(n)})_{i=1,\dots,n}$ may be considered as points in $\mathscr{B}_{E^{\star\star}}$, but the order of numeration has to be inversed. We put :

$$\eta_j^{(n)} = x_{n-j+1}^{(n)} \qquad j = 1,\dots,n ,$$

$$g_m^{(n)} = f_{n-m+1}^{(n)} \qquad m = 1,\dots,n .$$

Then $\eta_j^{(n)} \in \mathscr{B}_{E^{\star\star}}$ for $j = 1,\dots,n$, and $g_m^{(n)} \in \mathscr{B}_{E^\star}$ for $j = 1,\dots,n$. We have :

$$\begin{aligned} < \eta_j^{(n)} , g_m^{(n)} > &= < x_{n-j+1}^{(n)} , f_{n-m+1}^{(n)} > \\ &= \theta \quad \text{if } n-m+1 \leqslant n-j+1 \text{ , that is } j \leqslant m \\ &= 0 \quad \text{if } n-m+1 > n-j+1 \text{ , that is } j > m , \end{aligned}$$

and $E^\star$ is not super-reflexive.

2) Assume that $E^\star$ is not super-reflexive.

Then there are, for every $n \geqslant 1$, every θ, $0 < \theta < 1$, elements $g_1^{(n)},\dots,g_n^{(n)}$ in $\mathscr{B}_{E^\star}$, $\eta_1^{(n)},\dots,\eta_n^{(n)}$ in $\mathscr{B}_{E^{\star\star}}$, satisfying 3) of the previous proposition. Let $\epsilon > 0$.

Since $\mathscr{B}_E$ is $\sigma(E^{\star\star}, E^\star)$ dense in $\mathscr{B}_{E^{\star\star}}$, we can find $x_1^{(n)},\dots,x_n^{(n)}$ in $\mathscr{B}_E$ with :

$$|< x_i^{(n)} , g_j^{(n)} > - < \eta_i^{(n)} , g_j^{(n)} >| < \epsilon \quad \text{for } j = 1,\dots,n,$$

for $i = 1,\dots,n$.

This gives :

$$\begin{cases} |< x_i^{(n)} , g_j^{(n)} >| > \theta - \epsilon & \text{if } i \geqslant j \\ |< x_i^{(n)} , g_j^{(n)} >| \leqslant \epsilon & \text{if } j > i \end{cases}$$

Therefore, for every k , $1 \leqslant k \leqslant n$, every sequence of positive numbers $a_1, \ldots, a_k$, $a_{k+1}, \ldots, a_n$, with $\sum_1^k a_i = \sum_{k+1}^n a_i = 1$, we obtain :

$$\left\| \sum_1^k a_i x_i^{(n)} - \sum_{k+1}^n a_i x_i^{(n)} \right\| \geqslant \left| g_k^{(n)}\left(\sum_1^k a_i x_i^{(n)} - \sum_{k+1}^n a_i x_i^{(n)} \right) \right|$$

$$\geqslant g_k^{(n)}\left(\sum_{k+1}^n a_i x_i^{(n)} \right) - g_k^{(n)}\left(\sum_1^k a_i x_i^{(n)} \right)$$

$$\geqslant \theta - \epsilon - \epsilon = \theta - 2\epsilon ,$$

and therefore

$$\mathrm{dist}(\mathrm{conv}(x_1^{(n)}, \ldots, x_k^{(n)}), \mathrm{conv}(x_{k+1}^{(n)}, \ldots, x_n^{(n)})) \geqslant \theta - 2\epsilon$$

which proves our proposition.

PROPOSITION 9. - *Every subspace, every quotient, of a super-reflexive space are super-reflexive.*

PROOF. - For subspaces, apply theorem 2. For quotients, apply duality and corollary 8.

To end this chapter devoted to finite representability, we mention without proof Dvoretzky's Theorem, which is one of the deepest results in Banach Space Theory.

DVORETZKY's THEOREM. - *The space* ℓ_2 *is finitely representable in every Banach space.*

This theorem is very important for the study of the so called "local structure" of Banach spaces. But also it gives informations about the matters studied in the previous chapters. For example, it implies that the modulus of convexity of every Banach space is dominated by the modulus of convexity of the Hilbert space, which means

$$\delta_E(\epsilon) \leqslant C\,\epsilon^2 .$$

EXERCISES ON CHAPTER I.

EXERCISE 1. - Let $(x_n(t))_{n\in\mathbb{N}}$ be a Schauder basis of $\mathscr{C}([0,1])$. We call E_n the space spanned by $x_1,\dots,x_n$. Let $E = (\prod_{n\geqslant 1} E_n)_2$.

- Show that $\mathscr{C}([0,1])$ is finitely representable in E.

- Deduce from first part, chap. III, exercise 3, that E is separable and reflexive.

- Using Banach-Mazur Theorem (second part, chap. IV, § 2, prop. 1), show that every separable Banach space is finitely representable in E.

EXERCISE 2. - Show that the following are equivalent, for a Banach space E.

1) E is not reflexive.

2) For some θ, $0<\theta<1$, there are sequences $(x_n)_{n\in\mathbb{N}}$ in $\mathscr{B}_E$, $(f_n)_{n\in\mathbb{N}}$ in $\mathscr{B}_{E^\star}$, with :

$$f_n(x_k)\geqslant\theta \quad \text{if} \quad n\leqslant k$$

$$f_n(x_k)\leqslant\frac{\theta}{2} \quad \text{if} \quad n>k$$

give the corresponding proposition for super-reflexivity.

EXERCISE 3. - Let E and F be Banach spaces, and $\mathscr{U}$ a non-trivial ultrafilter on $\mathbb{N}$. Put $\tilde{E} = E^{\mathbb{N}}/\mathscr{U}$, $\tilde{F} = F^{\mathbb{N}}/\mathscr{U}$. For every bounded sequence $(x_n)_{n\in\mathbb{N}}$ in E, put :

$$\tilde{T}x = y,$$

if $x\in\tilde{E}$ is the class of $(x_n)_{n\in\mathbb{N}}$ and y is the class, in $\tilde{F}$, of $(Tx_n)_{n\in\mathbb{N}}$.

Show that $\tilde{T}$ is a continuous operator from $\tilde{E}$ into $\tilde{F}$ with $\|\tilde{T}\| = \|T\|$. Show that, if E is embedded in $\tilde{E}$, the restriction of $\tilde{T}$ to E is T.

EXERCISE 4. - Let E be a Banach space, and T an operator from E into itself. Show that, for every $\lambda \in \mathbb{C}$ in the boundary of the spectrum of T, one can find a sequence of norm-one vectors $(x_n)_{n \in \mathbb{N}}$ in E such that

$$Tx_n - \lambda x_n \xrightarrow[n \to +\infty]{} 0 .$$

Deduce that every such λ, is an eigenvalue of $\widetilde{T}$ (defined in exercise 3).

EXERCISE 5. - Let E, F, E_1, F_1 be Banach spaces, T, T_1 operators from E into F, from E_1 into F_1 respectively.

We say that T_1 is *finitely representable* in T if, for every $\epsilon > 0$, every finite-dimensional subspace E_1° of E_1, one can find a finite dimensional subspace E° of E, and (setting $F^\circ = TE^\circ$, $F_1^\circ = T_1 E_1^\circ$), two isomorphisms U and V, from E_1° onto E° and from F_1° onto F° respectively, with, on E_1° : $TU = VT_1$, and :

$$\begin{cases} \|U\| \cdot \|U^{-1}\| \leqslant 1 + \epsilon \\ \|V\| \cdot \|V^{-1}\| \leqslant 1 + \epsilon . \end{cases}$$

Show that this definition is transitive

Show that $\widetilde{T}$, from $\widetilde{E}$ into $\widetilde{F}$, if finitely representable in T.

EXERCISE 6. - An operator T, from E into F, is said to have the *Finite Tree Property* if, for some $\epsilon > 0$, for all $n \geqslant 1$, one can find in the unit ball $\mathscr{B}_E$, 2^n points $x_1, \ldots, x_{2^n}$ such that the images $Tx_1, \ldots, Tx_{2^n}$ form an (n, ϵ)-branch in F.

Prove that an operator T does not have the Finite Tree Property if and only if every operator T_1, finitely representable in T, is weakly compact. (Use exercise 8, first part, chapter III). Such an operator is called *uniformly convexifying*.

Show that if T, from E into F, is uniformly convexifying, and if S_1, from X into E, and S_2, from F into Y, are any operators, then $S_2 \circ T \circ S_1$ is uniformly convexifying, from X into Y.

REFERENCES ON CHAPTER I.

The notion of finite-representability was introduced by R.C. JAMES [27]. The connection with Ultrapowers was developed later by J.L. KRIVINE (see J. STERN [48]). The notions of Finite and Infinite Trees are due to R.C. JAMES [28], who also gave theorem 2. Our proof uses the formalism of Ultrapowers, which, we think, is well adapted.

The reader may find an exposition of Dvoretzky's theorem and many related notions in the paper of T. FIGIEL - J. LINDENSTRAUSS - V. MILMAN [21].

The notion of finite representability for operators was introduced by the author [7], [8]. The exercises 5 and 6 come from our work [7].

COMPLEMENTS ON CHAPTER I.

The following result, due to J. LINDENSTRAUSS and H.P. ROSENTHAL [33] is known as "Principle of local reflexivity".

THEOREM. - *For every Banach space* E *, the bidual* E^{**} *is finitely representable in* E .

A proof of this result, using ultrapowers, was given by J. STERN in [48]. We refer the reader to it. In [48], J. STERN obtained also the following result :

THEOREM. - *For a Banach space* E *, the following are equivalent :*

1) E *is super-reflexive.*

2 *For every set* I *, every ultrafilter* $\mathcal{U}$ *on* I *,*

$$(E^I/\mathcal{U})^* = (E^*)^I/\mathcal{U} .$$

3) *There is a non-trivial ultrafilter* $\mathcal{U}$ *on* $\mathbb{N}$ *such that*

$$(E^{\mathbb{N}}/\mathcal{U})^* = (E^*)^{\mathbb{N}}/\mathcal{U} .$$

CHAPTER II

BASIC SEQUENCES IN SUPER-REFLEXIVE BANACH SPACES

This chapter is devoted to the proofs of two theorems of R.C. James concerning the existence of upper and lower estimates for basic sequences in a super-reflexive Banach space.

The first theorem is as follows :

THEOREM 1. - *Let* E *be a super-reflexive Banach space. For every* $C > 1$, *and every* $K \geq 1$, *there exists a number* s , *with* $1 < s < +\infty$, *such that, for every normalized basic sequence* $(x_n)_{n \in \mathbb{N}}$, *with basis constant* K , *for every finite sequence of scalars* (a_i) , *one has*

$$\| \sum_i a_i x_i \| \leq C(\sum_i |a_i|^s)^{1/s} .$$

(The basis constant of a basic sequence is defined p. 83).

PROOF. - Let $C > 1$, $K \geq 1$. We shall show that, if the conclusion fails, there is, for every $n \geq 1$, a $(n, \frac{1}{2(K+1)})$-branch in the unit ball of E : this will prove that E is not super-reflexive.

For this, we shall use the following lemma :

LEMMA 2. - *Let* $y_1,\dots,y_n$ *be a basic sequence, in the unit ball of* E , *with basis constant* K . *Assume that there is a number* $\alpha > 0$ *such that for every* y *in* $\text{conv}\{y_1,\dots,y_n\}$, *one has :*

$$\|y\| \geq \alpha .$$

Then, for every N *such that* $2^N \leq n$, *the points* $(y_1,\dots,y_{2^N})$ *form a* $(N, \frac{\alpha}{2K})$ *-branch in the unit ball of* E .

PROOF OF LEMMA 2. - By definition of the basis constant, we have, for every $k \leqslant 2^N$, every sequence $a_1,\dots,a_{2^N}$ of scalars :

$$\left\| \sum_1^k a_i\, y_i \right\| \leqslant K \left\| \sum_1^{2^N} a_i\, y_i \right\| ,$$

therefore, if $1 \leqslant k \leqslant k' \leqslant 2^N$,

$$\left\| \sum_k^{k'} a_i\, y_i \right\| \leqslant \left\| \sum_1^{k'} a_i\, y_i \right\| + \left\| \sum_1^{k} a_i\, y_i \right\| \leqslant 2K \left\| \sum_1^{2^N} a_i y_i \right\| .$$

So we obtain :

$$\left\| \frac{y_1+\dots+y_{2^{N-1}}}{2^{N-1}} - \frac{y_{2^{N-1}+1}+\dots+y_{2^N}}{2^{N-1}} \right\| \geqslant \frac{1}{2K} \left\| \frac{y_1+\dots+y_{2^{N-1}}}{2^{N-1}} \right\| \geqslant \frac{\alpha}{2K}$$

and, more generally, for $1 \leqslant j \leqslant N-1$,

$$\left\| \frac{y_1+\dots+y_{2^j}}{2^j} - \frac{y_{2^j+1}+\dots+y_{2^{j+1}}}{2^j} \right\| \geqslant \frac{1}{2K} \left\| \frac{y_1+\dots+y_{2^j}}{2^j} \right\| \geqslant \frac{\alpha}{2K} ,$$

$$\left\| \frac{y_{2^{j+1}+1}+\dots+y_{2^{j+2}}}{2^j} - \frac{y_{2^{j+2}+1}+\dots+y_{2^{j+3}}}{2^j} \right\| \geqslant \frac{1}{2K} \left\| \frac{y_{2^{j+1}+1}+\dots+y_{2^{j+2}}}{2^j} \right\| \geqslant \frac{\alpha}{2K} ,$$

and so on ; the lemma follows.

Let us turn now to the proof of the theorem.

Let $n \geqslant 1$. First, we choose θ with :

$$(1) \qquad 1 - \frac{1}{2n} < \theta < 1 .$$

Then we choose λ with $\theta^{1/2} < \lambda < 1$, $\lambda^2 C > 1$, and

$$(2) \qquad \frac{(C + 1)(1 - \lambda)}{\lambda^2 C - 1} < \frac{1}{n} (1 - \theta^{1/4}) .$$

Finally, we take $s > 1$ and close enough to 1 that $\lambda n < n^{1/s}$. Then, by Hölder's Inequality, if $\alpha,\beta \geqslant 0$:

$$(3) \qquad (\alpha + \beta)^{1/s} \geqslant \lambda(\alpha^{1/s} + \beta^{1/s}) ,$$

and, if $\beta_1,\ldots,\beta_n$ are positive numbers,

$$(4) \qquad \lambda n(\min_{1\leqslant i\leqslant n} \beta_i)^{1/s} \leqslant (\sum_1^n \beta_i)^{1/s} .$$

Since the conclusion fails, there is a normalized basic sequence $(x_i)_{i\in\mathbb{N}}$, with basis constant at most K, and a least positive integer m such that

$$(5) \qquad \sup_{(a_1,\ldots,a_m)} \frac{\left\| \sum_1^m a_i x_i \right\|}{(\sum_1^m |a_i|^s)^{1/s}} = M > C .$$

Observe that $\sum_1^m |a_i| \geqslant M(\sum_1^m |a_i|^s)^{1/s} \geqslant C(\sum_1^m |a_i|^s)^{1/s}$, and $\sum_1^m |a_i| \leqslant m^{1/s'}(\sum_1^m |a_i|^s)^{1/s}$ (with $\frac{1}{s} + \frac{1}{s'} = 1$), from which we obtain $m^{1/s'} \geqslant C$. But we have $\lambda n < n^{1/s}$, that is $\lambda < n^{-1/s'}$, and $C > \frac{1}{\lambda^2}$, or $C > n^{2/s'}$. So we obtain : $m \geqslant n^2$.

Now, the formula

$$\frac{\|x + y\|}{(\alpha + \beta)^{1/s}} \leqslant \frac{\|x\|}{\alpha^{1/s}} + \frac{\|y\|}{\beta^{1/s}} \qquad (\alpha,\beta \geqslant 0)$$

implies

$$\frac{\left\|\sum_1^{m-1} a_i x_i + a_m x_m\right\|}{(\sum_1^{m-1} |a_i|^s + |a_m|^s)^{1/s}} \leqslant \frac{\left\|\sum_1^{m-1} a_i x_i\right\|}{(\sum_1^{m-1} |a_i|^s)^{1/s}} + \frac{\|a_m x_m\|}{|a_m|} \leqslant C + 1 ,$$

and therefore $C < M \leqslant C + 1$. So we deduce from (2) that

$$(6) \qquad \left(\frac{M(1-\lambda)}{\lambda^2 M - 1}\right)^s < \frac{M(1-\lambda)}{\lambda^2 M - 1} < \frac{1}{n}\,(1 - \theta^{1/4}) .$$

We put $\theta' = \theta^{1/4}$. Let $(a_1,\dots,a_m)$ be a sequence of m scalars with $\left\| \sum_1^m a_i x_i \right\| = 1$, and :

$$(7) \qquad \frac{\left\| \sum_1^m a_i x_i \right\|}{\left(\sum_1^m |a_i|^s \right)^{1/s}} = \frac{1}{\left(\sum_1^m |a_i|^s \right)^{1/s}} = M$$

(the supremum in (5) is attained by compactness).

We shall show first that for each $k = 1,\dots,m$,

$$|a_k|^s \leqslant \frac{1}{n}\,(1 - \theta') \sum_1^m |a_i|^s .$$

By (3), we have :

$$\left(\sum_1^m |a_i|^s \right)^{1/s} = \lambda \left[|a_k| + \left(\sum_{i \neq k} |a_i|^s \right)^{1/s} \right]$$

and thus

$$\lambda M < \frac{|a_k| + \left\| \sum_{i\neq k} a_i x_i \right\|}{|a_k| + \left(\sum_{i \neq k} |a_i|^s \right)^{1/s}} \leqslant \frac{|a_k| + M\left(\sum_{i \neq k} |a_i|^s \right)^{1/s}}{|a_k| + \left(\sum_{i \neq k} |a_i|^s \right)^{1/s}}$$

and so :

$$|a_k| \leqslant \frac{M(1-\lambda)}{\lambda M - 1} \left(\sum_{i \neq k} |a_i|^s \right)^{1/s}$$

and

$$|a_k|^s \leqslant \left(\frac{M(1-\lambda)}{\lambda M - 1}\right)^s \sum_{i=1}^m |a_i|^s < \frac{1}{n}(1-\theta') \sum_{i=1}^m |a_i|^s ,$$

and our claim is proved.

We put

$$c_k = \frac{|a_k|^s}{\sum_{i=1}^m |a_i|^s} , \quad \text{for } k = 1,\dots,m .$$

We have $\sum_1^m c_k = 1$, and $c_k < \frac{1}{n}(1-\theta')$. Consider the n open intervals I_j $(j = 1,\dots,n-1)$ in $[0,1]$: $]\frac{j}{n} - \frac{1}{2n}(1-\theta') , \frac{j}{n} + \frac{1}{2n}(1-\theta')[$. For each j , there is an index m_j such that $\sum_1^{m_j} c_i \in I_j$. We take $m_o = 0$, $m_n = m$. So we have, for $j = 1,\dots,n$:

$$\left|\sum_1^{m_j} c_i - \frac{j}{n}\right| < \frac{1}{2n}(1-\theta') ,$$

and the sequence $(m_j)_{j=1,\dots,n}$ is strictly increasing (this construction is possible since $m \geqslant n^2 \geqslant n$).

So we obtain

$$\left|\sum_{m_{j-1}+1}^{m_j} c_i - \frac{1}{n}\right| = \left|\left(\sum_1^{m_j} c_i - \frac{j}{n}\right) - \left(\sum_1^{m_{j-1}} c_i - \frac{j-1}{n}\right)\right| < \frac{1}{n}(1-\theta') ,$$

which means that

$$\left|\sum_{m_{j-1}+1}^{m_j} |a_i|^s - \frac{1}{n}\sum_1^m |a_i|^s\right| < \frac{1}{n}(1-\theta') \sum_1^m |a_i|^s .$$

This implies that

$$\frac{1}{n}\theta' \cdot \sum_{1}^{m} |a_i|^s < \sum_{m_{j-1}+1}^{m_j} |a_i|^s < \frac{1}{n}(2-\theta') \sum_{1}^{m} |a_i|^s ,$$

so we obtain

$$\sum_{m_{j-1}+1}^{m_j} |a_i|^s < \frac{2-\theta'}{\theta'} \min \left\{ \sum_{m_{k-1}+1}^{m_k} |a_i|^s \quad ; \quad 1 \leqslant k \leqslant n \right\}$$

and since $\frac{2-\theta'}{\theta'} < \theta^{-1/2}$, we have :

$$\sum_{m_{j-1}+1}^{m_j} |a_i|^s < \theta^{-1/2} \min \left\{ \sum_{m_{k-1}+1}^{m_k} |a_i|^s \quad ; \quad 1 \leqslant k \leqslant n \right\} .$$

We put, for $j = 1,\ldots,n$, $u_j = \sum_{i=m_{j-1}+1}^{m_j} a_i e_i$. Then, we have :

$$\frac{1}{(\sum_{1}^{m} |a_i|^s)^{1/s}} = M \geqslant \frac{\|u_j\|}{(\sum_{m_{j-1}+1}^{m_j} |a_i|^s)^{1/s}} > \frac{\theta^{1/2s} \|u_j\|}{\min\left\{(\sum_{m_{k-1}+1}^{m_k} |a_i|^s)^{1/s} \; ; \; 1 \leqslant k \leqslant n\right\}}$$

$$> \frac{n\,\theta^{1/2s} \lambda \|u_j\|}{(\sum_{1}^{m} |a_i|^s)^{1/s}} \quad , \quad \text{by (4)}$$

$$> \frac{n\,\theta \|u_j\|}{(\sum_{1}^{m} |a_i|^s)^{1/s}} \quad , \quad \text{since} \quad \lambda > \theta^{1/2} .$$

Therefore, we have shown that :

$$\|u_j\| < \frac{1}{n\theta} \qquad \text{for} \quad j = 1,\ldots,n .$$

Put now $y_j = n\theta\, u_j$. We shall show that $(y_j)_{j=1,\dots,n}$ satisfies the assumptions of lemma 2, with $\alpha = \frac{1}{2}$.

If $\beta_1,\dots,\beta_n$ are positive numbers, with $\sum_1^n \beta_i = 1$, then

$$\left\| \sum_1^n \beta_i y_i \right\| \geqslant \left\| \sum_1^n y_i \right\| - \left\| \sum_1^n (1-\beta_i) y_i \right\| \geqslant n\theta \left\| \sum_1^n u_j \right\| - \sum_1^n (1-\beta_i) .$$

Since $\left\| \sum_1^n u_j \right\| = \left\| \sum_1^n a_i x_i \right\| = 1$, and $\theta > 1 - \frac{1}{2n}$, we obtain

$$\left\| \sum_1^n \beta_i y_i \right\| > n - \frac{1}{2} - (n-1) = \frac{1}{2} ,$$

and since the basis constant of the sequence $(y_j)_{j=1,\dots,n}$ is the same as the one of $(u_j)_{j=1,\dots,n}$, which, in turn, is smaller than the one of $(x_j)_{j=1,\dots,m}$, the assumptions of the lemma are satisfied, and the theorem follows.

If one is interested only in basic sequences with basis constant equal to one (that is, monotone basic sequences), one obtains easily the following weak result, for lower estimates :

PROPOSITION 3. - *Let* E *be a super-reflexive space. There exists two numbers* $C \geqslant 1$, $p > 1$ *such that, for every finite monotone basic sequence* (x_n) , *one has*

$$\left(\sum_n \|x_n\|^p \right)^{1/p} \leqslant C \left\| \sum_n x_n \right\| .$$

SKETCH OF THE PROOF. - For every $n \geqslant 1$, we define :

$$a_n = \inf \Big\{ \lambda \in \mathbb{R} \ ; \ \sup_{\substack{A \subset \mathbb{N},\\ |A| = n,\ A \supset \{1\}}} \ \inf_{i \in A} \|x_i\| \leqslant \lambda \Big\| \sum_{i \in I} x_i \Big\| \ , \text{ for every finite monotone basic sequence } (x_i)_{i \in I} \Big\} .$$

One checks that $0 \leqslant a_n \leqslant 1$, for every n , that the sequence $(a_n)_{n \geqslant 1}$ is decreasing, and submultiplicative :

$$a_{m.n} \leqslant a_m \cdot a_n \text{ , for every } m,n \geqslant 1 .$$

Then either $a_n = 1$ for every n , or there is some $p > 1$ such that $a_n \leqslant n^{-1/p}$. But the first case is impossible since E is super-reflexive. The result follows.

We now give the second James'Theorem, concerning lower estimates :

THEOREM 4. - *Let* E *be a super-reflexive space. For every* c *and every* K *satisfying* $0 < 2c < \frac{1}{K} < 1$, *there exists a number* r , *with* $1 < r < +\infty$, *such that, for every normalized basic sequence* $(x_n)_{n \in \mathbb{N}}$ *with basis constant* K , *for every finite sequence of scalars* (a_i) , *one has*

$$c\Big(\sum_i |a_i|^r\Big)^{1/r} \leqslant \Big\| \sum_i a_i x_i \Big\| .$$

PROOF. - Let c , K , with $0 < 2c < \frac{1}{K} < 1$. We shall show that, if the conclusion fails, the assumptions of the following lemma are satisfied with $\alpha' = 4K^2c^2$, $\beta' = K$.

LEMMA 5. - *Assume that there are positive numbers* α' , β' *such that, for every* $n \geqslant 1$, *one can find points* $y_1,\dots,y_n$ *in* E *such that, for all* k , $1 \leqslant k \leqslant n$, *all sequences of scalars* $(a_1,\dots,a_n)$:

$$\Big\| \sum_1^n a_i y_i \Big\| \geqslant \alpha' \max_{1 \leqslant i \leqslant n} |a_i| \quad , \quad \Big\| \sum_1^k y_i \Big\| \leqslant \beta' .$$

Then E *is not super-reflexive.*

PROOF OF LEMMA 5. - We put $z_k = \frac{1}{\beta'} \sum_1^k y_i$, for $k = 1,\dots,n$. Then $\|z_k\| \leqslant 1$. On $\operatorname{span}\{z_1,\dots,z_n\} = \operatorname{span}\{y_1,\dots,y_n\}$, we define, for $j = 1,\dots,n$, a linear functional g_j by $g_j(x_i) = \alpha'$ if $i = j$, 0 if $i \neq j$. Then

$$|g_j(\sum_1^n a_i y_i)| = \alpha' |a_j| \leqslant \| \sum_1^n a_i y_i \| ,$$

for every sequence $(a_1,\dots,a_n)$.

Therefore g_j can be extended to the whole space, with $\|g_j\| \leqslant 1$. But we have

$$\begin{aligned} g_j(z_i) &= \frac{\alpha'}{\beta'} \quad \text{if} \quad j \leqslant i , \\ &= 0 \quad \text{if} \quad i < j , \end{aligned}$$

and by chap. I, prop. 7, the lemma follows.

Let us come back to the proof of theorem 4.

Let $n \geqslant 1$. We first choose λ with

$$(1) \qquad 2Kc < \lambda < 1$$

and r large enough that

$$(2) \qquad \lambda^r < \frac{1}{n}(1 - \lambda) .$$

Then, if $\beta_1,\dots,\beta_n$ are positive numbers,

$$(3) \qquad (\sum_1^n \beta_i)^{1/r} \leqslant n^{1/r} (\max_{1 \leqslant i \leqslant n} \beta_i)^{1/r} \leqslant \frac{1}{\lambda} (\max_{1 \leqslant i \leqslant n} \beta_i)^{1/r} .$$

Since there is a basic sequence $(x_n)_{n \in \mathbb{N}}$, with basis constant K , for which the conclusion fails, there is a least integer $m \geqslant 1$ for which

$$(4) \qquad \inf_{a_1,\dots,a_m} \frac{\| \sum_1^m a_i x_i \|}{(\sum_1^m |a_i|^r)^{1/r}} = \delta < c .$$

We observe that :

$$\frac{\left\| \sum_1^n a_i x_i \right\|}{\left(\sum_1^n |a_i|^r \right)^{1/r}} \geqslant \frac{1}{2K} \frac{\max_{1 \leqslant i \leqslant n} |a_i|}{\left(\sum |a_i|^r \right)^{1/r}} \geqslant \frac{\lambda}{2K} > c$$

(we have used the fact that, since $(x_n)_{n \in \mathbb{N}}$ has basis constant K,

$$\left\| \sum_i a_i x_i \right\| > \frac{1}{2K} \max_i |a_i|)$$

and therefore $m > n$.

Now, we choose $a_1, \ldots, a_m$ with $\left\| \sum_{i=1}^m a_i x_i \right\| = 1$, and

$$\frac{1}{\left(\sum_1^m |a_i|^r \right)^{1/r}} = \frac{\left\| \sum_{i=1}^m a_i x_i \right\|}{\left(\sum_{i=1}^m |a_i|^r \right)^{1/r}} = \delta$$

(The Infimum in (4) is attained by compactness).

For each $k = 1, \ldots, m$, we have :

$$|a_k| \leqslant 2K \left\| \sum_1^m a_i x_i \right\| = 2K\delta \left(\sum_1^m |a_i|^r \right)^{1/r} < 2Kc \left(\sum_1^m |a_i|^r \right)^{1/r}$$

and thus

$$|a_k|^r \leqslant (2Kc)^r \sum_1^m |a_i|^r < \lambda^r \sum_1^m |a_i|^r < \frac{1}{n} (1 - \lambda) \sum_1^m |a_i|^r .$$

As in the proof of the previous theorem, we deduce that there exists a sequence $m_1 < \ldots < m_n = m$, such that, for every $j = 1, \ldots, n$:

$$\left| \sum_{i=1}^{m_j} |a_i|^r - \frac{j}{n} \sum_1^m |a_i|^r \right| < \frac{1}{2n} (1 - \lambda) \sum_1^m |a_i|^r$$

which implies (with $m_o = 0$) :

$$\left| \sum_{m_{j-1}+1}^{m_j} |a_i|^r - \frac{1}{n} \sum_1^m |a_i|^r \right| < \frac{1}{n} (1 - \lambda) \sum_1^m |a_i|^r$$

and

$$\frac{\lambda}{n}\sum_1^m |a_i|^r < \sum_{m_{j-1}+1}^{m_j} |a_i|^r < \frac{1}{n}(2-\lambda)\sum_1^m |a_i|^r < \frac{1}{n\lambda}\sum_1^m |a_i|^r .$$

So we obtain :

$$(5) \qquad \sum_{m_{j-1}+1}^{m_j} |a_i|^r \geqslant \lambda^2 \max \left\{ \sum_{m_{k-1}+1}^{m_k} |a_i|^r \ ; \quad 1 \leqslant k \leqslant n \right\} .$$

We now define, for $j = 1,\ldots,r$:

$$u_j = \sum_{m_{j-1}+1}^{m_j} a_i x_i .$$

We have :

$$\frac{1}{\left(\sum_1^m |a_i|^r\right)^{1/r}} = \delta \leqslant \frac{\|u_j\|}{\left(\sum_{m_{j-1}+1}^{m_j} |a_i|^r\right)^{1/r}}$$

(since m is the least integer in (4)),

$$< \frac{\|u_j\|}{\lambda^{2/r} \max\limits_{1\leqslant k\leqslant n} \left(\sum_{m_{k-1}+1}^{m_k} |a_i|^r\right)^{1/r}} \qquad \text{by (5)}$$

$$< \frac{\|u_j\|}{\lambda^3 \left(\sum_1^m |a_i|^r\right)^{1/r}} , \qquad \text{by (3), and since } r > 1 .$$

So $\|u_j\| > \lambda^3$.

We obtain, for every k , $1 \leqslant k \leqslant n$:

$$\left\| \sum_1^n a_j u_j \right\| \geqslant \frac{1}{2K} \|a_k u_k\| \geqslant \frac{\lambda^3}{2K} |a_k| > \frac{(2Kc)^3}{2K} |a_k| = 4K^2c^2 |a_k|$$

And we have, finally, for every $k = 1,\dots,n$:

$$1 = \left\| \sum_1^m a_i x_i \right\| = \left\| \sum_1^n u_j \right\| \geq \frac{1}{K} \left\| \sum_1^k u_j \right\| ,$$

and the assumptions of the lemma are fulfilled with $\alpha' = 4K^2c^2$, and $\beta' = K$. Thus theorem 4 is proved.

REFERENCES ON CHAPTER II.

As we already mentioned, both theorems proved in this chapter are due to R.C. JAMES [27], and we have just reproduced James'proofs, with only a few notational changes, necessary to make them consistent with the previous chapters. Proposition 3 was suggested by B. MAUREY.

CHAPTER III

UNIFORMLY NON-SQUARE AND J-CONVEX BANACH SPACES

In this chapter, we continue our study of super-reflexive Banach spaces. But, whereas the results of the previous chapter were clearly invariant under isomorphisms, we shall now introduce some metric notions.

§ 1. UNIFORMLY NON-SQUARE BANACH SPACES.

A Banach space E is said to be *uniformly non-square* if there exists a positive number δ such that, for all x, y in the unit ball, the conditions

$$\left\| \frac{1}{2}(x+y) \right\| > 1 - \delta \quad \text{and} \quad \left\| \frac{1}{2}(x-y) \right\| > 1 - \delta$$

cannot hold simultaneously, that is,

$$\text{if} \quad \left\| \frac{x-y}{2} \right\| > 1 - \delta , \quad \text{then} \quad \left\| \frac{x+y}{2} \right\| \leqslant 1 - \delta .$$

This condition is of the same kind as uniform convexity, but involves only points which are at large distances from each other : in the definition of uniform convexity, we know that $\left\| \frac{x+y}{2} \right\| \leqslant 1 - \delta$ if $\|x-y\| > \epsilon$, here we have the same conclusion only for $\epsilon = 2(1 - \delta)$.

On the contrary, two points x, y , with $\|x\| \leqslant 1$, $\|y\| \leqslant 1$, satisfying for some $\delta > 0$ the two conditions $\left\| \frac{x-y}{2} \right\| \geqslant 1 - \delta$, $\left\| \frac{x+y}{2} \right\| \geqslant 1 - \delta$ will define a δ-*square*.

Clearly, the property of uniform non-squareness is a metric one, and may disappear if the norm is replaced by an equivalent one.

The main result of this paragraph is due to R.C. JAMES [24] :

THEOREM 1. - *A uniformly non-square Banach space is reflexive.*

PROOF. - We assume on the contrary that E is not reflexive. We recall that, for every θ , $0<\theta<1$, we can find a sequence $(x_n)_{n\in\mathbb{N}}$ in the unit ball of E , a sequence $(f_n)_{n\in\mathbb{N}}$ in the unit ball of $E^\star$, such that

$$(1)\quad \begin{cases} f_n(x_k) = \theta & n\leqslant k \\ \quad\quad\;\; = 0 & n>k\ . \end{cases}$$

(See first part, chap. III, § 1, th. 6).

For every sequence of norm-one linear functionals, $(g_j)_{j\in\mathbb{N}}$, every $n\geqslant 1$, every strictly increasing sequence of integers $p_1<p_2<\dots<p_{2n}$, we define :

$$S\{p_1,\dots,p_{2n}\ ;\ (g_j)_{j\in\mathbb{N}}\} =$$

$$\{x\in E\ ;\ \tfrac{3}{4}\leqslant(-1)^{i-1}\,g_k(x)\leqslant 1,\ \text{ for }\ p_{2i-1}<k\leqslant p_{2i}\ ,\ \ i=1,\dots,n\}\ ,$$

and :

$$K(n,\ (g_j)_{j\in\mathbb{N}}) =$$

$$\liminf_{p_1\to+\infty}\dots\liminf_{p_{2n}\to+\infty}\ \inf\{\|z\|,\ z\in S(p_1,\dots,p_{2n}\ ;\ (g_j)_{j\in\mathbb{N}})\}$$

and finally

$$K_n = \mathrm{Inf}\{K(n,\ (g_j)\ ;\ \ \|g_j\| = 1\ \text{ for every }\ j\in\mathbb{N}\}.$$

These definitions make sense in every Banach space, but K_n may be $+\infty$ identically. However, this is not the case if E is non-reflexive :

LEMMA 2. - *If* E *is non-reflexive,* $K_n\leqslant 2n$ *for every* $n\geqslant 1$.

PROOF OF LEMMA 2. - Let f_n , x_k satisfying (1), for some $\theta > \frac{3}{4}$. For every $n \geq 1$, every $p_1 < \dots < p_{2n}$, we put

$$w = \sum_{j=1}^{n} (-1)^{j-1}(-x_{p_{2j-1}} + x_{p_{2j}}) .$$

Thus, if $p_{2i-1} < k \leq p_{2i}$, we have, by (1) :

- if $j < i$, $\quad f_k(-x_{p_{2j-1}} + x_{p_{2j}}) = -\theta + \theta = 0$,

- if $j > i$, $\quad f_k(x_{p_{2j-1}}) = f_k(x_{p_{2j}}) = 0$

- for $j = i$, $\quad f_k(-x_{p_{2i-1}} + x_{p_{2i}}) = \theta$.

So we get :

$$(-1)^{i-1} f_k(w) = \theta ,$$

and $w \in S(p_1,\dots,p_{2n} ; (f_j)_{j\in\mathbb{N}})$. Since $\|w\| \leq 2n$, we get $K_n \leq 2n$, and the lemma is proved.

We observe also that (K_n) is positive and increasing. Therefore, $\lim_{n\to+\infty} K_n$ exists (finite or infinite). Also, if $x \in S(p_1,\dots,p_{2n} ; (g_j)_{j\in\mathbb{N}})$, then $\|x\| \geq \frac{3}{4}$, and consequently $K_n \geq \frac{3}{4}$ for every $n \geq 1$.

Let $\delta > 0$, and r with $1 > r > 1 - \delta$. Then, for every $\epsilon < \frac{3}{4}\frac{1-r}{2r+1}$ we have, for every $n \geq 1$,

$$K_n \geq \frac{3}{4} > \frac{(2r+1)\epsilon}{1-r} ,$$

and thus

$$\frac{K_n - \epsilon}{K_n + 2\epsilon} > r , \quad \text{for every} \quad n \geq 1 .$$

Since $K_n \leqslant 2n$, we have $\operatorname{Lim\,inf}_{n \to +\infty} \frac{K_n}{K_{n-1}} = 1$, both when $\operatorname{Lim}_{n \to +\infty} K_n < +\infty$ and when $\operatorname{Lim}_{n \to +\infty} K_n = +\infty$. Therefore, there is a $m \geqslant 1$ such that

$$\frac{K_{m-1} - \epsilon}{K_m + 2\epsilon} > 1 - \delta .$$

Consequently, by the definition of K_m, there is a sequence $(g_j)_{j \in \mathbb{N}}$ of norm-one linear functionals such that

$$K(m, (g_j)) < K_m + \epsilon .$$

We can now find $4m$ integers

$$(2) \quad p_1 < q_1 < p_2 < p_3 < q_2 < q_3 < p_4 < p_5 < q_4 < q_5 < \ldots < q_{2m-2} < q_{2m-1} < p_{2m} < q_{2m} ,$$

such that :

$$(3) \begin{cases} \operatorname{Inf}\{\|z\| \; ; \; z \in S(p_1,\ldots,p_{2m} \; ; \; (g_j)_{j \in \mathbb{N}})\} < K(m, (g_j)_{j \in \mathbb{N}}) + \epsilon \\ \operatorname{Inf}\{\|z\| \; ; \; z \in S(q_1,\ldots,q_{2m} \; ; \; (g_j)_{j \in \mathbb{N}})\} < K(m, (g_j)_{j \in \mathbb{N}}) + \epsilon \\ \operatorname{Inf}\{\|z\| \; ; \; z \in S(q_1,p_2,q_3,p_4,\ldots,q_{2m-1},p_{2m} \; ; \; (g_j)_{j \in \mathbb{N}})\} > \\ \qquad > K(m, (g_j)_{j \in \mathbb{N}}) - \epsilon \\ \operatorname{Inf}\{\|z\| \; ; \; z \in S(p_3,q_2,p_5,q_4,\ldots,p_{2m-1},q_{2m-2} \; ; \; (g_j)_{j \in \mathbb{N}})\} > \\ \qquad > K(m-1, (g_j)_{j \in \mathbb{N}}) - \epsilon \end{cases}$$

Therefore we can find $u \in S(p_1,\ldots,p_{2m} \; ; \; (g_j)_{j \in \mathbb{N}})$ with $\|u\| < K(m, (g_j)_{j \in \mathbb{N}}) + \epsilon$, and we can find $v \in S(q_1,\ldots,q_{2m} \; ; \; (g_j)_{j \in \mathbb{N}})$ with $\|v\| < K(m, (g_j)_{j \in \mathbb{N}}) + \epsilon$.

Then, by definition,

$$(4) \quad \text{and} \quad \begin{cases} \frac{3}{4} \leqslant (-1)^{i-1} g_k(u) \leqslant 1 , & p_{2i-1} < k \leqslant p_{2i} , \quad i = 1,\dots,m , \\ \frac{3}{4} \leqslant (-1)^{i-1} g_k(v) \leqslant 1 , & q_{2i-1} < k \leqslant q_{2i} , \quad i = 1,\dots,m . \end{cases}$$

Then, taking into account the order (2), we obtain

$$\begin{cases} \frac{3}{4} \leqslant (-1)^{i-1} g_k(u) \leqslant 1 \\ \frac{3}{4} \leqslant (-1)^{i-1} g_k(v) \leqslant 1 \end{cases} \quad \text{if} \quad q_{2i-1} < k \leqslant p_{2i} , \quad i = 1,\dots,m$$

that is

$$\frac{3}{4} \leqslant (-1)^{i-1} g_k\left(\frac{u + v}{2} \right) \leqslant 1 ,$$

and this means that

$$\frac{u + v}{2} \in S(q_1, p_2, q_3, p_4, \dots, q_{2m-1}, p_{2m} ; (g_j)_{j \in \mathbb{N}})$$

which implies, by (3) :

$$\left\| \frac{u + v}{2} \right\| \geqslant K(m, (g_j)_{j \in \mathbb{N}}) - \epsilon .$$

Also, we have :

$$\begin{cases} \frac{3}{4} \leqslant (-1)^{i} g_k(u) \leqslant 1 , & p_{2i+1} < k \leqslant p_{2i+2} , \quad i = 1,\dots,m-1 \\ \frac{3}{4} \leqslant (-1)^{i-1} g_k(v) \leqslant 1 , & q_{2i-1} < k \leqslant q_{2i} , \quad i = 1,\dots,m . \end{cases}$$

Which gives :

$$\frac{3}{4} \leqslant (-1)^{i-1} g_k\left(\frac{v - u}{2} \right) \leqslant 1 , \qquad p_{2i+1} < k \leqslant q_{2i} , \qquad i = 1,\dots,m-1$$

and

$$\frac{v - u}{2} \in S(p_3, q_2, p_5, q_4, \dots, p_{2m-1}, q_{2m-2} ; (g_j)_{j \in \mathbb{N}})$$

which implies, by (3) :

$$\left\| \frac{u - v}{2} \right\| \geqslant K(m - 1, (g_j)_{j \in \mathbb{N}}) - \epsilon .$$

After normalization, the points u and v will give the required δ-square. We put :

$$x = \frac{u}{K_m + 2\epsilon} \quad , \qquad y = \frac{v}{K_m + 2\epsilon}$$

then $\|x\| < 1$, $\|y\| < 1$,

$$\left\| \frac{x + y}{2} \right\| = \frac{1}{K_m + 2\epsilon} \left\| \frac{u + v}{2} \right\| \geqslant \frac{1}{K_m + 2\epsilon} (K(m, (g_j)_{j \in \mathbb{N}}) - \epsilon)$$

$$\geqslant \frac{K_m - \epsilon}{K_m + 2\epsilon} > 1 - \delta$$

and

$$\left\| \frac{x - y}{2} \right\| = \frac{1}{K_m + 2\epsilon} \left\| \frac{u - v}{2} \right\| \geqslant \frac{1}{K_m + 2\epsilon} (K(m - 1, (g_j)_{j \in \mathbb{N}}) - \epsilon)$$

$$\geqslant \frac{K_{m-1} - \epsilon}{K_m + 2\epsilon} > 1 - \delta ,$$

which proves the theorem.

We now observe that uniform non-squareness is a super-property :

LEMMA 3. - *If* F *is finitely representable in* E , *and has a* δ-*square, then* E *has a* δ'-*square, for every* $\delta' > \delta$.

PROOF. - Let $x, y \in F$ forming a δ-square. For every $\epsilon > 0$, we can find in E two points x' , y' and an isomorphism T , from span{x , y} onto span{x' , y'} , with $\|T\| \leqslant 1$, $\|T^{-1}\| \leqslant 1 + \epsilon$. This gives :

$$\|x'\| \leqslant 1 , \quad \|y'\| \leqslant 1 , \qquad \left\| \frac{x' \pm y'}{2} \right\| \geqslant \frac{1}{1 + \epsilon} (1 - \delta) .$$

The conclusion follows.

So we obtain :

COROLLARY 4. - *A uniformly non-square Banach space is super-reflexive.*

But we have seen (chap. I, § 3, cor. 6) that super-reflexivity was invariant under isomorphisms. Therefore, we obtain the following stronger form of theorem 1 :

THEOREM 5. - *If a Banach space* E *admits an equivalent norm which is uniformly non-square, then* E *is super-reflexive.*

This can be formulated also :

THEOREME 5 bis. - *If a Banach space* E *is not super-reflexive, then, for every equivalent norm on* E *, for every* $\delta > 0$ *, one can find* δ*-squares in the unit ball of* E *, equipped with this norm.*

A super-reflexive space can very well have squares. This is the case, for example, of the space ℓ_2 , equipped with the norm :

$$\|(x_n)_{n\geqslant 1}\| = \max\left\{ |x_1| + |x_2| \ , \ (\sum_{i>2} |x_i|^2)^{1/2} \right\},$$

which is obviously equivalent to the ℓ_2-norm. In this space, the two points $x = (1,0,0,\ldots)$ and $y = (0,1,0,\ldots)$ satisfy

$$\|x\| = \|y\| = 1 \ , \qquad \|x \pm y\| = 2 \ .$$

§ 2. J-CONVEX BANACH SPACES.

We shall now extend the results of the previous paragraph, still using the same techniques. Previously, we have shown, in a non-reflexive space the existence of two points forming a square. We shall now show the existence of an arbitrary number of points having a specific geometric property.

THEOREM 1. - *If* E *is not reflexive, one can find, for every* $\delta > 0$ *, for every* $K \geq 1$ *, points* $x_1, \ldots, x_K$ *in the unit ball of* E *, such that, for all* $k \leq K$ *,*

$$\|x_1 + \ldots + x_k - (x_{k+1} + \ldots + x_K)\| \geq K\delta .$$

(for $k = K$ *, this is intended as* $\|x_1 + \ldots + x_K\|$ *).*

PROOF. - We use the same notations as in the proof of theorem 1. The number $\delta > 0$ being given, r , ϵ , m are chosen as previously, and we have

$$\frac{K_{m-1} - \epsilon}{K_m + 2\epsilon} > 1 - \delta .$$

We also take a sequence $(g_j)_{j \in \mathbb{N}}$ of norm-one linear functionals such that

$$K(m, (g_j)_{j \in \mathbb{N}}) < K_m + \epsilon .$$

We now find $2Km$ integers $p_j^{(k)}$, $k = 1,\ldots,K$, $j = 1,\ldots,2m$, in the following order :

$$p_1^{(1)} < p_1^{(2)} < p_1^{(3)} < \ldots < p_1^{(K)} < p_2^{(1)} < p_3^{(1)} < p_2^{(2)} < p_3^{(2)} < \ldots$$

$$\ldots < p_2^{(k)} < p_3^{(k)} < \ldots < p_2^{(K)} < p_3^{(K)} < \ldots$$

$$\ldots < p_{2j-4}^{(1)} < p_{2j-3}^{(1)} < p_{2j-4}^{(2)} < p_{2j-3}^{(2)} < \ldots < p_{2j-4}^{(K-1)} < p_{2j-3}^{(K-1)} < p_{2j-4}^{(K)} < p_{2j-3}^{(K)} < \ldots$$

$$\ldots < p_{2j-2}^{(1)} < p_{2j-1}^{(1)} < \ldots\ldots\ldots\ldots < p_{2j-2}^{(K-1)} < p_{2j-1}^{(K-1)} < p_{2j-2}^{(K)} < p_{2j-1}^{(K)} < \ldots$$

$$\ldots < p_{2m-2}^{(1)} < p_{2m-1}^{(1)} < \ldots\ldots\ldots\ldots < p_{2m-2}^{(K-1)} < p_{2m-1}^{(K-1)} < p_{2m-2}^{(K)} < p_{2m-1}^{(K)} < \ldots$$

$$\ldots < p_{2m}^{(1)} < p_{2m}^{(2)} < \ldots < p_{2m}^{(K)} ,$$

such that :

a) for $k = 1,\ldots,K$:

$$\mathrm{Inf}\{\|z\| \; ; \; z \in S(p_1^{(k)},\dots,p_{2m}^{(k)} \; ; \; (g_j)_{j\in\mathbb{N}})\} < K(m, (g_j)_{j\in\mathbb{N}}) + \epsilon \; .$$

b) for $k = 1,\dots,K$:

$$\mathrm{Inf}\{\|z\| \; ; \; z \in S(p_3^{(k)},p_2^{(k+1)},\dots,p_{2i-1}^{(k)},p_{2i-2}^{(k+1)},\dots,p_{2m-1}^{(k)},p_{2m-2}^{(k+1)} \; ; \; (g_j)_{j\in\mathbb{N}})\}$$

$$> K(m-1, (g_j)_{j\in\mathbb{N}}) - \epsilon \; .$$

c)

$$\mathrm{Inf}\{\|z\| \; ; \; z \in S(p_1^{(K)},p_2^{(1)},p_3^{(K)},p_4^{(1)},\dots,p_{2i-1}^{(K)},p_{2i}^{(1)},\dots,p_{2m-1}^{(K)},\dots p_{2m}^{(1)})\}$$

$$> K(m, (g_j)_{j\in\mathbb{N}}) - \epsilon \; .$$

When this is done, we find, for $k = 1,\dots,K$, a point u_k in $S(p_1^{(k)},\dots,p_{2m}^{(k)} \; ; \; (g_j)_{j\in\mathbb{N}})$, with

$$\|u_k\| < K(m, (g_j)_{j\in\mathbb{N}}) + \epsilon \; .$$

Then, we have, for $k = 1,\dots,K$

$$\frac{3}{4} \leqslant (-1)^{i-1} g_h(u_k) \leqslant 1 \; , \quad \text{if} \quad p_{2i-1}^{(k)} < h \leqslant p_{2i}^{(k)} \; , \qquad i = 1,\dots,m \; ,$$

and therefore on the intersection $\bigcap_{k=1\dots K}]p_{2i-1}^{(k)},p_{2i}^{(k)}] =]p_{2i-1}^{(K)},p_{2i}^{(1)}]$, we have

$$\frac{3}{4} \leqslant (-1)^{i-1} g_h(u_k) \leqslant 1 \; , \text{ for every } k \, , \text{ if } \; p_{2i-1}^{(K)} < h < p_{2i}^{(1)} \; , \; i = 1,\dots,m.$$

Which implies

$$\frac{u_1 + \dots + u_K}{K} \in S(p_1^{(K)},p_2^{(1)},\dots,p_{2i-1}^{(K)},p_{2i}^{(1)},\dots,p_{2m-1}^{(K)},p_{2m}^{(1)} \; ; \; (g_j)_{j\in\mathbb{N}})$$

and, by c),

$$\left\| \frac{u_1 + \dots + u_K}{K} \right\| > K(m, (g_j)_{j\in\mathbb{N}}) - \epsilon \; .$$

The same way, on the intersection

$$]p_{2i-1}^{(1)},p_{2i}^{(1)}] \cap \ldots \cap]p_{2i-1}^{(k)},p_{2i}^{(k)}] \cap]p_{2i-3}^{(k+1)},p_{2i-2}^{(k+1)}] \cap \ldots \cap]p_{2i-3}^{(K)},p_{2i-2}^{(K)}] ,$$

we have

$$\frac{3}{4} \leqslant (-1)^{i-1} g_h(u_\ell) \leqslant 1 \qquad \text{if} \quad 1 \leqslant \ell \leqslant k$$

$$\frac{3}{4} \leqslant (-1)^{i-1} g_h(-u_\ell) \leqslant 1 \qquad \text{if} \quad k+1 \leqslant \ell \leqslant K ,$$

and this implies

$$\frac{u_1 + \ldots + u_k - (u_{k+1} + \ldots + u_K)}{K} \in$$

$$\in S(p_3^{(k)}, p_2^{(k+1)}, \ldots, p_{2i-1}^{(k)}, p_{2i-2}^{(k+1)}, \ldots, p_{2m-1}^{(k)}, p_{2m-2}^{(k+1)} \; ; \; (g_j)_{j \in \mathbb{N}})$$

and, by b)

$$\left\| \frac{u_1 + \ldots + u_k - (u_{k+1} + \ldots + u_K)}{K} \right\| \geqslant K(m, (g_j)_{j \in \mathbb{N}}) - \epsilon .$$

We put $x_j = \dfrac{u_j}{K_m + 2\epsilon}$, for $j = 1, \ldots, K$.

We get $\|x_j\| < 1$, and, for each $k < K$:

$$\left\| \frac{x_1 + \ldots + x_k - (x_{k+1} + \ldots + x_K)}{K} \right\| = \frac{1}{K_m + 2\epsilon} \left\| \frac{u_1 + \ldots + u_k - (u_{k+1} + \ldots + u_K)}{K} \right\|$$

$$\geqslant \frac{K(m, (g_j)_{j \in \mathbb{N}}) - \epsilon}{K_m + 2\epsilon} \geqslant \frac{K_m - \epsilon}{K_m + 2\epsilon} \geqslant 1 - \delta ,$$

and our theorem is proved.

This theorem contains of course the result of the previous paragraph. But we thought it preferable to develop first these notions in the case $K = 2$ before passing to the general case. Moreover, the notion of uniform non-squareness has its own interest.

We shall now give some definitions, which will allow us to give a different formulation of theorem 1.

A finite sequence of signs $\epsilon_1 \dots \epsilon_K$ will be called *admissible* if all + signs are before all - signs (so there are K different sequences of admissible signs of length K , counting the sequence $++,\dots,+$).

A Banach space is called J-*convex* if there is a number $K \geqslant 2$ and a number $\delta > 0$ such that, for every $x_1,\dots,x_K$ in the unit ball, there is an admissible choice of signs such that

$$\left\| \sum_1^K \epsilon_j x_j \right\| \leqslant (1 - \delta)K .$$

J-convexity is the negation of the property obtained in the theorem. Obviously, it is a super-property (same proof as in lemma 3, § 1). Therefore, we obtain :

THEOREM 2. - A *J-convex Banach space is super-reflexive.*

We have seen that uniform non-squareness was not equivalent to super-reflexivity. But now, this equivalence holds :

THEOREM 3. - A *Banach space is J-convex if and only if it is super-reflexive.*

PROOF. - Assume that E is not J-convex, that is, for every $K \geqslant 2$, every $\delta > 0$, there are points $x_1,\dots,x_K$ in the unit ball of E such that, for every admissible choice of signs,

$$\left\| \sum_1^K \epsilon_i x_i \right\| \geqslant (1 - \delta)K .$$

Let $K \geqslant 2$, and take $\delta < \frac{1}{K}$.

For every k , $1 \leqslant k < K$, let $\beta_1,\dots,\beta_k$, $\beta_{k+1},\dots,\beta_K$ be positive numbers, with

$$\sum_1^k \beta_i = \sum_{k+1}^K \beta_i = 1 .$$

Then we get :

$$\left\| \sum_1^k \beta_i x_i - \sum_{k+1}^K \beta_i x_i \right\| = \left\| \sum_1^k x_i - \sum_{k+1}^K x_i - \left(\sum_1^k (1-\beta_i) x_i - \sum_{k+1}^K (1-\beta_i) x_i \right) \right\|$$

$$\geqslant \left\| \sum_1^k x_i - \sum_{k+1}^K x_i \right\| - \left[\sum_1^k (1-\beta_i) + \sum_{k+1}^K (1-\beta_i) \right]$$

$$\geqslant K(1-\delta) - (k-1+K-k-1) = 2 - K\delta > 1 .$$

This proves that $\operatorname{dist}(\operatorname{conv}(x_1,\ldots,x_k) , \operatorname{conv}(x_{k+1},\ldots,x_K)) > 1$, and by chapter I, § 3, prop. 7, E is not super-reflexive. This ends the proof of our theorem.

We shall now give a homogeneous characterization of J-convexity, just as in third part, chapter II, we gave a homogeneous characterization of uniform convexity.

For a given $K \geqslant 2$, we call $\mathscr{E}_K$ the set of admissible sequences of signs, of length K . Thus $\operatorname{card} \mathscr{E}_K = K$.

PROPOSITION 4. - *For a Banach space* E , *the following are equivalent :*

1) E *is* J-*convex,*

2) *There exist an integer* $K \geqslant 2$, *a positive number* δ' , *such that, for every* $x_1,\ldots,x_K$ *in* E , *one can find an admissible sequence of signs* $\epsilon_1,\ldots,\epsilon_K$ *with :*

$$\left\| \sum_{j=1}^K \epsilon_j x_j \right\| \leqslant (1-\delta')\sqrt{K} \left(\sum_{j=1}^K \|x_j\|^2 \right)^{1/2} .$$

3) *There exist an integer* $K \geqslant 2$, *a positive number* δ'' *such that , for every* $x_1,\ldots,x_K$ *in* E ,

$$\frac{1}{K} \sum_{(\epsilon_j) \in \mathscr{E}_K} \left\| \sum_{j=1}^K \epsilon_j x_j \right\|^2 \leqslant (1-\delta'') K \sum_{j=1}^K \|x_j\|^2 .$$

PROOF. - We show first that 2) and 3) are equivalent.

2) ⇒ 3). We assume that, for an admissible sequence of signs $(\epsilon_j^\circ)_{j=1,\dots,K}$,

$$\left\| \sum_1^K \epsilon_j^\circ x_j \right\| \leqslant (1 - \delta') \sqrt{K} \left(\sum_1^K \|x_j\|^2 \right)^{1/2},$$

or

$$\left\| \sum_1^K \epsilon_j^\circ x_j \right\|^2 \leqslant (1 - \delta')^2 K \sum_1^K \|x_j\|^2 .$$

For the other admissible sequences of signs $(\epsilon_j)_{j=1,\dots,K}$, we just write :

$$\left\| \sum_1^K \epsilon_j x_j \right\|^2 \leqslant \left(\sum_{j=1}^K \|x_j\| \right)^2 \leqslant K \sum_1^K \|x_j\|^2 .$$

So we obtain :

$$\sum_{(\epsilon_j) \in \mathscr{E}} \left\| \sum_1^K \epsilon_j x_j \right\|^2 \leqslant K \sum_1^K \|x_j\|^2 \, [(1 - \delta')^2 + K - 1]$$

and

$$\frac{1}{K} \sum_{(\epsilon_j) \in \mathscr{E}} \left\| \sum_1^K \epsilon_j x_j \right\|^2 \leqslant (1 - \delta'') K \sum_1^K \|x_j\|^2$$

with $\delta'' = \frac{2\delta' - \delta'^2}{K} > 0$, and 3) is proved.

3) ⇒ 2). If

$$\frac{1}{K} \sum_{(\epsilon_j) \in \mathscr{E}_K} \left\| \sum_1^K \epsilon_j x_j \right\|^2 \leqslant (1 - \delta'') K \left(\sum_1^K \|x_j\|^2 \right),$$

then, for at least one admissible sequence of signs $(\epsilon_j^\circ)_{j=1,\dots,K}$, we get

$$\left\| \sum_1^K \epsilon_j^\circ x_j \right\|^2 \leqslant (1 - \delta'') K \sum_1^K \|x_j\|^2$$

or

$$\left\| \sum_1^K \epsilon_j^\circ x_j \right\| \leqslant \sqrt{1 - \delta''} \sqrt{K} \left(\sum_1^K \|x_j\|^2 \right)^{1/2}$$

and 2) follows.

2) ⇒ 1). Is obvious : one just takes $x_1, \dots, x_K$ in the unit ball. We shall now prove that 1) ⇒ 2).

So we assume there exist $K \geq 2$, $\delta > 0$ such that for every sequence $x_1,\dots,x_K$ with $\|x_j\| \leq 1$, there is an admissible sequence of signs $(\epsilon_j)_{j=1,\dots,K}$ such that :

$$(1) \qquad \left\| \sum_{j=1}^{K} \epsilon_j x_j \right\| \leq (1-\delta)K .$$

If $\|x_1\| = \dots = \|x_K\|$, we obtain

$$(2) \qquad \left\| \sum_{j=1}^{K} \epsilon_j x_j \right\| \leq (1-\delta) \sum_{1}^{K} \|x_j\| .$$

Now, take any $x_1,x_2,\dots,x_K$ in E , all different from 0 . We can write, for every $j = 1,\dots,K$, a decomposition

$$x_j = x_j' + x_j'' ,$$

where

$$x_j' = (1-\lambda_j)x_j$$

$$x_j'' = \lambda_j x_j$$

and

$$\lambda_j = \frac{\min_{1 \leq i \leq K} \|x_i\|}{\|x_j\|} .$$

We set

$$m = \min_{1 \leq i \leq K} \|x_i\| .$$

So we have $0 < \lambda_j \leq 1$, and

$$\|x_j''\| = m , \quad \text{for} \quad j = 1,\dots,K .$$

Using (2), we know that there is an admissible sequence of signs $(\epsilon_j)_{j=1,\dots,K}$, such that

$$\left\| \sum_{1}^{K} \epsilon_j x_j'' \right\| \leq (1-\delta) \sum_{1}^{K} \|x_j''\| .$$

So we obtain, for this sequence $(\epsilon_j)_{j=1,\dots,K}$:

$$\left\| \sum_1^K \epsilon_j x_j \right\| \leqslant \left\| \sum_1^K \epsilon_j x_j' \right\| + \left\| \sum_1^K \epsilon_j x_j'' \right\|$$

$$\leqslant \sum_1^K \|x_j'\| + (1-\delta) \sum_1^K \|x_j''\|$$

$$\leqslant \sum_1^K [(1-\lambda_j)\|x_j\| + (1-\delta)\lambda_j \|x_j\|]$$

$$\leqslant \sum_1^K (1-\delta\lambda_j)\|x_j\| .$$

By Cauchy-Schwarz inequality, we obtain :

$$\leqslant \left(\sum_1^K (1-\delta\lambda_j)^2\right)^{1/2} \left(\sum_1^K \|x_j\|^2\right)^{1/2} .$$

But for one j at least, we have $\lambda_j = 1$ (say $j = j_0$). For the other j 's , we have $\lambda_j \geqslant 0$. We obtain :

$$\sum_1^K (1-\delta\lambda_j)^2 \leqslant K - 1 + (1-\delta)^2 = K - 2\delta + \delta^2$$

and

$$\left[\sum_1^K (1-\delta\lambda_j)^2\right]^{1/2} \leqslant (K - 2\delta + \delta^2)^{1/2} \leqslant \sqrt{K}\,(1-\delta') ,$$

with

$$\delta' = 1 - \left(1 - \frac{2\delta - \delta^2}{K}\right)^{1/2} > 0 ,$$

and 2) is proved.

This proposition will be used to show that the space $L_2(\Omega,\mathscr{A},\mu\,;E)$ is super-reflexive if and only if E is super-reflexive. The next paragraph will be devoted to the study of this space.

We recall that $L_2(\Omega, \mathcal{A}, \mu ; E)$ is the space of measurable functions on $(\Omega, \mathcal{A})$, with values in E, such that

$$\left(\int \|f(t)\|^2 d\mu(t) \right)^{1/2} < +\infty .$$

PROPOSITION 5. - E *is super-reflexive if and only if* $L_2(\Omega, \mathcal{A}, \mu ; E)$ *is super-reflexive.*

PROOF. - First, E is isometric to a subspace of $L_2(\Omega, \mathcal{A}, \mu ; E)$: take $\Omega_0 \subset \Omega$, with $\mu(\Omega_0) = 1$; the application

$$x \in E \to x \cdot 1_{\Omega_0}$$

is the required isometry. So, if $L_2(\Omega, \mathcal{A}, \mu ; E)$ (in short, $L_2(\Omega ; E)$) is super-reflexive, E has the same property.

Now, assume E to be super-reflexive. Let $f_1, \dots, f_K \in L_2(\Omega ; E)$. For each $\omega \in \Omega$, we have, by 3), proposition 4 :

$$\frac{1}{K} \sum_{(\epsilon_j) \in \mathscr{E}_K} \left\| \sum_{j=1}^{K} \epsilon_j f_j(\omega) \right\|^2 \leqslant (1 - \delta'') K \sum_{j=1}^{K} \|f_j(\omega)\|^2$$

and so :

$$\frac{1}{K} \sum_{(\epsilon_j) \in \mathscr{E}_K} \int \left\| \sum_{j=1}^{K} \epsilon_j f_j(\omega) \right\|^2 d\mu(\omega) \leqslant (1 - \delta'') K \sum_{j=1}^{K} \int \|f_j(\omega)\|^2 d\mu(\omega)$$

and $L_2(\Omega ; E)$ is super-reflexive, by the same proposition.

Finally, we shall see that, if E is not super-reflexive, we can find finite sequences satisfying James'condition, not only for every θ, $0 < \theta < 1$, as announced in chapter I, § 3, proposition 7, but in fact for every θ, $0 < \theta < 2$:

PROPOSITION 6. - *If* E *is not super-reflexive, one can find, for every* θ, $0 < \theta < 2$, *a sequence* $x_1^{(n)}, \dots, x_n^{(n)}$ *in the unit ball of* E *such that for every* k, $1 \leqslant k \leqslant n$,

$$\operatorname{dist}(\operatorname{conv}(x_1^{(n)}, \dots, x_k^{(n)}), \operatorname{conv}(x_{k+1}^{(n)}, \dots, x_n^{(n)})) > \theta .$$

PROOF. - Let $n \geqslant 1$, $\theta > 0$. Put $\epsilon = 2 - \theta$, $\epsilon' = \frac{\epsilon}{n}$. Since E is not J-convex, there are points $x_1^{(n)},\ldots,x_n^{(n)}$ in the unit ball, such that, for every admissible choice of signs $(\epsilon_k)_{k \leqslant n}$,

$$\left\| \sum_{k=1}^{n} \epsilon_k x_k^{(n)} \right\| \geqslant n(1 - \epsilon') .$$

Now, let k , $1 \leqslant k \leqslant n$, $a_1,\ldots,a_k , a_{k+1},\ldots,a_n$ positive numbers with $\sum_1^k a_i = \sum_{k+1}^n a_i = 1$. We get

$$\left\| \sum_1^k a_i x_i^{(n)} - \sum_{k+1}^n a_i x_i^{(n)} \right\| \geqslant \left\| \sum_1^k x_i^{(n)} - \sum_{k+1}^n x_i^{(n)} \right\| - \sum_1^k (1 - a_i) - \sum_{k+1}^n (1 - a_i)$$

$$\geqslant n(1-\epsilon') - k + 1 - (n-k) + 1 = 2 - n\epsilon' = \theta ,$$

and the proposition is proved.

REFERENCES ON CHAPTER III.

The proof that a non-reflexive space has squares is taken from R.C. JAMES [26]. The statement is also given in [26] for three points ; for higher number of points, the adaptation is immediate ; it was made by J. SCHAFFER and SUNDARESAN in [44]. Proposition 4 (homogeneous characterization of J-convexity) is in the paper of G. PISIER [40] .

CHAPTER IV

RENORMING SUPER-REFLEXIVE BANACH SPACES

This chapter will be devoted to the proof of a renorming theorem for super-reflexive Banach spaces. The fact that every super-reflexive space admits an equivalent uniformly convex norm was first proved by P. ENFLO [20]. The result we present here is more precise and is due to G. PISIER [40].

THEOREM. - *Let* E *be a super-reflexive Banach space. There exists a* $p > 2$, *a* $C > 0$, *and an equivalent norm on* E , *which is uniformly convex and has a modulus of convexity* $\delta(\epsilon)$ *satisfying* :

$$\delta(\epsilon) \geqslant C\,\epsilon^{p} \qquad \text{for every } \epsilon > 0 .$$

The proof will be divided into several steps :

§ 1. BANACH-VALUED MARTINGALES.

In chapter I, we have explained how super-reflexivity was connected to the Finite Tree Property. In chapter II, we have given estimates (on both sides) for basic sequences. We shall now connect Finite Trees in E with basic sequences in a space $L_2(\Omega\,;\,E)$: this will allow us to use these estimates, since we know that $L_2(\Omega\,;\,E)$ is super-reflexive when E is.

We consider $\Omega = \{-1, +1\}^{\mathbb{N}}$. The σ-field $\mathscr{A}$ is $\mathscr{P}(\Omega)$, and the measure μ is the Haar measure P on Ω (Ω is a compact group). This measure is a probability, which can also be written

$$P = \left(\frac{\delta_{-1} + \delta_{1}}{2}\right)^{\otimes \mathbb{N}}$$

where δ_{-1} , δ_1 are the Dirac masses at -1 , $+1$ respectively. So we have for example

$$P\{(1,-1,\epsilon_3,\epsilon_4,\dots)\ ;\ \epsilon_k = \pm 1,\ k \geq 3\} = \frac{1}{4}\ .$$

In $\Omega = \{-1, +1\}^{\mathbb{N}}$, we may consider the following sequence of σ-fields (each of them has only finitely many atoms) :

$\mathscr{B}_0 = \{\phi\ ,\ \Omega\}$, and for every $n \geq 1$, $\mathscr{B}_n$ is the σ-field generated by the projection on the first n coordinates. That is :

$$\mathscr{B}_1 = \{\phi\ ,\ \Omega, \{(+1,\epsilon_2,\epsilon_3,\dots)\ ;\ \epsilon_k = \pm 1, k \geq 2\}, \{(-1,\epsilon_2,\epsilon_3,\dots)\ ;\ \epsilon_k = \pm 1,\ k \geq 2\}\}$$

and more generally, $\mathscr{B}_n$ $(n \geq 1)$ consists in ϕ , Ω , and 2^n atoms, which are :

$$B_{\epsilon_1,\dots,\epsilon_n} = \{(\epsilon_1,\dots,\epsilon_n,\ \epsilon_{n+1},\dots)\ ;\ \epsilon_k = \pm 1 \text{ for } k \geq n+1\}$$

for $\epsilon_1 = \pm 1,\dots,\epsilon_n = \pm 1$.

We also call Q_n the projection onto the first n coodinates, that is :

$$Q_n(\epsilon_1,\dots,\epsilon_n,\epsilon_{n+1},\dots) = (\epsilon_1,\dots,\epsilon_n)\ ,$$

and so

$$B_{\epsilon_1,\dots,\epsilon_n} = Q_n^{-1}\{(\epsilon_1,\dots,\epsilon_n)\}.$$

The sequence $(\mathscr{B}_n)_{n \geq 0}$ is an increasing sequence of σ-fields. Therefore, one can define the E-valued martingales corresponding to $(\Omega, (\mathscr{B}_n)_{n \in \mathbb{N}})$. By definition, such a martingale is a sequence $(M_n)_{n \in \mathbb{N}}$ of E-valued random variables with the following properties :

- M_n is $\mathscr{B}_n$-measurable for every $n \geq 0$,
- $E^{\mathscr{B}_n} M_{n+1} = M_n$, for every $n \geq 0$,

where $E^{\mathscr{B}_n}$ denotes the conditional expectation on the σ-field $\mathscr{B}_n$.

Here, due to the special form of Ω and $\mathscr{B}_n$'s , the corresponding martingales are very simple :

We call $(\epsilon_k)_{k\geq 1} \to x_{(\epsilon_k)_{k\geq 1}}$ a function from $\Omega = \{-1, +1\}^{\mathbb{N}}$ into E . The functions which we consider will depend only on a finite number of coordinates : for each such function $x_{(\epsilon_k)_{k\geq 1}}$, there is an index $k_o \geq 1$ such that

$$x_{\epsilon_1,\dots,\epsilon_{k_o},\epsilon_{k_o+1},\epsilon_{k_o+2},\dots} = x_{\epsilon_1,\dots,\epsilon_{k_o},\epsilon'_{k_o+1},\epsilon'_{k_o+2},\dots}$$

for every $\epsilon_{k_o+1},\epsilon'_{k_o+1},\epsilon_{k_o+2},\epsilon'_{k_o+2},\dots$.

A function with this property will be called *eventually constant*, or *stationary*. For every $n \geq 1$, we can define :

$$E^{\mathscr{B}_n}(x_{(\epsilon_k)_{k\geq 1}})(\epsilon_1,\dots,\epsilon_n) = x_{(\epsilon_k)_{k\geq 1}} \qquad \text{if } n \geq k_o$$

$$= \frac{1}{2^{k_o-n}} \sum_{\substack{\epsilon_{n+1} = \pm 1 \\ \vdots \\ \epsilon_{k_o} = \pm 1}} x_{\epsilon_1,\dots,\epsilon_n,\epsilon_{n+1},\dots,\epsilon_{k_o}} \quad \text{if } n < k_o$$

and we just call it $x_{\epsilon_1,\dots,\epsilon_n}$, since it depends only on the first n coordinates. We obtain a martingale, if we put :

$$M_o = \frac{x_1 + x_{-1}}{2} \qquad \text{(we call } x_o \text{ this point)} ,$$

and

$$M_n(\epsilon_1,\dots,\epsilon_n) = x_{\epsilon_1,\dots,\epsilon_n} .$$

Indeed, this function is $\mathscr{B}_n$-measurable, and

$$\begin{aligned}(E^{\mathscr{B}_n} M_{n+1})(\epsilon_1,\dots,\epsilon_n) &= \frac{1}{2}(x_{\epsilon_1,\dots,\epsilon_n,+1} + x_{\epsilon_1,\dots,\epsilon_n,-1}) \\ &= x_{\epsilon_1,\dots,\epsilon_n} \\ &= M_n ,\end{aligned}$$

and $E^{\mathscr{B}_o} M_1 = M_o$.

Conversely, every martingale which is stationary, that is, which satisfies :

"There is an index $k_o \geqslant 1$ such that $M_{k_o+1} = M_{k_o+2} = \dots$ " ,

is obtained this way : just put

$$x_{(\epsilon_k)_{k\geqslant 1}} = M_{k_o+1}((\epsilon_k)_{k\geqslant 1}) .$$

Geometrically, the connection with the Trees is obvious : the function M_n corresponds to the data of 2^n points $x_{\epsilon_1,\dots,\epsilon_n}$ $(\epsilon_i = \pm 1)$ (but no condition such as $\|x_{\epsilon_1,\dots,\epsilon_{n+1},+1} - x_{\epsilon_1,\dots,\epsilon_{n-1},-1}\| \geqslant \epsilon$ is required), and M_{n-1} corresponds to the associated (n-1)-branch.

The important thing is that the conditional expectations $E^{\mathcal{B}_n}$ are norm-one projections in $L_2(\Omega ; E)$, that is :

$$(1) \qquad \left(\int \|E^{\mathcal{B}_n} f\|^2 \, dP\right)^{1/2} \leqslant \left(\int \|f\|^2 \, dP\right)^{1/2} ,$$

for every $f \in L_2(\Omega ; E)$.

Here, this property is clear : it is simply a consequence of the convexity of $\|\cdot\|^2$ (that is, $\left\| \frac{x+y}{2} \right\|^2 \leqslant \frac{1}{2}(\|x\|^2 + \|y\|^2)$) , it means only that :

$$\|x_{\epsilon_1,\dots,\epsilon_n}\|^2 \leqslant \frac{1}{2^{k_o-n}} \sum_{\substack{\epsilon_{n+1} = \pm 1 \\ \vdots \\ \epsilon_{k_o} = \pm 1}} \|x_{\epsilon_1,\dots,\epsilon_n,\epsilon_{n+1},\dots,\epsilon_{k_o}}\|^2 .$$

To the martingale $(M_n)_{n\geqslant 0}$ we can associate the sequence of consecutive differences :

$$\Delta_o = M_o , \ \Delta_1 = M_1 - M_o , \ \Delta_2 = M_2 - M_1 , \dots , \ \Delta_{n+1} = M_{n+1} - M_n , \dots$$

Therefore $E^{\mathcal{B}_n} \Delta_{n+1} = 0$ for $n \geqslant 0$. A consequence is that the sequence $(\Delta_n)_{n\geqslant 0}$ is a monotone basic sequence in $L_2(\Omega ; E)$. Indeed, for every m , n with $m \leqslant n$, every sequence of scalars $a_o,\dots,a_n$, we have :

$$\left(\int \left\| \sum_{o}^{n} a_i \Delta_i \right\|^2 dP\right)^{1/2} \geqslant \left(\int \left\| E^{\mathscr{B}_m} \sum_{i=o}^{n} a_i \Delta_i \right\|^2 dP\right)^{1/2} \qquad \text{(by (1))}$$

$$\geqslant \left(\int \left\| \sum_{i=o}^{m} a_i \Delta_i \right\|^2 dP\right)^{1/2} ,$$

which proves that $(\Delta_n)_{n \geqslant 0}$ is basic, with basis constant 1 , in $L_2(\Omega ; E)$

For every $x \in E$, we call $\mathscr{M}(x)$ the set of martingales on $(\Omega , (\mathscr{B}_n)_{n \geqslant 0})$, with values in E , which are stationary, and which have x as "starting point", that is $\mathbb{E} M_n = x \quad (n \geqslant 0)$. ($\mathbb{E}$ is the expectation). This last conditions means also that

$$x = \frac{x_{+1} + x_{-1}}{2} .$$

We also call M_∞ the terminal value of the martingale $(M_n)_{n \geqslant 0}$; since it is stationary, we know that, for some $k_o \geqslant 1$,

$$M_\infty((\epsilon_k)_{k \geqslant 1}) = M_{k_o+1}((\epsilon_k)_{k \geqslant 1}) .$$

It should be observed that we have developed the language of vector-valued martingales since, as will be seen, it is quite well adapted to our aim, but this language is by no means necessary, and could be completely avoided, since the properties of martingales which we need are elementary, and can be established directly for trees.

§ 2. A SEQUENCE OF NORMS ON A SUPER-REFLEXIVE SPACE.

We have seen that, if E was super-reflexive, so was $L_2(\Omega ; E)$. Therefore, by chapter II, there is a constant $C \geqslant 1$, and a $p \geqslant 2$ such that, for every normalized monotone basic sequence (f_n) in $L_2(\Omega ; E)$, one has, for every finite sequence of scalars (a_n) :

$$\left(\Sigma \, |a_n|^p \right)^{1/p} \leqslant C \left\| \Sigma \, a_n f_n \right\|_{L_2(\Omega \, ; \, E)} .$$

This implies, for every monotone basic sequence $g_o,\ldots,g_N$ of length $N+1$:

$$\left(\sum_{n=o}^{N}\|g_n\|^p_{L_2(\Omega\ ;\ E)}\right)^{1/p}\leqslant C\left\|\sum_{n=o}^{N}g_n\right\|_{L_2(\Omega\ ;\ E)}.$$

By Hölder's inequality, we have :

$$\left(\sum_{n=o}^{N}\|g_n\|^2_{L_2(\Omega\ ;\ E)}\right)^{1/2}\leqslant (N+1)^{\frac{1}{2}-\frac{1}{p}}\left(\sum_{n=o}^{N}\|g_n\|^p_{L_2(\Omega\ ;\ E)}\right)^{1/p}.$$

And so we get :

$$(2)\qquad \left(\sum_{n=o}^{N}\|g_n\|^2_{L_2(\Omega\ ;\ E)}\right)^{1/2}\leqslant C(N+1)^{\frac{1}{2}-\frac{1}{p}}\left\|\sum_{n=o}^{N}g_n\right\|_{L_2(\Omega\ ;\ E)}.$$

We put

$$c_N=\frac{\sqrt{3}}{2}\,\frac{1}{C.(N+1)^{\frac{1}{2}-\frac{1}{p}}}.$$

For every $N\geqslant 1$, we define, if $x\in E$:

$$(3)\qquad \|x\|_N^2=\inf\left\{\|M_\infty\|^2_{L_2(\Omega\ ;\ E)}-c_N^2\sum_{j\in A}\|\Delta_j\|^2_{L_2(\Omega\ ;\ E)}\right.$$
$$\left.(M_n)_{n\geqslant o}\in\mathscr{M}(x)\ ,\ A\subset\mathbb{N}^\star\ ,\ |A|=N\right\}.$$

From the Tree - point of view, we know that M_n defines a n-branch $(x_{\epsilon_1,\ldots,\epsilon_n})$ $(\epsilon_i=\pm 1)$. We may consider $\mathscr{M}(x)$ as the set of all n-branches (for all $n\geqslant 1$) "starting" from x , that is, satisfying

$$x=\frac{1}{2^n}\sum_{\substack{\epsilon_1=\pm 1\\ \vdots\\ \epsilon_n=\pm 1}}x_{\epsilon_1,\ldots,\epsilon_n}.$$

If we set :

$$\delta_{+1} = x_{+1} - x \ , \quad \delta_{-1} = x_{-1} - x$$

and, for $n \geqslant 1$:

$$\delta_{\epsilon_1,\ldots,\epsilon_n} = x_{\epsilon_1,\ldots,\epsilon_n} - x_{\epsilon_1,\ldots,\epsilon_{n-1}}$$

(this corresponds to the differences Δ_n), we can also write (3) under the form :

$$\|x\|_N^2 = \inf \left\{ \frac{1}{2^n} \sum_{\substack{\epsilon_1 = \pm 1 \\ \vdots \\ \epsilon_n = \pm 1}} \|x_{\epsilon_1,\ldots,\epsilon_n}\|_E^2 - c_N^2 \sum_{j \in A} \frac{1}{2^j} \sum_{\substack{\epsilon_1 = \pm 1 \\ \vdots \\ \epsilon_j = \pm 1}} \|\delta_{\epsilon_1,\ldots,\epsilon_j}\|_E^2 \ ; \right.$$

all n , N with $n > N$, all $x_{\epsilon_1,\ldots,\epsilon_n}$, n-branches starting from x , all $A \subset \{1,\ldots,n\}$ with $|A| = N \Big\}$.

We now give some properties of $\|\cdot\|_N$.

a) If we take the martingale constantly equal to x , we get $\|x_N\| \leqslant \|x\|$.

b) If we apply (2), we obtain, if $A \subset \mathbb{N}^\star$ with $|A| = N$,

$$\sum_{j \in A} \|\Delta_j\|^2_{L_2(\Omega\ ;\ E)} \leqslant \sum_{j=o}^{N} \|\Delta_j\|^2_{L_2(\Omega\ ;\ E)}$$

$$\leqslant \left[C(N+1)^{\frac{1}{2} - \frac{1}{p}} \right]^2 \left\| \sum_{j=o}^{N} \Delta_j \right\|^2_{L_2(\Omega\ ;\ E)} \ ;$$

but since $(\Delta_j)_{j \geqslant 0}$ is monotone basic :

$$\left\| \sum_{j=o}^{N} \Delta_j \right\|_{L_2(\Omega\ ;\ E)} \leqslant \left\| \sum_{j=o}^{\infty} \Delta_j \right\|_{L_2(\Omega\ ;\ E)} = \|M_\infty\|_{L_2(\Omega\ ;\ E)} \ ,$$

and we obtain

$$(4) \qquad \sum_{j \in A} \|\Delta_j\|^2_{L_2(\Omega\,;\,E)} \leqslant C^2 (N+1)^{1-\frac{2}{p}} \|M_\infty\|^2_{L_2(\Omega\,;\,E)}$$

$$\leqslant \frac{3}{4c_N^2} \|M_\infty\|^2_{L_2(\Omega\,;\,E)} .$$

c) By definition of the norm $\|\cdot\|_N$, for every $x \in E$, every $\alpha > 0$, we can find a martingale $(M_n)_{n \in \mathbb{N}}$ in $\mathscr{M}(x)$ and a set $A \subset \mathbb{N}^\star$, with $|A| = N$, such that

$$\|x\|_N^2 \geqslant \|M_\infty\|^2_{L_2(\Omega\,;\,E)} - c_N^2 \sum_{j \in A} \|\Delta_j\|^2_{L_2(\Omega\,;\,E)} - \alpha .$$

Using (4), we get :

$$(5) \qquad \|x\|_N^2 \geqslant \frac{1}{4} \|M_\infty\|^2_{L_2(\Omega\,;\,E)} - \alpha .$$

Also, we have $x = \mathbb{E} M_\infty$, and therefore

$$\|x\| \leqslant \mathbb{E} \|M_\infty\|_E \leqslant (\mathbb{E} \|M_\infty\|_E^2)^{1/2} = \|M_\infty\|_{L_2(\Omega\,;\,E)} ,$$

which gives

$$\|x\|_N^2 \geqslant \frac{1}{4} \|x\|^2 - \alpha ,$$

and since this is true for every $\alpha > 0$, we obtain finally :

$$(6) \qquad \frac{1}{2}\|x\| \leqslant \|x\|_N \leqslant \|x\| .$$

We shall now show that $\|\cdot\|_N$ is a norm on E. This will be done in two steps :

LEMMA 2. - *For all* $x,y \in E$, $\left\| \frac{x+y}{2} \right\|_N^2 \leqslant \frac{1}{2} (\|x\|_N^2 + \|y\|_N^2)$.

PROOF OF LEMMA 2. - Let $x,y \in E$, $\alpha > 0$. We can find two martingales $(M_n)_{n \geqslant 1}$ and $(M'_n)_{n \geqslant 1}$, belonging respectively to $\mathscr{M}(x)$ and $\mathscr{M}(y)$, two sets A, A' contained in $\mathbb{N}^\star$, with $|A| = |A'| = N$, two numbers

$\beta > 0$, $\beta' > 0$, with $\beta < \alpha$, $\beta' < \alpha$, such that

$$\text{(a)} \quad \|x\|_N^2 = \|M_\infty\|^2_{L_2(\Omega\,;\,E)} - c_N^2 \sum_{j\in A} \|\Delta_j\|^2_{L_2(\Omega\,;\,E)} - \beta$$

(7)

$$\text{(b)} \quad \|y\|_N^2 = \|M'_\infty\|^2_{L_2(\Omega\,;\,E)} - c_N^2 \sum_{j\in A'} \|\Delta'_j\|^2_{L_2(\Omega\,;\,E)} - \beta' .$$

We shall first reduce to the case where $A = A'$ by "slowing down" the martingales $(M_n)_{n\geqslant 1}$ and $(M'_n)_{n\geqslant 1}$. This is done the following way :

Let k_1 be the first index which is, for example, in A and not in A' . Then we replace the martingale $(M'_n)_{n\geqslant 1}$ by the martingale $(\widetilde{M}'_n)_{n\geqslant 1}$, with

$\widetilde{M}'_n = M'_n$ for $n \leqslant k_1-1$

$\widetilde{M}'_{k_1} = M'_{k_1-1}$

$\widetilde{M}'_n = M'_{n-1}$ $n > k_1$.

Of course, $(\widetilde{M}'_n)_{n\geqslant 1}$ belongs to $\mathscr{M}(y)$, and since $\widetilde{\Delta}'_{k_1}$ is now equal to zero, since $M'_\infty = \widetilde{M}'_\infty$, $\Delta'_{j-1} = \widetilde{\Delta}'_j$ $j > k_1$, we have also :

$$\|y\|_N^2 = \|\widetilde{M}'_\infty\|^2_{L_2(\Omega\,;\,E)} - c_N^2 \sum_{j\in \widetilde{A}'} \|\widetilde{\Delta}'_j\|^2_{L_2(\Omega\,;\,E)} - \beta' ,$$

where $\widetilde{A}' = \{m \in \mathbb{N}^\star\,;\; m \in A' \text{ if } m < k_1 ,\; m-1 \in A' \text{ if } m > k_1\} \cup \{k_1\}$.

and now, we have a set $\widetilde{A}'$ which contains k_1 .

We go on the same way : if k_2 is the next integer which is in A but not in A' , or in A' but not in A , we replace M' or M by a martingale which satisfies $\widetilde{M}_j = M_j$ $j \leqslant k_2-1$, $\widetilde{M}_{k_2} = M_{k_2-1}$, $\widetilde{M}_j = M_{j-1}$ $j > k_2$. Since there are only finitely many points in A and A' , we shall finally obtain a set, which we call A , common for both. So we just take (7) (a) and (b) with $A' = A$.

We consider the shift S on $L_2(\Omega ; E)$, that is the application defined by

$$S(x_{(\epsilon_j)_{j\geqslant 1}}) = (x_{(\epsilon_j)_{j\geqslant 2}}) .$$

We now define a martingale $(M''_n)_{n\geqslant 1}$, using (7) (a) and (b), by :

$$M''_o = \frac{x+y}{2} ,$$

$$M''_1(\epsilon_1) = x \quad \text{if} \quad \epsilon_1 = +1$$

$$= y \quad \text{if} \quad \epsilon_1 = -1$$

and, if $n > 1$,

$$M''_n(\epsilon_1,\ldots,\epsilon_n) = (SM_{n-1})(\epsilon_1,\ldots,\epsilon_n) = M_{n-1}(\epsilon_2,\ldots,\epsilon_n) \quad \text{if} \quad \epsilon_1 = +1$$

$$= (SM'_{n-1})(\epsilon_1,\ldots,\epsilon_n) = M'_{n-1}(\epsilon_2,\ldots,\epsilon_n) \quad \text{if} \quad \epsilon_1 = -1$$

This construction can be more easily understood in terms of n-branches: if we start from $x_{\epsilon_1,\ldots,\epsilon_n}$ and $y_{\epsilon_1,\ldots,\epsilon_n}$, we define $z_{\epsilon_1,\ldots,\epsilon_n}$ by :

$$z_{+1} = x , \quad z_{-1} = y ,$$

and

$$z_{1,\epsilon_2,\ldots,\epsilon_n} = x_{\epsilon_2,\ldots,\epsilon_n} , \quad z_{-1,\epsilon_2,\ldots,\epsilon_n} = y_{\epsilon_2,\ldots,\epsilon_n} .$$

Obviously, we have :

$$\|M''_\infty\|^2_{L_2(\Omega ; E)} = \frac{1}{2}\left[\|M_\infty\|^2_{L_2(\Omega ; E)} + \|M'_\infty\|^2_{L_2(\Omega ; E)}\right]$$

and also, for $j > 1$:

$$\|\Delta''_j\|^2_{L_2(\Omega ; E)} = \frac{1}{2}\left[\|\Delta_{j-1}\|^2_{L_2(\Omega ; E)} + \|\Delta'_{j-1}\|^2_{L_2(\Omega ; E)}\right] .$$

For $j = 1$, we have

$$\|\Delta''_1\|^2_{L_2(\Omega\ ;\ E)} = \left\| \frac{x-y}{2} \right\|^2$$

and for $j = 0$,

$$\Delta''_o = \frac{x+y}{2} .$$

Obviously also, we have $(M''_n)_{n \geq 0} \in \mathcal{M}(\frac{x+y}{2})$. If we use this martingale, and the set $B = A + 1$, to estimate the norm $\left\| \frac{x+y}{2} \right\|^2_N$, we get :

$$\left\| \frac{x+y}{2} \right\|^2_N \leq \|M''_\infty\|^2_{L_2(\Omega\ ;\ E)} - c_N^2 \sum_{j \in B} \|\Delta''_j\|^2_{L_2(\Omega\ ;\ E)}$$

$$\leq \frac{1}{2} \left[\|M_\infty\|^2_{L_2(\Omega\ ;\ E)} + \|M'_\infty\|^2_{L_2(\Omega\ ;\ E)} \right]$$

$$- \frac{1}{2} c_N^2 \sum_{j \in A} \left[\|\Delta_j\|^2_{L_2(\Omega\ ;\ E)} + \|\Delta'_j\|^2_{L_2(\Omega\ ;\ E)} \right]$$

$$\leq \frac{1}{2} (\|x\|^2_N + \|y\|^2_N + \beta + \beta') \leq \frac{1}{2} (\|x\|^2_N + \|y\|^2_N) + \alpha$$

and since this holds for every $\alpha > 0$, the lemma is proved.

We shall now see that $\|\cdot\|_N$ is continuous on E .

LEMMA 3. - $\|\cdot\|_N$ *is continuous on* E .

PROOF OF LEMMA 3. - Let $x \in E$, $\alpha > 0$. Let $(M_n)_{n \geq 0} \in \mathcal{M}(x)$, and $A \in \mathbb{N}^\star$, with $|A| = N$, such that :

$$\|x\|^2_N \geq \|M_\infty\|^2_{L_2(\Omega\ ;\ E)} - c_N^2 \sum_{j \in A} \|\Delta_j\|^2_{L_2(\Omega\ ;\ E)} - \alpha .$$

For any $u \in E$, $(M_n + u)_{n \in \mathbb{N}}$ belongs to $\mathcal{M}(x+u)$, and the increments of $M_n + u$, for $n \geq 1$, are the same as the increments of M_n . Therefore, we get :

$$\begin{aligned}\|u + x\|_N^2 &\leq \mathbb{E}\,\|M_\infty + u\|_E^2 - c_N^2 \sum_{j\in A} \|\Delta_j\|^2_{L_2(\Omega\,;\,E)} \\ &\leq \mathbb{E}\,\|M_\infty\|_E^2 + 2\|u\|_E\ \mathbb{E}\,\|M_\infty\|_E + \|u\|_E^2 - c_N^2 \sum_{j\in A} \|\Delta_j\|^2_{L_2(\Omega\,;\,E)} \\ &\leq \|x\|_N^2 + \alpha + 2\|u\|_E\ \mathbb{E}\,\|M_\infty\|_E + \|u\|_E^2\ .\end{aligned}$$

But

$$\begin{aligned}\mathbb{E}\,\|M_\infty\|_E \leq (\mathbb{E}\,\|M_\infty\|_E^2)^{1/2} &\leq 4(\|x\|_N^2 + \alpha)\ , \qquad \text{by (5)} \\ &\leq 4(\|x\|^2 + \alpha)\ .\end{aligned}$$

So, since this is valid for every $\alpha > 0$, we get

$$\|x + u\|_N^2 - \|x\|_N^2 \leq 8\|u\|_E\ \|x\|_E^2 + \|u\|_E^2$$

which implies

$$\left|\,\|x + u\|_N^2 - \|x\|_N^2\right| \leq 8\|u\|_E\ \|x\|_E^2 + \|u\|_E^2\ ,$$

and the lemma is proved.

It is now clear that $\|\cdot\|_N$ is a norm on E : it is obviously positively homogeneous, and, from lemma 2, follows that

$$\left\|\frac{x + y}{2}\right\|_N \leq 1 \quad \text{if} \quad \|x\|_N = \|y\|_N = 1\ .$$

More generally,

$$\|\lambda x + (1 - \lambda)y\|_N \leq 1 \quad \text{if} \quad \|x\|_N = \|y\|_N = 1\ ,$$

and if λ can be written $\lambda = \sum_{j=1}^{n} a_j/2^j$, for some $n \geq 1$, $a_1,\ldots,a_n \in \mathbb{N}$. Since the set of such λ's is dense in $[0,1]$, and since $\|\cdot\|_N$ is continuous, the same relation holds for every $\lambda \in [0,1]$: this says that $\{x \in E\ ,\ \|x\|_N \leq 1\}$ is convex, and therefore $\|\cdot\|_N$ is a norm. By (6), this norm is equivalent to the norm of E. We shall now investigate its modulus of convexity, which we call δ_N.

LEMMA 4. - *If* x, y *are two points in* E *such that*

$$\frac{\|x - y\|}{\left[\frac{1}{2}(\|x\|_N^2 + \|y\|_N^2)\right]^{1/2}} \geq 8\sqrt{\frac{2}{3}}\, C\, N^{-\frac{1}{p}}$$

then

$$\frac{\left\| \frac{x + y}{2} \right\|_N^2}{\frac{1}{2}(\|x\|_N^2 + \|y\|_N^2)} \leq 1 - \frac{1}{N}$$

where C *is the constant in* (2).

PROOF OF LEMMA 4. - Let x, y in E, with $\frac{1}{2}(\|x\|_N^2 + \|y\|_N^2) = 1$. Let $\alpha > 0$. Let $(M_n)_{n \geq 0}$, $(M'_n)_{n \geq 0}$ be martingales in $\mathscr{M}(x)$, $\mathscr{M}(y)$ respectively, satisfying (7) (a) and (7) (b). Let $(M''_n)_{n \geq 0}$ be the martingale in $\mathscr{M}(\frac{x + y}{2})$ constructed in the proof of lemma 2.

We have seen that, for $j > 1$:

$$\|\Delta''_j\|^2_{L_2(\Omega\, ;\, E)} = \frac{1}{2}\left(\|\Delta_{j-1}\|^2_{L_2(\Omega\, ;\, E)} + \|\Delta'_{j-1}\|^2_{L_2(\Omega\, ;\, E)}\right).$$

Thus, if $|A| = N$

$$\begin{aligned} c_N^2 \sum_{j \in A} \|\Delta''_j\|^2_{L_2(\Omega\, ;\, E)} &= \frac{1}{2} c_N^2 \sum_{j \in A} \left(\|\Delta_{j-1}\|^2_{L_2(\Omega\, ;\, E)} + \|\Delta'_{j-1}\|^2_{L_2(\Omega\, ;\, E)}\right) \\ &\leq \frac{3}{8}\left(\|M_\infty\|^2_{L_2(\Omega\, ;\, E)} + \|M'_\infty\|^2_{L_2(\Omega\, ;\, E)}\right), \quad \text{by (4)}, \\ &\leq \frac{3}{2}\left(\|x\|_N^2 + \|y\|_N^2 + 2\alpha\right), \end{aligned}$$

that is :

$$c_N^2 \sum_{j \in A} \|\Delta''_j\|^2_{L_2(\Omega\, ;\, E)} \leq 3(1 + \alpha).$$

Therefore, there is a $j_o \in A$ such that

$$(8) \qquad c_N^2 \cdot \|\Delta''_{j_o}\| \leqslant \frac{3(1+\alpha)}{N} .$$

But also, we know that :

$$\|\Delta''_1\|^2 = \left\| \frac{x-y}{2} \right\|^2 .$$

Let us assume that

$$c_N^2 \left\| \frac{x-y}{2} \right\|^2 \geqslant \frac{4}{N} \quad ;$$

we obtain :

$$(9) \qquad \|\Delta''_1\|^2 \geqslant \frac{4}{N\, c_N^2} .$$

We define the set $B = ((A \setminus \{j_o\}) + 1) \cup \{1\}$. Then $|B| = N$, and we have :

$$c_N^2 \sum_{j \in B} \|\Delta''_j\|^2_{L_2(\Omega\,;\,E)} \geqslant c_N^2 \sum_{j \in A} \|\Delta''_{j+1}\|^2_{L_2(\Omega\,;\,E)} - \frac{3(1+\alpha)}{N} + \frac{4}{N}$$

$$\geqslant c_N^2 \sum_{j \in A} \|\Delta''_{j+1}\|^2_{L_2(\Omega\,;\,E)} + \frac{1-3\alpha}{N} .$$

Finally, we obtain :

$$\left\| \frac{x+y}{2} \right\|_N^2 \leqslant \|M''_\infty\|^2_{L_2(\Omega\,;\,E)} - c_N^2 \sum_{j \in B} \|\Delta''_j\|^2_{L_2(\Omega\,;\,E)}$$

$$\leqslant \frac{1}{2} \left(\|M_\infty\|^2_{L_2(\Omega\,;\,E)} + \|M'_\infty\|^2_{L_2(\Omega\,;\,E)}\right) - c_N^2 \sum_{j \in A} \|\Delta''_{j+1}\|^2_{L_2(\Omega\,;\,E)} - \frac{1-3\alpha}{N}$$

$$\leqslant \frac{1}{2} \left(\|M_\infty\|^2_{L_2(\Omega\,;\,E)} + \|M'_\infty\|^2_{L_2(\Omega\,;\,E)}\right)$$

$$- c_N^2 \sum_{j \in A} \frac{1}{2} \left(\|\Delta_j\|^2_{L_2(\Omega\,;\,E)} + \|\Delta'_j\|^2_{L_2(\Omega\,;\,E)}\right) - \frac{1-3\alpha}{N}$$

$$\leqslant \frac{\|x\|_N^2 + \|y\|_N^2 + 2\alpha}{2} - \frac{1-3\alpha}{N}$$

$$\leqslant 1 + \alpha + \frac{1-3\alpha}{N} .$$

Since this holds for every $\alpha > 0$, we obtain that, if $\|x\|_N^2 + \|y\|_N^2 = 2$ and $c_N^2 \left\| \frac{x-y}{2} \right\|^2 \geq \frac{4}{N}$, then

$$\left\| \frac{x+y}{2} \right\|_N^2 \leq 1 - \frac{1}{N} .$$

So if

$$\frac{\|x - y\|}{\left[\frac{1}{2} (\|x\|_N^2 + \|y\|_N^2)\right]^{1/2}} \geq 8\sqrt{\frac{2}{3}}\, C\, N^{-\frac{1}{p}} \geq \frac{4}{c_N \sqrt{N}} \tag{10}$$

then

$$\left\| \frac{x+y}{2} \right\|_N^2 \leq (1 - \frac{1}{N})\, \frac{1}{2}\, (\|x\|_N^2 + \|y\|_N^2) , \tag{11}$$

and the lemma is proved.

This is not yet uniform convexity of the norm $\|\cdot\|_N$, for two reasons : first, the distance in (10) is measured with the norm $\|\cdot\|$, and not $\|\cdot\|_N$ (but this is not a serious objection, because of the equivalence of the two norms), and, mainly, because we have a conclusion only for $8\sqrt{\frac{2}{3}}\, C\, N^{-\frac{1}{p}}$, and not for arbitrarily small $\epsilon > 0$, in the second member of (10). One can say that each of the norms $\|\cdot\|_N$ has the uniform convexity property only for points which are at distance at least ϵ_N from each other. Now, by mixing up the norms $\|\cdot\|_N$, we shall build a norm which is really uniformly convex.

LEMMA 5. - *Let* $\alpha > 1$, $\beta > 1$, *and let* $(\|\cdot\|_N)_{N \geq 1}$ *be a sequence of norms on* E , *with :*

1°) $\frac{1}{2}\|x\| \leq \|x\|_N \leq \|x\|$, *for all* $x \in E$, *all* $N \geq 1$,

2°) *for all* $x,y \in E$, *if*

$$\frac{\|x - y\|}{\frac{1}{2} (\|x\|_N^2 + \|y\|_N^2)^{1/2}} \geq \alpha N^{-\frac{1}{p}} , \textit{ then } \left\| \frac{x+y}{2} \right\|_N^2 \leq (1 - \frac{\beta}{N})\, \frac{1}{2}(\|x\|_N^2 + \|y\|_N^2)$$

then, for every $p' > p$ *there exists a constant* C' , *and a norm* $|\cdot|$ *on* E , *satisfying*

$$\frac{1}{2}\|x\| \leqslant |x| \leqslant \|x\| \qquad \textit{for all } x \in E ,$$

and which is uniformly convex, and has a modulus of convexity with $\delta(\epsilon) \geqslant C' \epsilon^{p'}$ *for all* $\epsilon > 0$.

PROOF OF LEMMA 5. - We put

$$|x|^2 = \frac{6}{\pi^2} \sum_{n=1}^{\infty} \frac{1}{n^2} \|x\|_{2^n}^2 \qquad \text{for all } x \in E .$$

This defines a norm on E , which satisfies obviously

$$(12) \qquad \frac{1}{2}\|x\| \leqslant |x| \leqslant \|x\| .$$

Take now $x, y \in E$ with $\frac{1}{2}(|x|^2 + |y|^2) = 1$, $|x - y| \geqslant \epsilon$. Then $\|x - y\| \geqslant \epsilon$, by (12). We have also, for all $N \geqslant 1$

$$\frac{1}{2}(\|x\|_N^2 + \|y\|_N^2) \leqslant \frac{1}{2}(\|x\|^2 + \|y\|^2) \leqslant 2(|x|^2 + |y|^2) = 4$$

and so :

$$\frac{\|x - y\|}{\left(\frac{1}{2}(\|x\|_N^2 + \|y\|_N^2)\right)^{1/2}} \geqslant \frac{\epsilon}{2} .$$

We take $N = 2^k$, where k is the smallest integer such that $\frac{\epsilon}{2} \geqslant \frac{\alpha}{2^{k/p}}$, that is

$$k = \left[\frac{p \operatorname{Log} \frac{2\alpha}{\epsilon}}{\operatorname{Log} 2}\right] + 1 \qquad \text{(where } [\cdot] \text{ is the entire part).}$$

Then we have

$$\frac{\|x - y\|}{\left(\frac{1}{2}(\|x\|_{2^k}^2 + \|y\|_{2^k}^2)\right)^{1/2}} \geqslant \frac{\alpha}{2^{k/p}} .$$

By assumption 2), this implies, for this k :

$$(13)\qquad \left\| \frac{x+y}{2} \right\|_{2^k}^2 \leq \left(1 - \frac{\beta}{2^k}\right) \frac{1}{2} \left(\|x\|_{2^k}^2 + \|y\|_{2^k}^2\right)$$

For every $n \geq 1$, we have, since $\|\cdot\|_{2^n}$ is a norm :

$$(14)\qquad \left\| \frac{x+y}{2} \right\|_{2^n}^2 \leq \frac{1}{2} \left(\|x\|_{2^n}^2 + \|y\|_{2^n}^2\right) .$$

So we obtain :

$$\left| \frac{x+y}{2} \right|^2 \leq \frac{6}{\pi^2} \sum_{n \neq k} \frac{1}{2n^2} \left(\|x\|_{2^n}^2 + \|y\|_{2^n}^2\right) + \frac{6}{\pi^2} \frac{1}{k^2} \left(1 - \frac{\beta}{2^k}\right) \frac{1}{2} \left(\|x\|_{2^k}^2 + \|y\|_{2^k}^2\right)$$

$$\leq \frac{1}{2} \left(|x|^2 + |y|^2\right) - \frac{6}{\pi^2 k^2} \frac{\beta}{2^k} \frac{1}{2} \left(\|x\|_{2^k}^2 + \|y\|_{2^k}^2\right)$$

but, using again 1°) and (12), we have :

$$\|x\|_{2^k}^2 + \|y\|_{2^k}^2 \geq \frac{1}{4} \left(|x|^2 + |y|^2\right)$$

and therefore :

$$\left| \frac{x+y}{2} \right|^2 \leq \frac{1}{2} \left(|x|^2 + |y|^2\right) - \frac{6\beta}{4\pi^2 k^2 2^{k+3}} \left(|x|^2 + |y|^2\right)$$

$$\leq 1 - \frac{3\beta}{\pi^2 k^2 2^{k+3}} \quad , \qquad \text{since} \quad |x|^2 + |y|^2 = 2 .$$

So we have obtained the following : for every $\epsilon > 0$, every $x,y \in E$ with $|x|^2 + |y|^2 = 2$, $|x - y| \geq \epsilon$, if we set

$$k = \left[\frac{p \operatorname{Log} \frac{2\alpha}{\epsilon}}{\operatorname{Log} 2} \right] + 1$$

then

$$\left| \frac{x+y}{2} \right|^2 \leq 1 - \frac{3\beta}{\pi^2 k^2 2^{k+3}} .$$

This shows indeed that the norm $|\cdot|$ is uniformly convex, but we still have to compute its modulus of convexity.

We have

$$2^k \leqslant 2\left(\frac{2\alpha}{\epsilon}\right)^p ,$$

and therefore

$$\delta(\epsilon) \geqslant \frac{3\beta}{\pi^2\left(\frac{p\,\mathrm{Log}(\frac{2\alpha}{\epsilon})}{\mathrm{Log}\,2}+1\right)^2 \; 2(\frac{2\alpha}{\epsilon})^p}$$

$$\geqslant \frac{3\beta(\mathrm{Log}\,2)^2}{2\pi^2\;(2\alpha)^p}\;\frac{\epsilon^p}{\mathrm{Log}(2(\frac{2\alpha}{\epsilon})^p\;)} .$$

If we put :

$$K = \frac{3\beta(\mathrm{Log}\,2)^2}{2\pi^2\;(2\alpha)^p}\cdot\frac{1}{p} ,$$

we have :

$$\liminf_{\epsilon\to 0} \frac{\delta(\epsilon)\;\mathrm{Log}^{1/\epsilon}}{K\,\epsilon^p} \geqslant 1$$

which implies, a fortiori, that for every $p' > p$, there exists a constant $C' > 0$ such that

$$\delta(\epsilon) \geqslant C\,\epsilon^{p'} , \quad \text{for every } \epsilon > 0 ,$$

and our theorem is proved.

So we have shown that every super-reflexive space could be endowed with an equivalent norm for which it was uniformly convex. Moreover, this norm has a modulus of convexity which is of "power-type", that is $\delta(\epsilon) \geqslant C\,\epsilon^q$, for some $q > 2$. The norm $|\cdot|$ which we have built satisfies

$$\frac{1}{2}\|x\| \leqslant |x| \leqslant \|x\| \qquad \text{for every} \quad x \in E ,$$

but we can replace it by another one, closer to $\|\cdot\|$:

PROPOSITION 6. - *Let* E *be a super-reflexive Banach space,* $\|\cdot\|$ *its norm. Let* $|\cdot|$ *be a uniformly convex norm on* E *, with*

$$\frac{1}{2}\|x\| \leqslant |x| \leqslant \|x\| \qquad \textit{for every} \quad x \in E .$$

Then, for each γ *,* $0 < \gamma \leqslant 1$ *, the norm :*

$$|x|_\gamma = (\|x\|^2 + \gamma|x|^2)^{1/2}$$

satisfies

$$\text{(a)} \qquad \sqrt{1 + \frac{\gamma}{4}}\ \|x\| \leqslant |x|_\gamma \leqslant \sqrt{1+\gamma}\ \|x\|$$

and its modulus of convexity, $\overset{(\gamma)}{\delta}(\epsilon)$ *, satisfies :*

$$\text{(b)} \qquad \overset{(\gamma)}{\delta}(\epsilon) \geqslant \frac{\gamma}{8(1+\gamma)}\ \delta\left(\frac{\epsilon}{2}\sqrt{\frac{1+\frac{\gamma}{4}}{1+\gamma}}\right) .$$

PROOF. - The inequalities (a) are obvious, by (12) and the definition of $|\cdot|_\gamma$. For (b), let $x,y \in E$, with $|x|_\gamma^2 + |y|_\gamma^2 = 2$, $|x-y|_\gamma \geqslant \epsilon$. Then we have :

$$\frac{1}{2}(|x|^2 + |y|^2) \leqslant \frac{1}{2}(\|x\|^2 + \|y\|^2) \leqslant \frac{1}{2}\left(1 + \frac{\gamma}{4}\right)^{-1}(|x|_\gamma^2 + |y|_\gamma^2) \leqslant \frac{1}{1+\frac{\gamma}{4}} .$$

Also :

$$|x-y| \geqslant \frac{1}{2}\|x-y\| \geqslant \frac{1}{2\sqrt{1+\gamma}}|x-y|_\gamma = \frac{\epsilon}{2\sqrt{1+\gamma}} ,$$

which gives

$$\frac{|x-y|}{\left[\frac{1}{2}(|x|^2+|y|^2)\right]^{1/2}} \geqslant \frac{\epsilon}{2}\sqrt{\frac{1+\frac{\gamma}{4}}{1+\gamma}} .$$

By uniform convexity, we obtain, putting $\epsilon' = \frac{\epsilon}{2}\sqrt{\frac{1+\frac{\gamma}{4}}{1+\gamma}}$.

$$\left| \frac{x+y}{2} \right|^2 \leqslant (1 - \delta(\epsilon'))\ \frac{1}{2}\ (|x|^2 + |y|^2)\ .$$

Since we have

$$\left\| \frac{x+y}{2} \right\|^2 \leqslant \frac{1}{2}\ (\|x\|^2 + \|y\|^2)\ ,$$

we get

$$\left| \frac{x+y}{2} \right|_\gamma^2 \leqslant \frac{1}{2}\ (\|x\|^2 + \|y\|^2) + \frac{\gamma}{2}\ (|x|^2 + |y|^2) - \frac{\gamma}{2}\ (|x|^2 + |y|^2)\delta(\epsilon')$$

$$\leqslant \frac{1}{2}\ (|x|_\gamma^2 + |y|_\gamma^2) - \frac{\gamma}{2}\ (|x|^2 + |y|^2)\delta(\epsilon')\ .$$

But

$$|x|^2 + |y|^2 \geqslant \frac{1}{4}\ (\|x\|^2 + \|y\|^2) \geqslant \frac{1}{4(1+\gamma)}\ (|x|_\gamma^2 + |y|_\gamma^2)$$

and so finally

$$\left| \frac{x+y}{2} \right|_\gamma^2 \leqslant 1 - \frac{\gamma}{8(1+\gamma)}\ \delta(\epsilon')\ ,$$

and the proposition is proved.

This proposition shows that if $\delta(\epsilon) \underset{\epsilon \to 0}{\sim} C\,\epsilon^q$, then $\overset{(\gamma)}{\delta}(\epsilon) \underset{\epsilon \to 0}{\sim} C_\gamma\,\epsilon^q$, for the same exponent q . But of course the constant C_γ depends on γ , and $C_\gamma \xrightarrow[\gamma \to 0]{} 0$, if the original norm $\|\cdot\|$ on E was not already uniformly convex.

PROPOSITION 7. - *Every super-reflexive space admits an equivalent uniformly smooth norm* $|\cdot|'$, *with a modulus of smoothness* $\rho(\tau)$ *satisfying, for some* $C'' > 0$, *some* r , $1 < r < 2$,

$$\rho(\tau) \leqslant C''\,\tau^r\ .$$

PROOF. - We know that if E is super-reflexive, so is $E^\star$ (chapter I, § 3, corollary 8). So, by the previous result, $E^\star$ admits an equivalent norm, $|\cdot|^{\star}$, with

$$\frac{1}{2}\,\|\xi\|_{E^{\star}} \leqslant |\xi|^{\star} \leqslant \|\xi\|_{E^{\star}} \quad , \quad \text{for all } \xi \in E^{\star}$$

and the modulus of convexity of $|\cdot|^{\star}$, which we call $\delta^{\star}$, satisfies

$$\delta^{\star}(\epsilon) \geqslant C\,\epsilon^{q} \text{ , for some } C > 1 \text{ , some } q > 2 \text{ .}$$

On E , we define a new norm by

$$|x| = \sup_{|\xi|^{\star} \leqslant 1} |\xi(x)| \text{ ,}$$

then we get

$$\|x\| \leqslant |x| \leqslant 2\|x\| \qquad \text{for all } x \in E \text{ .}$$

Let us now compute the modulus of smoothness of $|\cdot|$. For this purpose, we use the second duality formula we met in third part, chapter II :

$$\rho_{E}(\tau) = \sup_{0 \leqslant \epsilon \leqslant 2} \left(\frac{\tau\epsilon}{2} - \delta^{\star}(\epsilon) \right) \text{ ,}$$

for E equipped with $|\cdot|$.

If $\delta^{\star}(\epsilon) \geqslant C\,\epsilon^{q}$, then :

$$\rho(\tau) \leqslant \sup_{0 \leqslant \epsilon \leqslant 2} \left(\frac{\tau\epsilon}{2} - C\,\epsilon^{q} \right) \text{ .}$$

The maximum of $\frac{\tau\epsilon}{2} - C\,\epsilon^{q}$ is obtained for $\epsilon = \left(\frac{\tau}{2qC}\right)^{1/q-1}$, its value is

$$\frac{\tau^{r}}{r2^{r}C^{r-1}q^{r-1}} \quad , \quad \text{if } \frac{1}{r} + \frac{1}{q} = 1 \text{ ,}$$

and so,

$$\rho(\tau) \leqslant C''\,\tau^{r} \text{ ,}$$

as announced.

Therefore, on a super-reflexive space, we can construct two equivalent norms : one which is uniformly convex, and with $\delta(\epsilon) \geqslant C_1 \epsilon^q$, $q \geqslant 2$, and one which is uniformly smooth, with $\rho(\tau) \leqslant C_2 \tau^r$, $1 < r \leqslant 2$. It was shown by E. ASPLUND [2] that one can find a third, equivalent, norm which is at the same time both uniformly convex and uniformly smooth (see also [6]), but it is not known if, for this new norm, one can keep the estimates $\delta(\epsilon) \geqslant C_1' \epsilon^q$, $\rho(\tau) \leqslant C_2' \tau^r$ (C_1 and C_2 may change, but q and r being conserved) ; the known proofs give, for this third norm, $q' > q$ and a $r' < r$: these numbers q' and r' are not therefore directly related to the exponents in James' estimates for the basic sequences.

EXERCISES ON CHAPTER IV.

EXERCISE 1. - Let E be a Banach space and q a continuous semi-norm on E : there is a constant $C > 0$ such that

$$q(x) \leqslant C\|x\| \qquad \text{for all } x \text{ in } E .$$

We say that two points x_1, x_{-1} form a $(1, \epsilon)$ q-branch in E if $q(x_1 - x_{-1}) \geqslant \epsilon$.

Assume that a $(n-1, \epsilon)$ q-branch has been defined. We say that the 2^n points $(x_{\epsilon_1, \ldots, \epsilon_n})_{\epsilon_i = \pm 1}$ form a (n, ϵ) q-branch if

$$q(x_{\epsilon_1, \ldots, \epsilon_{n-1}, +1} - x_{\epsilon_1, \ldots, \epsilon_{n-1}, -1}) \geqslant \epsilon \qquad \text{for all } \epsilon_1, \ldots, \epsilon_{n-1} = \pm 1 ,$$

and if the 2^{n-1} points

$$\left(\tfrac{1}{2}\,(x_{\epsilon_1, \ldots, \epsilon_{n-1}, +1} + x_{\epsilon_1, \ldots, \epsilon_{n-1}, -1})\right)_{\epsilon_i = \pm 1}$$

form a $(n-1, \epsilon)$ q-branch.

If, for some $\epsilon > 0$, there is, for every $n \geqslant 1$, a (n, ϵ) q-branch in the unit ball of E, we say that E has the finite q-Tree property.

Let F be another Banach space, and T an operator from E into F. Put $q(x) = \|Tx\|$. Show that T is uniformly convexifying (see chapter I, exercise 6) if and only if E does not have the finite q-Tree property.

EXERCISE 2. - Let q be a continuous semi-norm, as in exercise 1. Assume that there exists on E an equivalent norm, $|\cdot|$, satisfying the following property :

$$(\star)\quad \begin{cases} \text{For every } \epsilon > 0 \text{, there is a } \delta > 0 \text{ such that, for all } x,y \in E, \\ \text{if} \\ |x| \leqslant 1 , \quad |y| \leqslant 1 , \quad q(x - y) \geqslant \epsilon , \\ \text{then} \\ \left| \dfrac{x + y}{2} \right| \leqslant 1 - \delta . \end{cases}$$

Show that E cannot have the finite q-Tree property.

EXERCISE 3. - Let $\epsilon > 0$, and $z \in E$. We say that the couple (x_{+1}, x_{-1}) is a $(1, \epsilon)$ q-partition of z if

$$x_{+1} + x_{-1} = z , \qquad \|x_{+1}\| = \|x_{-1}\| ,$$

and

$$q\left(\frac{x_{+1}}{\|x_{+1}\|} - \frac{x_{-1}}{\|x_{-1}\|} \right) \geqslant \epsilon .$$

Assume that a $(n-1, \epsilon)$ q-partition of z has been defined. We say that $(x_{\epsilon_1,\dots,\epsilon_n})_{\epsilon_i = \pm 1}$ is a (n, ϵ) q-partition of z if

- $$\|x_{\epsilon_1,\dots,\epsilon_{n-1},+1}\| = \|x_{\epsilon_1,\dots,\epsilon_{n-1},-1}\| , \quad \epsilon_1,\dots,\epsilon_{n-1} = \pm 1$$

- $$q\left(\frac{x_{\epsilon_1,\dots,\epsilon_{n-1},+1}}{\|x_{\epsilon_1,\dots,\epsilon_{n-1},+1}\|} - \frac{x_{\epsilon_1,\dots,\epsilon_{n-1},-1}}{\|x_{\epsilon_1,\dots,\epsilon_{n-1},-1}\|} \right) \geqslant \epsilon , \quad \epsilon_1,\dots,\epsilon_{n-1} = \pm 1$$

- the 2^{n-1} points

  $$\left(\frac{1}{2} (x_{\epsilon_1,\dots,\epsilon_{n-1},+1} + x_{\epsilon_1,\dots,\epsilon_{n-1},-1}) \right) , \quad \epsilon_1,\dots,\epsilon_{n-1} = \pm 1$$

  form a $(n-1, \epsilon)$ q-partition of z.

We assume that E does not have the finite q-Tree property .

1°) Show that there exist an integer $n > 0$ and a $\delta > 0$ such that, for every $z \in E$, every (n , ϵ) q-partition of z , $(x_{\epsilon_1,\ldots,\epsilon_n})_{\epsilon_i = \pm 1}$, one has

$$\sum_{\epsilon_i = \pm 1} \|x_{\epsilon_1,\ldots,\epsilon_n}\| \geqslant (1 + \delta)\|z\| .$$

2°) Let $\epsilon > 0$, and n , δ given by 1°). Assume $0 < \delta < \epsilon < \frac{1}{8}$.

Put, for $m \geqslant 1$, $c_m = (1 + \frac{\delta}{2}(1 + \frac{1}{4} + \ldots + \frac{1}{4^m}))^{-1}$, and define, for all $x \in E$:

$$|x| = \inf c_m \sum_{\epsilon_i = \pm 1} \|u_{\epsilon_1,\ldots,\epsilon_m}\| ,$$

the infimum being taken over all $m \geqslant 1$, and all (m , ϵ) q-partitions, $(u_{\epsilon_1,\ldots,\epsilon_m})_{\epsilon_i = \pm 1}$, of x .

Show that :

a) $(1 - \delta)\|x\| \leqslant |x| \leqslant \|x\|$, for all $x \in E$.

b) $|\alpha x| = |\alpha|\ |x|$, $\alpha \in \mathbb{R}$, $x \in E$.

c) If $\|x\| = \|y\| = 1$ and $q(x - y) \geqslant \epsilon$, then

$$|x + y| \leqslant |x| + |y| - \frac{\delta}{2\cdot 4^{n+2}} .$$

(For c), follow the same ideas as in the proof of the theorem, in the last chapter).

3°) Define, for $\epsilon > 0$:

$$|x|_\epsilon = \inf \sum_{k \geqslant 1} |x_k - x_{k-1}| ,$$

the infimum being taken over all finite sequences $x_o,\ldots,x_K$, with $x_o = 0$ and $x_K = x$.

Show that $|\cdot|_\epsilon$ is a norm on E , satisfying :

a) $(1 - \frac{\epsilon}{4})\|x\| \leqslant |x|_\epsilon \leqslant \|x\|$, for all x in E ,

b) For all x , y in E , if

$\|x\| = \|y\| = 1$, $q(x - y) \geqslant \epsilon$,

then $|x + y|_\epsilon \leqslant |x|_\epsilon + |y|_\epsilon - \delta'$, with $\delta' = \dfrac{\epsilon\,\delta}{10 \cdot 4^{n+2}}$.

4°) Deduce from 3°) that, for all x , y in E ,

if $|x|_\epsilon = |y|_\epsilon = 1$, $q(x - y) \geqslant 4\epsilon$,

then

$$|x + y|_\epsilon \leqslant |x|_\epsilon + |y|_\epsilon - \delta' .$$

5°) Show that there exists a $\delta'' > 0$ such that, for all x , y in E , if

$$q(x - y) \geqslant 8\epsilon \max(|x|_\epsilon , |y|_\epsilon)$$

then

$$\left| \frac{x + y}{2} \right|^2 \leqslant \frac{1}{2} (1 - \delta'')(|x|_\epsilon^2 + |y|_\epsilon^2) .$$

6°) Use lemma 5 (p. 287) to establish the following theorem :

THEOREM. - *If* E *does not have the finite* q-*Tree property, one can find on* E *a new norm,* $|\cdot|$, *equivalent to the original one, satisfying* (☆) *of exercise* 2.

(The converse implication was proved in exercise 2).

Deduce the following result :

THEOREM. - *An operator* T , *from* E *into* F , *is uniformly convexifying if and only if there exists on* E *an equivalent norm,* $|\cdot|$, *such that :*

For all $\epsilon > 0$, *there is a* $\delta_T > 0$, *such that for all* x , y *in* E , *if*

$\|x\| \leqslant 1$, $\|y\| \leqslant 1$, $\|Tx - Ty\| \geqslant \epsilon$,

then

$$\left\| \frac{x + y}{2} \right\| \leqslant 1 - \delta_T .$$

REFERENCES ON CHAPTER IV.

As we said before the statement of the theorem, the fact that every super-reflexive space admits an equivalent uniformly convex norm is due to P. ENFLO [20]. But Enflo's proof did not give any estimate of $\delta(\epsilon)$ like $\delta(\epsilon) \geqslant C\,\epsilon^p$. These estimates were given by G. PISIER [40], who first used martingales instead of trees, and could make use of James' estimates for basic sequences in $L_2(\Omega\,;E)$. Pisier's proof depended upon deep properties of martingales, much more than the proof we present here, which is, from this point of view, quite elementary. The present proof is due to B. MAUREY, and is reproduced here with his kind permission (it has not been published elsewhere). The definition in terms of martingales allows to make use of James' estimates. The other ideas of the proof (like "slowing down" the martingales, mixing the sequence of norms, and so on) were already in ENFLO's paper [20], and are just adapted from [20]. It seems to us that the present proof is "minimal", in the sense that one cannot avoid the different steps it contains.

Super-reflexivity and uniform convexity can also be defined for operators between Banach spaces : this was done by the author in [7], [8], and the exercises in this chapter are taken from this work. So, some of the results of chapters I, II, III, IV can be extended to this new frame, but James' estimates for basic sequences do not hold anymore. There is, however, a renorming theorem (exercise 3 above, taken from [7]), which gives a new norm on the space on which the operator is defined, but the modulus of convexity does not satisfy in general $\delta_T(\epsilon) \geqslant C\,\epsilon^p$, but can be, for example like $\delta_T(\epsilon) \underset{\epsilon\to 0}{\sim} C\,e^{-1/\epsilon}$.

We shall finish with some complements about Banach-valued martingales.

COMPLEMENTS ON CHAPTER IV.

The martingales used in this chapter are of very special type, because of the very peculiar definition of Ω , $\mathscr{B}_n$, P . They are called "Walsh-Paley martingales" by G. PISIER in [40].

More generally, if $(\mathscr{B}_\alpha)_{\alpha\in I}$ is a monotone increasing net of sub σ-fields of $\mathscr{A}$, one says that a family $(f_\alpha)_{\alpha\in I}$ of functions defined on $(\Omega, \mathscr{A}, P)$, with values in E , is a martingale if each f_α is

$\mathscr{B}_\alpha$-measurable (see second part, chapter VI, § 3), and if $E^{\mathscr{B}_\beta} f_\alpha = f_\beta$, for $\beta \leqslant \alpha$ ($E^{\mathscr{B}_\beta}$ is the conditional expectation on $\mathscr{B}_\beta$). This means also that $\int_A f_\alpha dP = \int_A f_\beta dP$, if $A \in \mathscr{B}_\beta$, $\beta \leqslant \alpha$.

The martingale is called L_1-*bounded* if all the f_α's are integrable, that is $f_\alpha \in L_1(\Omega, \mathscr{B}_\alpha , P ; E)$, and if

$$\sup_\alpha \int_\Omega \|f_\alpha\| dP < +\infty \ .$$

An L_1-bounded martingale is called *uniformly integrable* if

$$\lim_{P(A) \to 0} \sup_\alpha \int_A \|f_\alpha\| dP = 0 \ .$$

The convergence of such martingales in $L_1(\Omega, \mathscr{A}, P ; E)$ is equivalent to the Radon-Nikodym property for E , as the following theorem shows : (This theorem is due to S. CHATTERJI, see [15] or [46] for a proof and for complements).

THEOREM. - *A Banach space* E *has R.N.P. if and only if for every probability space* $(\Omega, \mathscr{A}, P)$, *every* L_1-*bounded, uniformly integrable martingale converges in* $L_1(\Omega, \mathscr{A}, P ; E)$.

If E has the infinite tree property (see chapter I), this tree defines (with the same notations as in the chapter above) a martingale $(f_n)_{n \in \mathbb{N}}$ on $(\Omega, (\mathscr{B}_n), P)$, which is L_1-bounded and uniformly integrable. This martingale fails to converge, since $\int \|f_n - f_{n+1}\| dP \geqslant \frac{\epsilon}{2}$, for all n . This means, as we already noticed, that a space with the infinite tree property cannot have R.N.P.

☆
☆ ☆
☆

..... suadentque cadentia sidera somnos.

BIBLIOGRAPHY

[1] ANDO, T. - Contractive Projections in L_p-spaces. Pacific J. of Maths 17 (1966), 391-405.

[2] ASPLUND, E. - Averaged norms. Israël J. of Maths, 5 (1967), 227-233.

[3] ASPLUND, E. - Fréchet differentiability of convex functions. Acta Math. 121 (1968), 31-47.

[4] ASPLUND, E. and NAMIOKA, I. - A geometric proof of Ryll-Nardzewski's fixed point theorem. Bull. A.M.S., 73 (1967), 443-445.

[5] BANACH, S. - Théorie des opérations linéaires, Varsovie, 1932.

[6] BEAUZAMY, B. - Espaces d'Interpolation réels : Topologie et Géométrie. Lecture Notes n° 666. Springer-Verlag, (1978).

[7] BEAUZAMY, B. - Opérateurs uniformément convexifiants. Studia Math. 57 (1976), 103-139.

[8] BEAUZAMY, B. - Quelques propriétés des opérateurs uniformément convexifiants. Studia Math., 60 (1977), 211-222.

[9] BEAUZAMY, B. et MAUREY, B. - Points minimaux et ensembles optimaux dans les Espaces de Banach. Journal of Functional Analysis, 24-2 (1977), 107-139.

[10] BOURBAKI N. - Eléments de Mathématiques : Topologie Générale. Hermann, Paris.

[11] BOURBAKI N. - Eléments de Mathématiques : Espaces vectoriels topologiques. Hermann, Paris.

[12] CHOQUET, G. - Lectures on Analysis. Vol. I, II, III. Mathematics Lecture Notes Series. Benjamin.

[13] DAY, M.M. - Normed linear spaces, Springer-Verlag.

[14] DIESTEL, J. - Geometry of Banach spaces : Selected Topics. Lecture Notes n° 485. Springer-Verlag.

[15] DIESTEL, J. and UHL, J.J. - Vector Measures. Mathematical surveys of the A.M.S. Vol. 15, 1977.

[16] DUNFORD, N. and SCHWARTZ, J.T. - Linear Operators. Vol. I, II, III. Pure and Applied Mathematics, Interscience Publishers. 1963.

[17] EKELAND, I. - Cours d'Analyse, Ecole Polytechnique, Palaiseau, France.

[18] EKELAND, I. and TEMAM, R. - Convex Analysis and Variational Problems. North Holland.

[19] ENFLO, P. - A counter-example to the approximation problem in Banach spaces. Acta Math., 130 (1973), 309-317.

[20] ENFLO, P. - Banach spaces which can be given an equivalent uniformly convex norm. Israël J. of Maths, 13 (1972), 281-288.

[21] FIGIEL, T. - LINDENSTRAUSS, J. and MILMAN, V.D. - The dimension of almost spherical sections of convex sets. Acta Math., 139 (1977), 53-94.

[22] GURARII, V.I. and GURARII, N.I. - On bases in uniformly convex and uniformly smooth Banach spaces. Izvestia Akad. Nauk SSSR, Ser. Mat. 35 (1971), 210-215.

[23] JAMES, R.C. - Reflexivity and the supremum of linear functionals. Ann. of Maths, 66 (1957), 159-169.

[24] JAMES, R.C. - Bases and reflexivity of Banach spaces. Ann. of Maths, 52 (1950), 518-527.

[25] JAMES, R.C. - Weak compactness and reflexivity. Israël J. of Maths, 2 (1964), 101-119.

[26] JAMES, R.C. - Uniformly non square Banach spaces. Ann. of Maths, 80 (1964), 542-550.

[27] JAMES, R.C. - Super-reflexive spaces with bases. Pacific J. of Maths, 41-2 (1972), 409-419.

[28] JAMES, R.C. - Some self-dual properties of normed linear spaces. Symposium on Infinite Dimensional Topology, Annals of Math. Studies, 69 (1972), 159-175.

[29] JAMES, R.C. - Super-reflexive Banach spaces. Canadian J. of Maths, 24 (1972), 896-904.

[30] KADEC, M.I. and PEŁCZYŃSKI, A. - Bases, lacunary sequences and complemented subspaces in the space L_p. Studia Math., 21 (1962), 161-176.

[31] KÖTHE, G. - Topological Vector Spaces. Springer-Verlag, 1969.

[32] LACEY, E. - The isometric theory of classical Banach spaces. Springer-Verlag, 1974.

[33] LINDENSTRAUSS, J. and ROSENTHAL, H.P. - The $\mathscr{L}_p$-spaces, Israël J. of Maths, 7 (1969), 325-349.

[34] LINDENSTRAUSS, J. and TZAFRIRI, L. - Classical Banach Spaces, vol. I and II, Springer-Verlag.

[35] MAUREY, B. - Le système de Haar. Exposés 1 et 2, Séminaire Maurey-Schwartz, 1974-1975. Ecole Polytechnique. Palaiseau, France.

[36] MILUTIN, A.A. - Isomorphisms of spaces of continuous functions on compacta of power continuum. Tieoriea Funct. (Kharkow), 2 (1966), 150-156 (Russian).

[37] NACHBIN, L. - A theorem of Hahn-Banach type for linear transformations. Transactions A.M.S., 68 (1950), 28-46.

[38] NEVEU, J. - Notions élémentaires de la théorie des probabilités. Masson.

[39] PISIER, G. - Bases, lacunary sequences and complemented subspaces in the space L_p, d'après Kadec-Pełczyński. Exposé n° 3, Séminaire Maurey-Schwartz, 1972-1973, Ecole Polytechnique. Palaiseau. France.

[40] PISIER, G. - Martingales with values in uniformly convex spaces. Israël J. of Maths, 20-3.4 (1975), 326-350.

[41] ROSENTHAL, H.P. - On subspaces of L_p. Ann. of Maths, 97 (1973), 344-373.

[42] ROSENTHAL, H.P. - On factors of $\mathscr{C}([0,1])$ with non-separable dual. Israël J. of Maths, 13 (1972), 361-378.

[43] RUDIN, W. - Real and Complex Analysis. Tata Mc Graw Hill.

[44] SCHÄFFER, J.J. and SUNDARESAN, K.S. - Reflexivity and the girth of spheres. Math. Ann. 184 (1970), 163-168.

[45] SCHWARTZ, L. - Cours d'Analyse. Hermann. Paris.

[46] SCHWARTZ, L. - Fonctions mesurables et ★-scalairement mesurables, propriété de Radon-Nikodym. Exposés 4, 5 et 6, Séminaire Maurey-Schwartz, 1974-1975. Ecole Polytechnique, Palaiseau. France.

[47] SINGER, I. - Bases in Banach Spaces, I . Springer-Verlag, 1970.

[48] STERN, J. Propriétés locales et ultrapuissances d'espaces de Banach. Exposés 7 et 8, Séminaire Maurey-Schwartz, 1974-1975. Ecole Polytechnique, Palaiseau. France.

[49] SZANKOWSKI, A. - $B(H)$ does not have the approximation property. To appear.

[50] YOSHIDA, K. - Functional Analysis, Springer-Verlag.

[51] ZYGMUND, A. - Trigonometric series. Vol. I and II. Cambridge University Press.

INDEX